# Uniscience Series on Fine Particle Science and Technology

Editor-in-Chief

**John Keith Beddow, Ph.D.**
Professor of Chemicals and Materials Engineering
Division of Materials Engineering
University of Iowa
Iowa City, Iowa

**Advanced Particulate Morphology**
J. K. Beddow and T. P. Meloy

**Separation of Particles From Air and Gases, Volumes I and II**
Akira Ogawa

**Particle Characterization in Technology**
Volume I: Applications and Microanalysis
Volume II: Morphological Analysis
J. K. Beddow, Editor

**Vortex Flow**
Akira Ogawa

# Particle Characterization in Technology

## Volume II
## Morphological Analysis

Editor
**John Keith Beddow, Ph.D.**
Professor of Chemical and Materials Engineering
College of Engineering
Division of Materials Engineering
The University of Iowa
Iowa City, Iowa

CRC Series on Fine Particle Science and Technology

Editor-in-Chief
**John Keith Beddow**

CRC Press
Taylor & Francis Group
Boca Raton London New York

CRC Press is an imprint of the
Taylor & Francis Group, an informa business

First published 1984 by CRC Press
Taylor & Francis Group
6000 Broken Sound Parkway NW, Suite 300
Boca Raton, FL 33487-2742

Reissued 2018 by CRC Press



**Library of Congress Cataloging in Publication Data**
Main entry under title:

Particle characterization in technology.

(CRC series on fine particle science and technology)
Bibliography.
Includes index.
1. Particles. 2. Bulk solids. 3. Particle size determination. I. Beddow, John K. II. Series.
TA418.78.P36 1984 620'.43 83-14363
ISBN 0-8493-5784-5 (v. 1)
ISBN 0-8493-5785-3 (v. 2)

A Library of Congress record exists under LC control number: 83014363

Publisher's Note
The publisher has gone to great lengths to ensure the quality of this reprint but points out that some imperfections in the original copies may be apparent.

Disclaimer
The publisher has made every effort to trace copyright holders and welcomes correspondence from those they have been unable to contact.

ISBN 13: 978-1-315-89626-7 (hbk)
ISBN 13: 978-1-351-07536-7 (ebk)

Visit the Taylor & Francis Web site at http://www.taylorandfrancis.com and the
CRC Press Web site at http://www.crcpress.com

# INTRODUCTION

The problem of particle characterization is one of the central themes of particulate technology. It has as its ultimate goal the development of the ability to predict from a knowledge of the particle characteristics, the properties and behavior of particles individually and also en masse. Our present stage of knowledge indicates that our goal should be judged as a high ideal — we are indeed a long way from our ultimate goal. However, there is clearly no cause for despair. This is because of the great progress that has been made in the last few years. Some of the more important of these are the subject of the chapters of these two volumes.

Not so very long ago we had to be satisfied with a low power look at particulate materials. There is now available a remarkable set of techniques under the collective title of microanalysis. Some of the more significant of these methods are reviewed in Volume I by Gavrilovic (SIMS), Kaufmann and Wieser (LAMMA), and Davidson (electron microscopy and spectroscopy).

The remainder of Volume I presents an important exposition of some of the most significant areas where particle characterization is applied. The technological fields include pharmaceutical materials by Hersey, bulk solids by Smith and Lohnes, and explosions by Palmer. In each of these fields the authors describe a great deal of work that has to be done to be able to state the properties of the bulk particulate material with quantitative exactness. Some information to relate these properties to particle characteristics is already available and reported in this and Volume II. The remainder will be generated in the future. The other fields of application reported out in this volume include applications of fractal analysis by Kaye, fine particle characterization methods in liquids by Lieberman, practical aspects of electrozone size analysis by Karuhn and Berg, and finally, a brief description of the well-known COULTER® Counter by Kinsman.

Morphological analysis has been with us since the early 1970s, but it made its great leap after the 1977 research workshop at the University of Iowa, which was funded by the National Science Foundation, Particulates Processing and Multiphase Flow Program under Dr. Morris Ojalvo. The reader may compare the proceedings of that workshop (*Advanced Particulate Morphology,* Beddow and Meloy, Eds., CRC Press, 1980) with this volume in order to grasp the giant steps forward that have been made in a very few years. The first section of Volume II deals with both theory and methods of morphological analysis. In particular there are reviews by Luerkens, Beddow, Vetter, Gotoh, and Meloy. PIAS instrumentation is described by Chang. For the moment, the general consensus favors the use of the Fourier series in morphological analysis. The use of the Fast Fourier Transform would speed up the rate of data acquisition. The FFT is fully discussed by Kanatani (Chapter 4). The task of data analysis is of primary importance, and Inoue has written a chapter on the potential use of a relatively new computer language called FORTH. It is believed that its adoption at this time in place of FORTRAN and the like may help reduce some of our current difficulties. To date, morphological analysis has been concerned exclusively with the profile of the particle. In addition, there is much information to be obtained from within the particle and also from within clusters and agglomerates. There is therefore an important need to extend the technique so that we may deal with textural features of particles. Leong's chapter on color as a morphic feature represents a beginning step in this direction.

Data analysis is to morphological analysis as the identification of chemical elements is to chemical analysis. Various topics are discussed in the second volume including a multivariate pattern recognition approach by Ehrlich and Full. This deals with ways in which morphic mixtures may be analyzed for their end members. The Fourier series is an infinite series. How far one should go before making the decision to truncate is dealt with by Lenth in his chapter. Bezdek and Solomon demonstrate further use of fuzzy sets in morphological analysis. (The reader is referred to Bezdek's recent book, *Pattern Recognition With Fuzzy Objective Function Algorithms,* Plenum Press, New York, 1981 for additional information.) The chapter by Hua (Chapter 10) underscores the need for introducing texture as a feature in morphological analysis. In this chapter Hua shows that it is not possible to differentiate between alumina clusters (designated O and N) merely on the basis of the information in their outer profiles. This section of the book is brought to a conclusion by the chapter dealing with data presentation in the form of histograms by Full, Ehrlich, and Kennedy.

Applications of morphological analysis include a review by Beddow, relation of the size and shape of abrasive particles to the rate of wear that they cause in the case of steel samples by Vetter, and identification of the quantitative shape characteristics associated with particles produced by different processes by Guo and Ramakrishnan. The study reported by Carmichael shows that not only can three-dimensional morphic features account for the sedimentation behavior of particles of different regular geometries, but their orientations are also accounted for. Sieving or screening are normally perceived as methods for separating particles on a basis of their size differences. Rajpal et al. reports that this technique may also be used for shape separation of particles. Chang et al. describes correlations obtained between bulk properties (including flow-time, Hausner ratio, and angle of internal friction) and the morphic features of a number of materials. Hua et al. (Chapter 18) discusses the modeling of crystallization using morphological analysis.

## THE EDITOR-IN-CHIEF

**John Keith Beddow** received his Ph.D. in Metallurgy from Cambridge University, Cambridge, England, in 1959. Currently Secretary of the Fine Particle Society, he is a member of the Faculty at the University of Iowa, where he heads a small research group in fine particle science with emphasis on morphological analysis. Dr. Beddow is an active lecturer and author. He has also been active as a Consultant in metallurgy, powder metallurgy, and powder technology for numerous corporations. He is also president of Shape Technology, Ltd. His present research activities are in particle morphological analysis. Dr. Beddow is married, with four daughters, and has resided in the U.S. since 1966.

## THE EDITOR

**John Keith Beddow** received his Ph.D. in Metallurgy from Cambridge University, Cambridge, England, in 1959. Currently Secretary of the Fine Particle Society, he is a member of the Faculty at the University of Iowa, where he heads a small research group in fine particle science with emphasis on morphological analysis. Dr. Beddow is an active lecturer and author. He has also been active as a Consultant in metallurgy, powder metallurgy, and powder technology for numerous corporations. He is also President of Shape Technology, Ltd. His present research activities are in particle morphological analysis. Dr. Beddow is married, with four daughters, and has resided in the U.S. since 1966.

## CONTRIBUTORS


**John Keith Beddow, Ph.D.**
Professor of Chemical and Materials Engineering
College of Engineering
Division of Materials Engineering
The University of Iowa
Iowa City, Iowa

**Robert H. Berg, M.S., P.E.**
President
Particle Data, Inc.
Particle Data Laboratories, Ltd.
Elmhurst, Illinois

**James C. Bezdek, Ph.D.**
Head
Department of Computer Science
University of South Carolina
Columbia, South Carolina

**Gordon Butters, Ph.D.**
BP Chemicals, Ltd.
Research and Development Department
South Wales Division
Sully, South Glamorgan, Wales

**Gregory Carmichael, Ph.D.**
Associate Professor
Chemical and Materials Engineering Program
College of Engineering
Division of Materials Engineering
The University of Iowa
Iowa City, Iowa

**Chi-Ren Chang, Ph.D.**
College of Engineering
Division of Materials Engineering
The University of Iowa
Iowa City, Iowa

**S. Collins**
Image Processing Laboratory
Information Engineering
The University of Iowa
Iowa City, Iowa

**David L. Davidson, Ph.D.**
Institute Scientist
Southwest Research Institute
San Antonio, Texas

**Robert Ehrlich, Ph.D.**
Department of Geology
University of South Carolina
Columbia, South Carolina

**William E. Full, Ph.D.**
Department of Geology and Geography
Wichita State University
Wichita, Kansas

**John Gavrilovic, Ph.D.**
Senior Research Scientist
Walter C. McCrone Associates, Inc.
Chicago, Illinois

**Keishi Gotoh, Ph.D.**
Professor
Department of Energy Engineering
Toyohashi University of Technology
Toyohashi, Japan

**Andy Guo**
Fine Particle Research Group
College of Engineering
Division of Materials Engineering
The University of Iowa
Iowa City, Iowa

**John A. Hersey, Ph.D. (deceased)**
Sigma Industrial Professor of Pharmaceuticals
Victorian College of Pharmacy
Melbourne, Australia

**Louise Hua**
Fine Particle Research Group
College of Engineering
Division of Materials Engineering
The University of Iowa
Iowa City, Iowa

**Toshio Inoue, Ph.D.**
Professor
Department of Mineral Development Engineering
Faculty of Engineering
University of Tokyo
Tokyo, Japan

**Ken-ichi Kanatani, Ph.D.**
Associate Professor
Department of Computer Science
Gunma University
Gunma, Japan

**Richard F. Karuhn, B.S.**
Director
Particle Data Laboratories, Ltd.
Elmhurst, Illinois

**Prof. Dr. Raimund Kaufmann**
Lehrstuhl für Klinische Physiologie der Universität
Düsseldorf, West Germany

**Brian H. Kaye, Ph.D.**
Professor of Physics
Director
Fineparticle Research Institute
Laurentian University
Sudbury, Ontario, Canada

**Stephen Kennedy, Ph.D.**
Postdoctoral Fellow
Department of Geology
University of South Carolina
Columbia, South Carolina

**Shepard Kinsman**
909 Eastridge Village Drive
Miami, Florida

**Mitsuyoshi Kumamoto**
Graduate Student
Department of Energy Engineering
Toyohashi University of Technology
Toyohashi, Japan

**Russell V. Lenth, Ph.D.**
Associate Professor
Department of Statistics and Actuarial Science
The University of Iowa
Iowa City, Iowa

**Poh-Leng Leong, S.M.**
Operations Research
Massachusetts Institute of Technology
Cambridge, Massachusetts

**Alvin Lieberman, M.S.**
Technical Specialist
Particle Measuring Systems, Inc.
Boulder, Colorado

**Robert A. Lohnes, Ph.D.**
Professor
Civil Engineering Department
Iowa State University
Ames, Iowa

**David W. Luerkens, Ph.D.**
Research Engineer
E. I. DuPont de Nemours & Company
Atomic Energy Division
Savannah River Laboratory
Aiken, South Carolina

**Thomas P. Meloy, Ph.D.**
Particle Analysis Center
West Virginia University
Morgantown, West Virginia

**K. N. Palmer**
Head of Fire Protection Division
Fire Research Station
Borehamwood, England

**Samir Rajpal**
Fine Particle Research Group
College of Engineering
Division of Materials Engineering
The University of Iowa
Iowa City, Iowa

**Prof. Dr. P. Ramakrishnan**
Department of Metallurgical Engineering
Indian Institute of Technology
Bombay, India

**C. Seemuth**
Audio Visual Center
The University of Iowa
Iowa City, Iowa

**David L. O. Smith, Ph.D.**
Assistant Professor
Civil Engineering Department
Iowa State University
Ames, Iowa

**Kenneth Solomon, Ph.D.**
Research Associate
U.S. Salinity Laboratory
Riverside, California

**Paul Swanson, Ph.D.**
John Deere & Company
Moline, Illinois

**Arthur F. Vetter, M.S.**
Department of Chemical and Materials Engineering
College of Engineering
Division of Materials Engineering
The University of Iowa
Iowa City, Iowa

**Prof. Dr. Paul Wieser**
Institut für Physik
Universität Hohenheim
Stuttgart, West Germany

**Ming-Jau Yin**
College of Engineering
Division of Materials Engineering
The University of Iowa
Iowa City, Iowa

## ACKNOWLEDGMENTS

As Editor of these two volumes, I would like to extend my sincere thanks to the many authors who have contributed so splendidly to the final outcome. Particulate technology is such a diverse field and its activists are so very busy. This is the main reason why there are so few books and why most of us have to go into the original literature if we wish to obtain information. I hope that the contributing authors to this book have by their outstanding efforts given the reader invaluable new information and ideas and that in the future, these will contribute to substantial programs and applications. If this is so, then they will have earned our gratitude.

# DEDICATION

This book is dedicated to John Arthur Hersey who was Sigma Industrial Professor of Pharmaceutics at the Victorian College of Pharmacy in Melbourne, Australia. He died unexpectedly in Melbourne on February 14, 1982. He was well known in North America, Europe, and the Pacific Region. He will best be remembered to particle technologists for his brilliant work on ordered mixing. His chapter in Volume I is entitled, "Characterization of Pharmaceutical Particulate Materials".

# TABLE OF CONTENTS

Volume I

Volume II

THEORY AND METHODS



DATA ANALYSIS

APPLICATIONS

# *Part I*
# *Theory and Methods*

Chapter 1

# THEORY OF MORPHOLOGICAL ANALYSIS

**David W. Luerkens, John Keith Beddow, and Arthur F. Vetter**

## TABLE OF CONTENTS

# I. ABSTRACT

The objectives and mathematical principles of morphological analysis are presented leading to the development of invariant Fourier descriptors via the (R,θ) method. For particles with reentrant profiles, the (ϕ,ℓ) method is discussed. Finally, a generalized (R,S) method is presented.

# II. PARTICLE MORPHOLOGY

Particle morphology may be defined as that part of fine particle science and technology concerned with the form of finely divided matter and the structures, homologies, and processes of change that govern that form. Fine particles may be described in terms of four general characteristics: size, shape, chemistry, and physics.

Physical properties include electrical and thermal conductivities, optical behavior, density, magnetic properties, etc. Chemical properties include gross as well as point-to-point chemical analysis. Chemical kinetics may also be included here. The properties of size and shape may be collected together as particle geometry.

In this chapter, three methods of morphological analysis are discussed: (1) (R,θ), (2) (ϕ,ℓ), and (3) (R,S).

## A. Historical Development

There are two stages in the historical development of morphological analysis. In the first stage, advantage was taken of the incredible human ability to observe, recognize, and codify particles of differing shapes. This culminated in the development of the *Particle Atlas,*[1] which is a unique and valuable guide to the identification of fine particles using the power of human observation in conjunction with some very fine science. The second stage began with the developments of Beddow and Philip,[2] Ehrlich and Weinberg,[3] and Meloy.[4] More recently, quantitative three-dimensional analysis is being studied.[5] Gotoh[6] reports on the recent contributions from Japan.

## B. Morphological Analysis Research

In general, morphological research covers three broad, interconnected areas:

1. Fundamental research to improve morphological analysis and procedures.
2. The study of the relations between morphic properties and the mode of origin of the particles.
3. The relation of the morphic properties of the particles to physical-chemical and bulk properties.

## C. The (R,Θ) Method

Particle shape has been defined as "the pattern of all the points on the boundary or surface of a particle profile."[2]

In order to measure the shape of a particle profile, the sample set of coordinate points satisfying the above definition must be obtained. This data set must then be condensed to a minimum while retaining the original pattern of the data.

The procedure is as follows:

1. An image of the particle is prepared, either photographically or electronically.[7]
2. The profile is digitized into a set of (x,y) coordinates.
3. The (x,y) pairs may be treated in a number of ways. The way used here is to convert the (x,y) pairs to polar coordinates (R,Θ), using the center gravity as the origin.

4. The polar pairs are analyzed to form a Fourier equation of the form

$$R(\theta) = a_0 + \sum_{n=1}^{\infty} (a_n \cos n\theta + b_n \sin n\theta) \quad (1)$$

5. The following minimum of invariant Fourier descriptors is obtained:

$$R_0 = A_0^2 + \frac{1}{2} \sum_{n=1} A_n^2 \quad (2)$$

where

$$A_n^2 = a_n^2 + b_n^2 \quad (3)$$

and

$$A_0 = a_0 \quad (4)$$

$R_0$ is termed "equivalent radius" and is the radius of a circle having the same area as that of the particle profile, and n is the order of the coefficient, $n + m = 2,3,4, \ldots N$.

$$L_0 = \frac{A_0}{R_0} \quad (5)$$

$$L_1(n) = 0 \text{ for all N} \quad (6)$$

$$L_2(n) = \frac{1}{2R_0^2} A_n^2 \quad (7)$$

$$L_3(m,n) = \frac{3}{4R_0^3} [A_m A_n A_{m+n} \cos (\alpha_{n+m} - \alpha_m - \alpha_n)] \quad (8)$$

in which

$$\alpha_n = \tan^{-1} \frac{b_n}{a_n} \quad (9)$$

The size and shape terms are directly related to the mean and moments about the mean of the particle profile radial distribution as follows:

$$\mu_0 = L_0 R_0 \text{ (mean radius)} \quad (10)$$

$$\mu_1 = R_0 \sum_{n=1}^{\infty} L_1(n) = 0 \quad (11)$$

$$\mu_2 = R_0^2 \sum_{n=1} L_2(n) \quad (12)$$

$$\mu_3 = R_0^3 \sum_{m=1} \sum_{n=1} L_3(m,n) \quad (13)$$

in which $\mu_0$ is the mean of the radial distribution and $\mu_1$, $\mu_2$, and $\mu_3$ are the first, second, and third moments about the mean of the radial distribution, respectively.

There are three major types of information available from this analysis: the individual terms $L_2(n)$, $L_3(m,n)$ describe the shape of the profile of the object. The terms then represent shape features, which may be used as identifiers in classifiers.[8] The $R_0$ term defines the size of the profile. It, too, may be used for classification. The third type of information, the radial moment, can be used in a variety of applications including crystallization, modeling, sedimentation studies, abrasive wear studies, and chemical corrosive attacks.

### D. Previous Practice

It should be noted that the development of the invariant descriptors represents a significant advance over previous practice. In the past, the following form of the Fourier equation was used:

$$f(\theta) = A_0 + \sum_{n=1}^{\infty} A_n \cos (n\theta - \alpha_n) \tag{14}$$

in which

$$A_0 = a_0$$

$$A_n = a_n^2 + b_n^2$$

$$\alpha_n = \tan^{-1}\left(\frac{b_n}{a_n}\right)$$

In the past, the coefficients $A_n$ were used in morphological analysis studies. Insofar as $A_n$ is an invariant feature, this is satisfactory. However, $\alpha_n$ is not invariant and for this reason, in the past, $\alpha_n$ was ignored and omitted from consideration. This problem has been overcome by a recent development.[9]

## III. RELATIONSHIPS BETWEEN INVARIANT TERMS AND RADIAL DISTRIBUTION

The size and shape terms are related to the mean and the moments about the mean of the profile radial distribution. This is discussed and demonstrated as follows:

### A. The Size Term $R_0$

The area of the particle profile may be stated in terms of the Fourier coefficients:

$$\text{Area} = \int_0^{2\pi}\int_0^{R(\theta)} r dr d\theta = \pi\left[a_0^2 + \frac{1}{2}\sum_{n=1}^{\infty}(a_n^2 + b_n^2)\right] \tag{15}$$

Keeping in mind that a circle of radius $R_0$ has an area of $\pi R_0^2$, the terms are equated by

$$\pi R_0^2 = \pi\left[a_0^2 + \frac{1}{2}\sum_{n=1}^{\infty}(a_n^2 + b_n^2)\right] \tag{16}$$

This leads directly to Equation 2.

### B. The Shape Term $a_0$

From the calculus, it is known that the average value of a function, $y = f(x)$ for $a \leq y \leq b$, is given by

$$\overline{Y_{ave}} = \frac{1}{b-a}\int_a^b f(x)dx$$

By analogy:

$$\overline{R(\theta)} = \frac{1}{2\pi}\int_0^{2\pi} rR(\theta)d\theta = a_0 \tag{17}$$

Therefore, $a_0$ is the average or mean radius.

**C. The Other Shape Terms**

Similarly,

$$\overline{R(\theta)^2} = \frac{1}{2\pi}\int_0^{2\pi} R(\theta)^2 d\theta = R_0^2 \tag{18}$$

So,

$$R_0^2 - a_0^2 = \overline{R(\theta)^2} - \overline{R(\theta)}^2 = \overline{[R(\theta) - a_0]^2}$$

Therefore, from Equation 18:

$$\overline{[R(\theta) - a_0]^2} = \frac{1}{2}\sum_{n=1}^{\infty} (a_n^2 + b_n^2) \tag{19}$$

Equation 19 is the second moment about the mean of the radial distribution of the particle profile. Making use of the general form of the moment equation,

$$\mu_n = \frac{1}{2\pi}\int_0^{2\pi} [R(\theta) - a_0]^n \, d\theta = \overline{[R(\theta) - a_0]^n}$$

the higher moments can also be expressed as functions of the Fourier coefficients, $a_n$ and $b_n$.

The third moment about the mean is derived in a similar fashion, and is

$$\begin{aligned}\mu_3 &= \frac{1}{2\pi}\int_0^{2\pi} [R(\theta) - a_0]^3 \, d\theta \\ &= \frac{3}{4}\sum_{n=1}^{\infty} (a_n^2 a_{2n} - b_n^2 a_{2n} + 2a_n b_n b_{2n})\end{aligned} \tag{20}$$

This leads directly to Equation 8.

To summarize, the size and shape terms are directly related to the mean and moments about the mean of the particle profile radial distribution as follows:

$$\mu_0 = L_0 R_0 \text{ (mean radius)} \tag{21}$$

$$\mu_1 = R_0 \sum_{n=1}^{\infty} L_{1,n} = 0, \quad L_{1,n} = 0 \text{ for all n} \tag{22}$$

$$\mu_2 = R_0^2 \sum_{n=1}^{\infty} L_{2,n} \tag{23}$$

$$\mu_3 = R_0^3 \sum_{n=1}^{\infty} L_{3,n} \tag{24}$$

in which $\mu_0$ is the mean of the radial distribution and $\mu_1$, $\mu_2$, and $\mu_3$ are the first, second, and third moments about the mean of the radial distribution, respectively.

## IV. ROTATIONAL AND SIZE INVARIANCE

Consider a particle with coefficients $a_0$, $a_n$, and $b_n$. Let the particle be rotated at an angle $\theta'$ such that the new coefficients are $a_0'$, $a_n'$, $b_n'$. In this case, the original particle representation $R(\theta)$ is related to the new, rotated particle representation by the relationship

$$R'(\theta) = R(\theta + \theta') \tag{25}$$

The required rotational transformation equations are obtained from Equation 25 simply by expanding the trigonometric terms on the right-hand side of Equation 25. The coefficients for the nth cos nθ term and the coefficients for the nth sin nθ term are collected together. The coefficients for these terms on the left- and right-hand side of Equation 25 are then equated. This procedure leads to the following set of relationships:

$$a_0' = a_0 \tag{26}$$

$$a_n' = a_n \cos n\theta' + b_n \sin n\theta' \tag{27}$$

$$b_n' = b_n \cos n\theta' - a_n \sin n\theta' \tag{28}$$

These are the rotational transformation equations for an (R,θ) particle. (Note that each sin nθ term and each cos nθ term is independent on the interval $0 - 2\pi$.)

When Equations 26 to 28 are substituted into Equations 5 to 8 it is found that

$$a_0' = a_0 \tag{29}$$

$$(a_n')^2 + (b_n')^2 = a_n^2 + b_n^2 \tag{30}$$

$$(a_n')^2 a_{2n}' - (b_n')^2 a_{2n}' + 2a_n' b_n' b_{2n}'$$

$$= a_n^2 a_{2n} - b_n^2 a_{2n} + 2a_n b_n b_{2n} \tag{31}$$

and

$$R_0' = R_0$$

That is, the size and shape terms are rotationally invariant. It follows then that $\mu_0$, $\mu_1$, $\mu_2$, and $\mu_3$ are also rotationally invariant.

## V. SIZE CHANGE INVARIANCE

When an image is either magnified or reduced in size during the morphological analysis process it is especially important that the morphic descriptors remain invariant. For example, let both the x-axis and the y-axis be scaled by a factor α. It will be shown that the magnitude of the radii change, but that the angular relationships do not change.

Definitions:

$$r = \sqrt{x^2 + y^2}$$

$$\theta \equiv \tan^{-1} \frac{y}{x} \tag{32}$$

Also by definition:

$$x' = \alpha x$$

and

$$y' = \alpha y \tag{33}$$

in which "′" indicates the new scale. Then,

$$r_s = \sqrt{(\alpha x)^2 + (\alpha y)^2} = \alpha\sqrt{x^2 + y^2} = \alpha r \tag{34}$$

and

$$\theta_s = \tan^{-1}\frac{\alpha y}{\alpha x} = \tan^{-1}\frac{y}{x} = \theta \tag{35}$$

in which the subscript, s, means scaled. It follows that

$$R_s(\theta) = \alpha R(\theta) \tag{36}$$

In Equation 36, $R_s(\theta)$ is the scaled Fourier series. Using this equation, a scaled equivalent radius is determined:

$$R_{0s} = \alpha R_0 \tag{37}$$

And, from Equation 2,

$$R_{0s}^2 = \alpha^2 R_0^2 = (\alpha a_0)^2 + \frac{1}{2}\sum_{n=1}^{\infty} (\alpha a_n)^2 + (\alpha b_n)^2 \tag{38}$$

Dividing through by $(\alpha R_0)^2$ eliminates the scaling factor, $\alpha$, completely. This procedure ensures that the descriptors are normalized with respect to the equivalent radius $R_0$. This is shown in Equations 5 to 8.

## VI. THE $(\phi,\ell)$ METHOD

Many particles possess relatively simple $(R,\theta)$ shapes, but many others have much more complex outlines which contain reentrants. A reentrant produces multiple values of R at certain values of the angle $\theta$, and the Fourier transformation cannot handle multivalues. Therefore, a more generalized method of analysis is required.

The $(\phi,\ell)$ method, which can also analyze complex shapes, has been reported in the literature.[10] It has a disadvantage because it does not yield an expression for the area. However, it does have the capability for analyzing complex shapes.

The profile is parameterized by its arc length $\ell$ and the change of slope $\phi(\ell)$ from its starting point, i.e., $(x,y)\rightarrow(\ell,\phi(\ell))$. Total arc length (or perimeters) are normalized by defining:

$$t = \left(\frac{\ell}{L}\right) 2\pi$$

$$\phi^*(t) = \phi\left(\frac{\ell}{L} 2\pi\right) + t$$

The interval of interest is $t = [0,2\pi]$ for $\ell = [0,L]$.

The function $\phi^*(t)$ is expanded into a Fourier series:

$$\phi^*(t) = \mu_0 + \sum_{k=1}^{\infty} (A_k \cos kt + B_k \sin kt) \tag{39}$$

$$\mu_0 = \frac{1}{2\pi}\int_0^{2\pi} \phi^*(t)dt \tag{40}$$

$$A_k = \frac{1}{\pi}\int_0^{2\pi} \phi^*(t) \cos kt \, dt \tag{41}$$

$$B_k = \frac{1}{\pi}\int_0^{2\pi} \phi^*(t) \sin kt \, dt \tag{42}$$

The morphological information is contained in the coefficients $(A_k,B_k)$. As the ordered pair, $(\ell,\phi(\ell))$ is unique for every point on the profile, multivaluedness no longer appears. The profile is fully represented by $(A_k,B_k)$, conducted in an earlier paper by Beddow and Philip.[2] However, the interpretations are quite different and this is illustrated in the section comparing the R,θ and R,S methods, for example.

## VII. THE (R,S) METHOD[11]

This is a general method of analysis which, in some respects, is analogous to the (R,θ) method. It is, however, fundamentally different. Being a general method, reentrant profiles can be analyzed. The (x,y) coordinates of the profile are transformed to the polar coordinate (r,θ). A Fourier series expansion is used to represent R as a function of s, the normalized arc length. Another Fourier series is used to represent θ′(s) as a function of s:

$$R(s) = a_0 + \sum_{n=1}^{N} (a_n \cos ns + b_n \sin\theta\ ns) \tag{43}$$

and

$$\theta'(s) = c_0 + \sum_{n=1}^{N} (c_n \cos ns + d_n \sin ns) \tag{44}$$

in which

$$s = \left(\frac{\ell}{L}\right) 2\pi,\ \ell_i = \sum_{i=1}^{k} \sqrt{\Delta x_i^2 + \Delta y_i^2},\ L = \sum_i \ell_i$$

and

$$\theta'(s) = \frac{d}{ds}(\theta(s))$$

θ′(s) can be formally integrated to find θ(s), in order to regenerate the particle profile. The function θ′(s) is used directly to calculate the area of the profile.

The function θ′(s) can be used to discriminate between reentrant and nonreentrant profiles. Thus, profiles with reentrants will have negative values of θ′(s) at some s whereas profiles without reentrants will always have positive values of θ′(s) for all s.

## VIII. $R_0$, THE EQUIVALENT RADIUS

This is defined from the area of the particle profile. Under the transformation,

$$x(s) = r(s) \cos \theta(s) \tag{45}$$

and

$$y(s) = r(s) \sin \theta(s) \tag{46}$$

The area of the particle profile is given by

$$\text{Area} = \iint dxdy = \iint \left| J\left(\frac{x,y}{r,s}\right) \right| drds \tag{47}$$

in which

$$\left|J\left(\frac{x,y}{r,s}\right)\right| \equiv \begin{vmatrix} \frac{\partial x}{\partial r} & \frac{\partial x}{\partial s} \\ \frac{\partial y}{\partial r} & \frac{\partial y}{\partial s} \end{vmatrix} \quad (48)$$

is the Jacobian of x and y with respect to r and s. Then

$$\left|J\left(\frac{x,y}{r,s}\right)\right| \equiv \begin{vmatrix} \cos\theta(s) & \cos\theta(s)\frac{dr}{ds} - r\sin\theta(s)\frac{d\theta}{ds} \\ \sin\theta(s) & r\cos\theta(s)\frac{d\theta}{ds} + \sin\theta(s)\frac{dr}{ds} \end{vmatrix}$$
$$= r(s)\frac{d\theta}{ds} \quad (49)$$

The equivalent radius, $R_0$, is then *defined* as

$$R_0^2 \equiv \frac{1}{\pi}\int_0^{2\pi}\int_0^{R(s)} r\left(\frac{d\theta}{ds}\right) drds$$
$$= \frac{1}{2\pi}\int_0^{2\pi} R(s)^2\theta'(s)ds \quad (50)$$

## IX. SHAPE TERMS

The shape terms for the R(S) expansion are

$$L_0 = \frac{A_0}{R_0} \quad (51)$$

$$L_2(n) = \frac{A_n^2}{2R_0^2} \quad (52)$$

$$L_3(m,n) = \frac{3}{4R_0^3}[A_mA_nA_{m+n}\cos(\alpha_{n+m} - \alpha_n - \alpha_m)] \quad (53)$$

The shape terms for the $\theta'(s)$ expansion are

$$\hat{\ell}_0 = \frac{B_0}{R_0} \quad (54)$$

$$\hat{\ell}_2(n) = \frac{B_n^2}{2R_0^2} \quad (55)$$

$$\hat{\ell}_3(m,n) = \frac{3}{4R_0^3}[B_nB_mB_{n+m}\cos(\gamma_{m+n} - \gamma_n - \gamma_m)] \quad (56)$$

in which

$$B_n^2 = C_n^2 + D_n^2$$

and

$$\gamma_n = \tan^{-1}\frac{D_n}{C_n}$$

Note that all of the shape terms are rotationally invariant. A rotation of a profile may be characterized by a translation in s along the boundary of the profile. This translation produces a phase shift in the R(s) and θ′(s) curves and it has been shown earlier (see Section VIII) that the forms in Equations 51 to 53 are independent of any phase shift.

## X. CONTRAST BETWEEN THE (R,Θ) AND (R,S) METHODS

The R(s) curve in Equation 43 is not the same as the R(θ) curve in Equation 1. This is because geometrically θ and s are not the same. In general, the magnitude of the radii may be different when θ is numerically equal to s. For this reason, the shape terms in the (R,S) method will have a different geometrical interpretation than those in the (R,θ) method.

This is best illustrated in the case of the average value of y with respect to x, defined to be

$$\overline{(y_{ave})_x} = \overline{b-a} \int_a^b f(x)dx \tag{57}$$

where

$$y = f(x), \ a \leqslant x \leqslant b$$

It may be possible to express y as a function of x, or, alternatively, as a function of u. The $\overline{(y_{ave})_x}$ and $\overline{(y_{ave})_u}$ need not be equal. This is the case, in physics, for a freely falling body starting from rest.

$$s = \frac{1}{2} gt^2, \qquad v = gt, \qquad v = \sqrt{2gs}$$

where s is the distance, t is time, and v is the velocity. Then

$$\overline{(v_{ave})_t} = \frac{1}{2} v_2, \qquad \overline{(v_{ave})_s} = \frac{2}{3} v_2$$

where $v_2$ is the velocity at $t = t_2$ (note $s_2 = {}^1/_2 g\, t_2^2$). Thus,

$$\overline{(v_{ave})_t} \neq \overline{(v_{ave})_s} \tag{58}$$

By analogy $L_0$ in the (R,θ) method may be interpreted as the size-normalized mean radius with respect to θ, and the $L_0$ term in the (R,S) method may be interpreted as the size-normalized mean radius with respect to s. Similar analogies for the other shape terms, $L_2(n)$ and $L_3(m,n)$ can also be shown.

The $L_2(n)$ term in the (R,S) method, then, are the size-normalized inner products of the m-th and n-th harmonic components of the R(s) function with respect to s, and the $L_3(m,n)$ terms are the size-normalized inner products of the m-th and n-th harmonic components of the R(s) function with respect to s. It follows, then, that the (R,θ) and (R,S) methods are geometrically different. As with the (R,θ) method, the original profile may be represented from the morphic terms.

## XI. CONCLUSIONS

The above discussion on three different methods of morphological analysis makes available a set of techniques which enables the investigator to analyze any required profile.

This analysis deals with the two-dimensional profile of the particle. The feasibility of a shadowing technique to obtain a three-angled projection of the particle has been demonstrated.[15] From time to time, a three-dimensional holographic technique has been suggested for morphological analysis of fine particles.

The current approach to the problem of three-dimensionality appears to be that of sampling particle profiles:

1. In the case of a classification problem which has been demonstrated for coal macerals,[12] it appears that the analysis of the two-dimensional profile provides a large amount of information such that a satisfactory classification can be achieved. This is undoubtedly connected with the fact that at least 200 coordinate points on the profile of each particle are measured. This represents a large amount of information which facilitates the classification process.
2. In relating morphic features of particles to the behavior of individual particles, for example, as in the case of the sedimentation studies, a satisfactory correlation was obtained in using transverse and normal projections for the profile analysis.[13]
3. In the case of larger sets of particles, particularly small particles which have to be mounted, it yet remains to be seen whether sampling of profiles will enable the observer to develop satisfactory correlations. The problem here hinges on the fact that a section in one plane cuts through a set of particles that reveals an appropriate set of profiles. A section in another orientation through the same mass of particles will reveal a different set of profiles, and these two sets of profiles do not possess a one-to-one correspondence.

## LIST OF SYMBOLS

| | |
|---|---|
| $R$ = | Radius vector |
| $\ell$ = | Arc length |
| $s$ = | Normalized arc length |
| $R_0$ = | Equivalent radius |
| $a_0$, $a_n$, $b_n$ = | Fourier coefficients |
| $L_0$, $L_{2(n)}$, $L_{3(m,n)}$ = | Morphic features |
| $\mu_1$, $\mu_2$, $\mu_3$ = | First, second, and third moment of radial distribution |
| $\mu_0$ = | Mean of the radial distribution |
| $A_n$, $A_m$ = | Fourier coefficients |
| $\sigma$ = | Angle |
| $\phi(\ell)$ = | Change of slope of tangent |

## REFERENCES


1. **McCrone, W.,** *Particle Atlas,* Ann Arbor Science Publ., Ann Arbor, Mich., 1979.
2. **Beddow, J. K. and Philip, G. C.,** as cited in **Jsubari, J. and Jimbo, G.,** *Powder Technol.,* 21, 161, 1979.
3. **Ehrlich, R. and Weinberg, B.,** *J. Sediment. Petrol.,* 40, 205, 1970.
4. **Beddow, J. K. and Meloy, T. P.,** *Testing and Characterization of Powders and Fine Particles,* Heyden & Son, London, 1980, 1.
5. **Weichert, R.,** Three-dimensional shape analysis of particles, Int. Powder and Bulk Solids Conf., Rosemont, Ill., May 1980.
6. **Gotoh, K.,** Review of Recent Morphological Analysis Research in Japan, Int. Symp. Powder Technol., Kyoto, Japan, September 1981.

7. **Lenth, R., Beddow, J. K., Chang, C. R., and Vetter, A. F.,** Particle image analyzing system, *Particle Technology: Processes and Fundamentals*, Hemisphere Publishing, New York, 1983.
8. **Beddow, J. K. and Vetter, A. F.,** *J. Powder and Bulk Solids Technol.*, 1, 42, 1977.
9. **Luerkens, D. W., Beddow, J. K., and Vetter, A. F.,** Morphological Fourier descriptors, *Powder Technol.*, 31(2), 209, 1981.
10. **Fong, S.-T., Beddow, J. K., and Vetter, A. F.,** Refined method of particle shape representation, *Powder Technol.*, 22, 1, 1979.
11. **Luerkens, D. W., Beddow, J. K., and Vetter, A. F.,** The (R,S) method, *Powder Technol.*, 31(2), 217, 1981.
12. **Beddow, J. K. et al.,** Principles and Applications of Morphological Analysis, Int. Symp. Powder Technol., Kyoto, Japan, September 1981, 10.
13. **Carmichael, G.,** Effect of particle shape on solids flow, *I & EC Process Design Dev.*, 21, 401, 1982.

Chapter 2

# PARTICLE SHAPE CHARACTERIZATION FROM PACKING DENSITY

**Keishi Gotoh, Mitsuyoshi Kumamoto, and T. P. Meloy**

## TABLE OF CONTENTS

## I. ABSTRACT

The equation of state for rigid particles of arbitrary shape, available in the statistical thermodynamics, is applied to characterize particle shape from the packing density. Although the theory is valid for relatively low concentrations, one can discuss the effect of the particle shape on the packing density. For given compressibilities, one obtains a relation between the shape parameter and the packing density. Oblate and prolate ellipsoids with axial ratio k ($\leq 1$), and a rectangular parallelopiped with sides $1 \times 1 \times k$, are chosen as the reference standard shape whose shape parameter is calculated from k. For a random packing situation, one can therefore estimate from the packing density the size ratio k of the granular particle.

## II. INTRODUCTION

Shape characterization is one of the major research subjects in particulate science and technology. It is important not only for the specification itself, but also for elucidating the relationship of the physical and chemical properties of a single particle and/or the assembly. This research has culminated in recent progress in shape characterization by the Fourier analysis method.[1,2] However, there is little knowledge thus far about the relationship between the shape characteristics and the physical/chemical properties. Of course, one should note that particle-fluid interactions have been investigated most extensively in this respect.[3] Beddow[4] applied morphological analysis to express the rate of crystal growth. He also proposed the shape characterization of powders from the packing density.[5] It can be said that there are already various methods for characterizing the shape of particulates. However, few theoretical relationships have been derived between the shape characteristics and the physical/chemical properties of powders.

The object of this study is to discuss the effect of particle shape on the packing density and to present a method for characterizing particle shape from the packing density. The equation of state for rigid particles of arbitrary shape, available in the statistical thermodynamics, is used for these purposes.

## III. OUTLINE OF THE EQUATION OF STATE

This section outlines the equation of state for rigid particles of arbitrary shape derived from the scaled particle theory in the statistical thermodynamics.[6-8]

The argument begins with the work required to add a single particle to a particle assembly. The work is calculated from the probability of finding the fluid particle in the volume excluding the presence of other particles. It is assumed that the work is expressed by a cubic polynomial in the particle size. Then the thermodynamic relation for pressure yields the equation of state for the rigid particle system. For systems of particles with the same shape, one obtains the following equation:[7-9]

$$Z = \frac{P}{\rho k_B T} = \frac{1}{1 - \phi} + \frac{A\phi}{(1 - \phi)^2} + \frac{(A\phi)^2}{3(1 - \phi)^3}$$

$$A = RS/V \tag{1}$$

where P is the pressure, $\rho$ is the number density, $k_B$ is Boltzmann's constant, T is the absolute temperature, $\phi$ is the bulk-mean particle volume fraction, R is the particle radius averaged over all orientations, and S and V are the surface area and volume of the single particle, respectively; $A = 3$ for spherical particles. Although Equation 1 is not rigorous in the sense that it cannot predict occurrence of the thermodynamic phase change in the high-concentration region, one can use it to estimate the packing density for the present engineering purpose. Moreover, Equation 1 is the only one available for expressing the effect of the particle shape on the packing density.

## IV. SHAPE CHARACTERIZATION FROM PACKING DENSITY

This section discusses application of Equation 1 to the shape characterization of particles. Under a constant temperature, if one compresses the system to the number density $\rho$ with pressure P, the packing density $\phi$ can be determined from Equation 1 for known shape parameter A. The left-hand side, Z, of Equation 1 denotes the work required for the mechanical compression of one particle at temperature T. For given Z values, one can obtain the relation between $\phi(A)$ and $\phi(A = 3)$ as depicted in Figure 1, where $\phi(A)$ denotes the packing density $\phi$ as the function of A; A = 3 for spherical particles. Accordingly, one can obtain the shape parameter A from the measurements of $\phi(A)$ and $\phi(A = 3)$ at the same Z condition.

Although Equation 1 is derived for the molecular particles, let us now apply it to granular materials. It is assumed that, as far as the packing structure of the granular particles is uniformly random, Z value, which is the work per one particle required for making the packing in units of hypothetical thermal energy for keeping the random arrangement, becomes constant. In a common sense, hypothetical thermal vibration is improbable in the packing of the granular particles. Mechanical vibration cannot always correspond to thermal vibration because the particle arrangement is apt to become regular in the field of vibration. Accordingly, as mentioned before, it is assumed that when the packing structure of granular particles is uniformly random, the left-hand side of Equation 1 becomes constant.

The shape parameter, A in Equation 1, can be obtained without difficulty.[10,11]

For oblate and prolate ellipsoids,

$$A = \frac{3}{4\sqrt{1 - e^2}}\left(\sqrt{1 - e^2} + \frac{1}{e}\sin^{-1}e\right)\left(1 + \frac{1 - e^2}{2e}\ln\frac{1 + e}{1 - e}\right)$$

$$e \equiv \sqrt{1 - k^2} \tag{2}$$

where k is the axial ratio ($k \leq 1$).

For a rectangular parallelopiped with sides $1 \times 1 \times k$,

$$A = (2 + k)(1 + 2k)/(2k) \tag{3}$$

For a cylinder with k = diameter/length,

$$A = (2 + \pi k)(2 + k)/(4k) \tag{4}$$

In this paper, the ellipsoids and rectangular parallelopiped are chosen as the reference standard shape. Figures 2 and 3 depict Equations 2 and 3, respectively.

Table 1 lists granular particles tested (see also Figure 4). In preliminary experiments, the particle beds exhibited regular arrangement in the field of vibration. In order to make it random, the beds are made by pouring particles into a cylinder (diameter = 5 cm), the bulk-mean particle volume fraction being listed in Table 1. Each material has no distribution in particle size. All the packed beds are considered as having been made under the same conditions. The particle arrangement can be regarded as uniformly random. As one can read from Table 1, $\phi = 0.617$ for spherical particles, from which one can obtain the shape parameter A from Figure 1, and then the size ratio k from Figure 2 or 3 as listed in Table 1; k values are in agreement with the direct measurement.

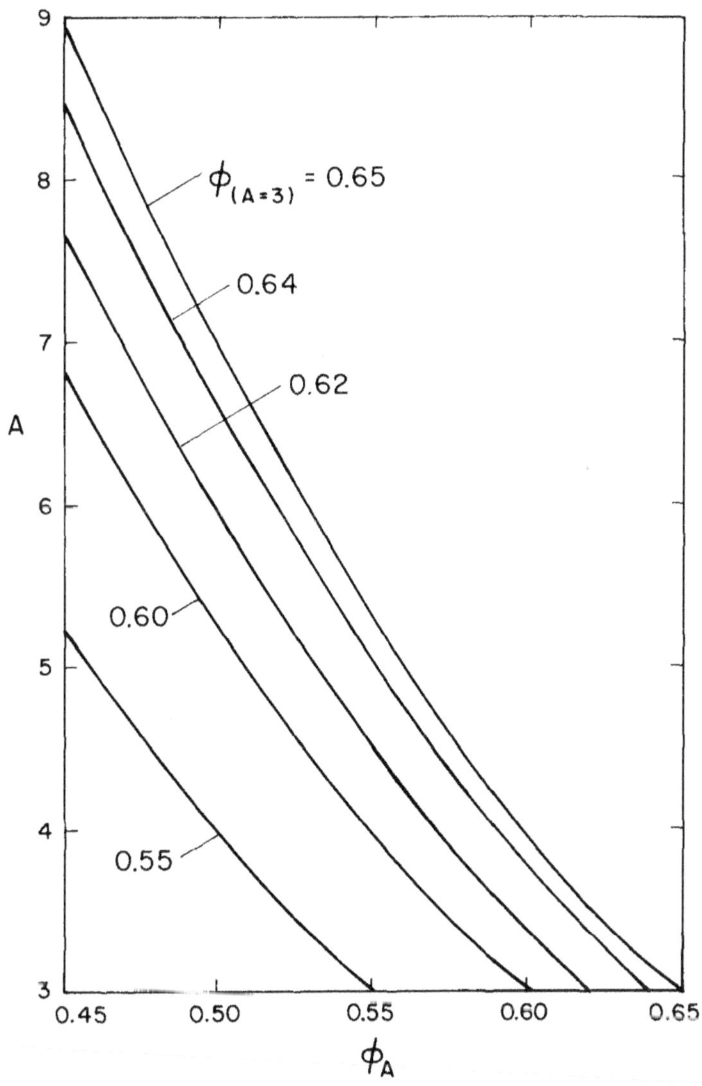


FIGURE 1. Relation between the shape parameter A and the bulk-mean particle volume fraction $\phi(A)$.

As Beddow et al.[5] reviewed, several methods such as the ratio of apparent/tap density and the porosity of a powder mass under a given degree of compaction have been proposed for characterizing the shape of particulates. However, no theoretical basis is available so far for this purpose. The present study may be the only one of the examples which deals with the particle shape explicitly.

## V. CONCLUSION

The effect of particle shape on the packing density is discussed from the equation of state for rigid molecular particles in the statistical thermodynamics. It is assumed that when the particle arrangement in the packed bed is regarded as being uniformly random, the compressibility, Z, becomes constant. Hence, for given compressibilities one can obtain from Equation 1 the relationship between the shape parameter A and the packing density $\phi$, as

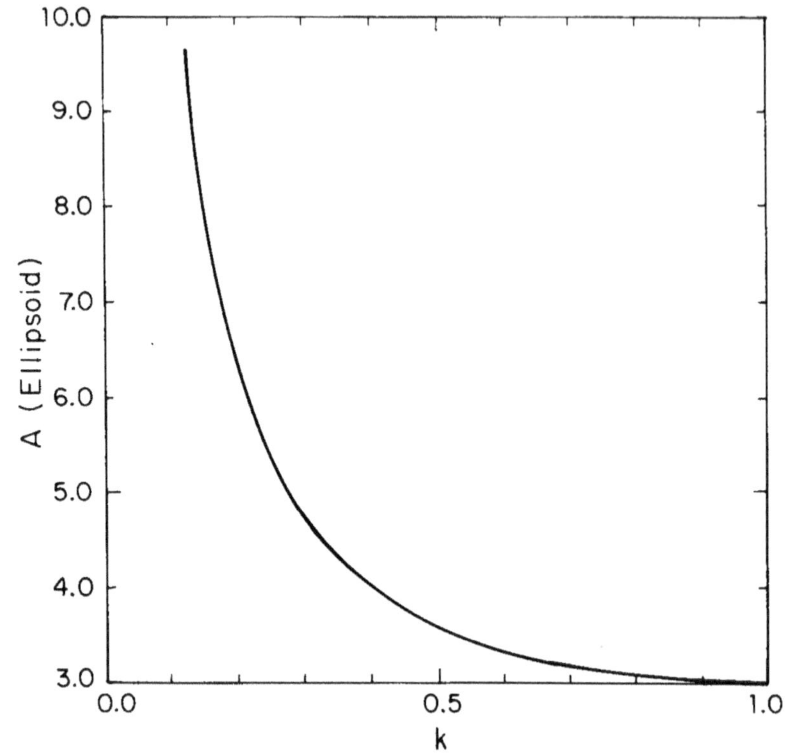


FIGURE 2. Shape parameter A of ellipsoids with axial ratio k(≤1).

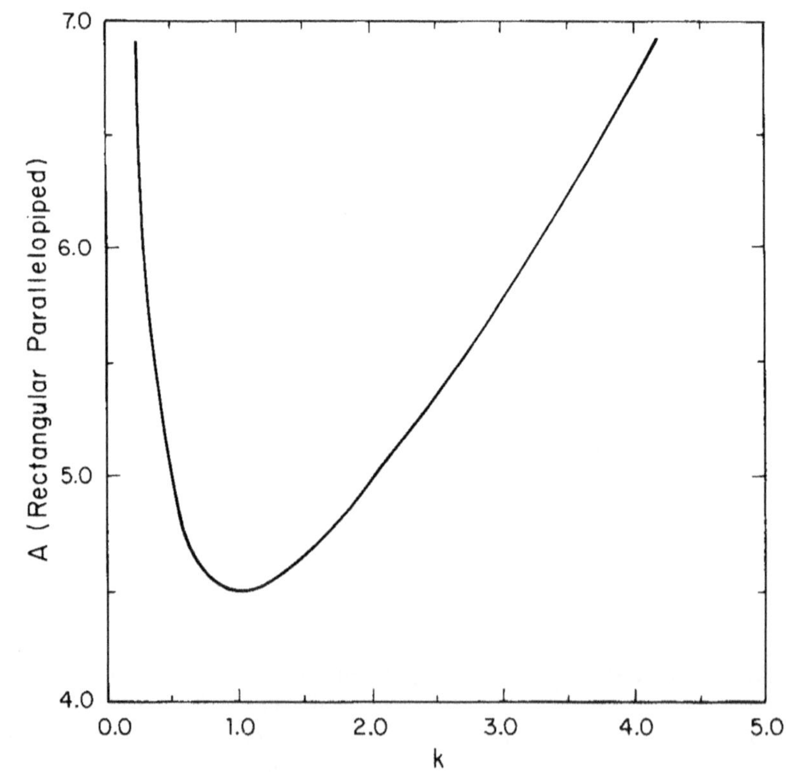


FIGURE 3. Shape parameter A of rectangular parallelopiped with sides $1 \times 1 \times k$.

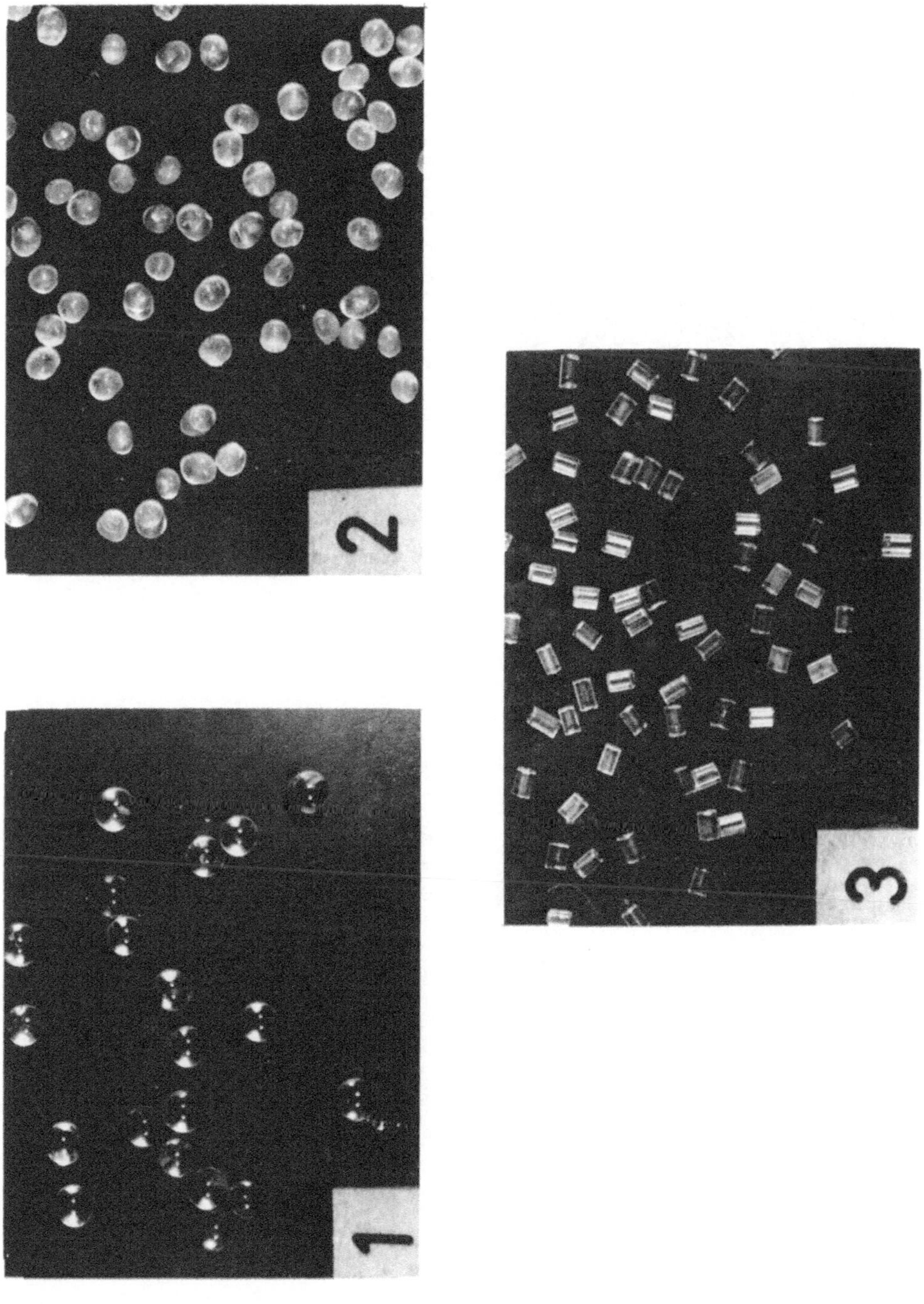
2
1
3

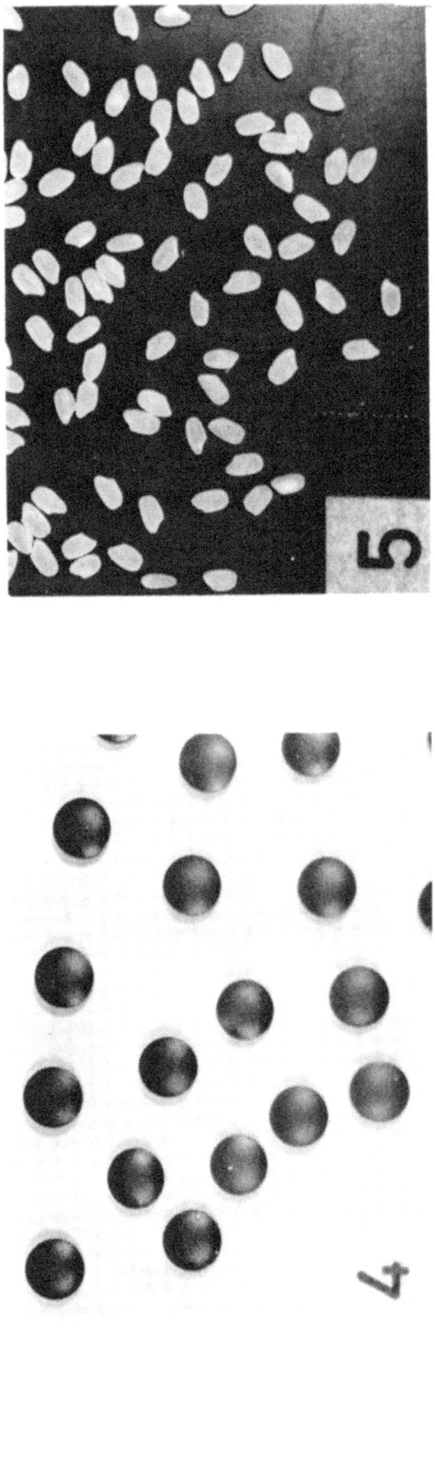


FIGURE 4. Photographs of granular particles. Arabic numbers are written in 8 mm size for comparison, and correspond to those in Table 1.

**Table 1**
**GRANULAR PARTICLES AND THEIR SIZE RATIOS**

| Material | Shape | Packing density (ϕ) | Shape parameter (A) | Size ratio (k) | |
|---|---|---|---|---|---|
| | | | | Calcd. | Measured |
| 1. Glass beads | Spherical | 0.617 | 3.00 | — | 1.00 |
| 2. Polyethylene pellets | Prolate ellipsoidal | 0.611 | 3.20 | 0.66 | 0.63 |
| 3. Acrylic resin chips | Cylindrical | 0.605 | 3.30 | 0.61 | 0.53 |
| 4. Tablets | Oblate ellipsoidal | 0.570 | 4.00 | 0.40 | 0.40 |
| 5. Rice grain | Prolate ellipsoidal | 0.526 | 5.15 | 0.27 | 0.38 |

depicted in Figure 1. The oblate and prolate ellipsoids with the axial ratio k ($\leq 1$), and the rectangular parallelopiped with the sides $1 \times 1 \times k$, are chosen as the reference standard shape whose shape parameter A is obtained from k as shown in Figures 2 and 3. Since the particle arrangement formed by pouring granular particles into the cylindrical container is regarded as being uniformly random, its bulk-mean particle volume fraction can be measured, from which the estimated shape parameter is in good agreement with the direct measurement as listed in Table 1.

## LIST OF SYMBOLS

| | |
|---|---|
| A | Shape parameter in Equation 1 |
| k | Size ratio of a particle |
| $\phi, \phi(A)$ | Packing density, i.e., bulk-mean particle volume fraction |

## REFERENCES


1. **Beddow, J. K.,** Particulate science and technology, *Fine Particle Characterization,* Chemical Publ. Co., New York, 1980, chap. 6.
2. **Beddow, J. K. and Meloy, T. P., Eds.,** *Advanced Particulate Morphology,* CRC Press, Boca Raton, Fla., 1980.
3. **Davies, C. N.,** Particle-fluid interaction, *J. Aerosol Sci.,* 10, 477, 1979.
4. **Beddow, J. K.,** Particle shape characterization, Int. Symp. Powder Technol., Kyoto, Jpn., September 1981.
5. **Beddow, J. K., Vetter, A. F., and Sisson, K.,** Powder metallurgy review, particle shape analysis, *Powder Metall. Int.,* 8, 69, 1976.
6. **Lebowitz, J. L., Helfand, E., and Praestgaard, E.,** Scaled particle theory of fluid mixtures, *J. Chem. Phys.,* 43, 774, 1965.
7. **Gibbons, R. M.,** The scaled particle theory for particles of arbitrary shape, *Mol. Phys.,* 17, 81, 1969.
8. **Gibbons, R. M.,** The scaled particle theory for mixtures of hard convex particles, *Mol. Phys.,* 18, 809, 1970.
9. **Mohanty, K. K. and Davis, H. T.,** An equation of state for polyatomic fluids, *AIChE J.,* 25, 701, 1979.
10. **Ishihara, A. and Hayashida, T.,** Theory of high polymer solutions. I and II, *J. Phys. Soc. Jpn.,* 6, 40, 46, 1951.
11. **Kihara, T.,** Virial coefficients and models of molecules in gases, *Rev. Mod. Phys.,* 25, 831, 1953.

Chapter 3

# A RESEARCH MODE PARTICLE IMAGE ANALYZING SYSTEM

**Russell V. Lenth, Chi-Ren Chang, John Keith Beddow, and Arthur F. Vetter**

## TABLE OF CONTENTS

## I. INTRODUCTION

The shape descriptors, discussed in other chapters of this book, are obtained through the following procedure: start with the (x,y) coordinates, convert the (x,y) data set to other coordinate systems, and then calculate the descriptors. Therefore, obtaining (x,y) coordinates — a procedure which is called digitizing — is basically the starting point of analysis of any particle shape.

A traditional way to digitize particle shape includes the following steps:

1. Take a picture of a particle through a microscope
2. Enlarge the picture to the desired size
3. Trace this enlargement on a digitizer, e.g., a Grafpen® or a Tablet

Using the Grafpen® or similar technique, it takes about 15 min to trace a particle (including tracing the edge and storing the digitized coordinates on a tape). Digitizing in this way is very inefficient. A more efficient piece of equipment or system is desired, especially when tens of thousands of particles are to be digitized. This sparks the idea of developing a high-speed, automated particle digitizing system. The particle image analyzing system (PIAS), to be described in this chapter, was developed in light of the above considerations. Using PIAS, an average rate of over 200 particles per hour is possible.

The (x,y) data obtained from PIAS is then fed into a sequence of computer programs in order to obtain a set of morphological descriptors for size and shape.

It was decided to build up PIAS piece by piece in cooperation with the Image Analyzing Laboratory of the University of Iowa. A TV camera, an A/D convertor, a monitor, and an interface between the convertor and the computer were purchased and connected with the already-available PDP-11/34 computer. Later a Ramtek device for image manipulation was added. Finally, a quality microscope was purchased and installed in order that the particles may be digitized directly, without the need for an intervening photographic process.

In this chapter, discussions will be focused on the design and components of the system and the algorithm used to search the edge of a particle profile.

## II. SYSTEM DESIGN AND COMPONENTS

A picture of the complete PIAS is shown in Figure 1. This system includes a camera head, a camera control unit, a TV monitor, a Ramtek, and a host computer. The system works in such a manner that the scene is first observed by the camera head, from which analog video signals are sent to an A/D convertor where the analog signals are converted to digital signals. These digital signals are then transferred to the host computer which immediately transfers these video data to the Ramtek which displays the image. The Ramtek is a digital color image display device which has its own memory and therefore is readable and writeable. After all the video data have been transferred to the Ramtek (i.e., when the Ramtek shows the complete picture), an edge-tracing routine is evoked to trace the edge, and the edge being traced is displayed on the Ramtek superimposed on the original particle image. A schematic illustration of the PIAS is shown in Figure 2.

The camera provides several scanning modes such that choices of total effective scanning lines of 256, 512, or 1024 per screen are available. In addition, a choice of one or four data points per each horizontal scanning line is available. (In our work, 1 datum per each horizontal scanning line and a total effective number of 256 scanning lines are used.) For a given desired sampling line (X-value), the system will send the gray level of each point (0 — darkest, 255 — brightest), one by one, along the desired sampling line in the direction from the top of the particle image divergence. That is, the first data sent out represent the gray level of the point (X, 255), the second data represent the gray level of the point (X, 254),

FIGURE 1. Particle image analyzing system (PIAS).

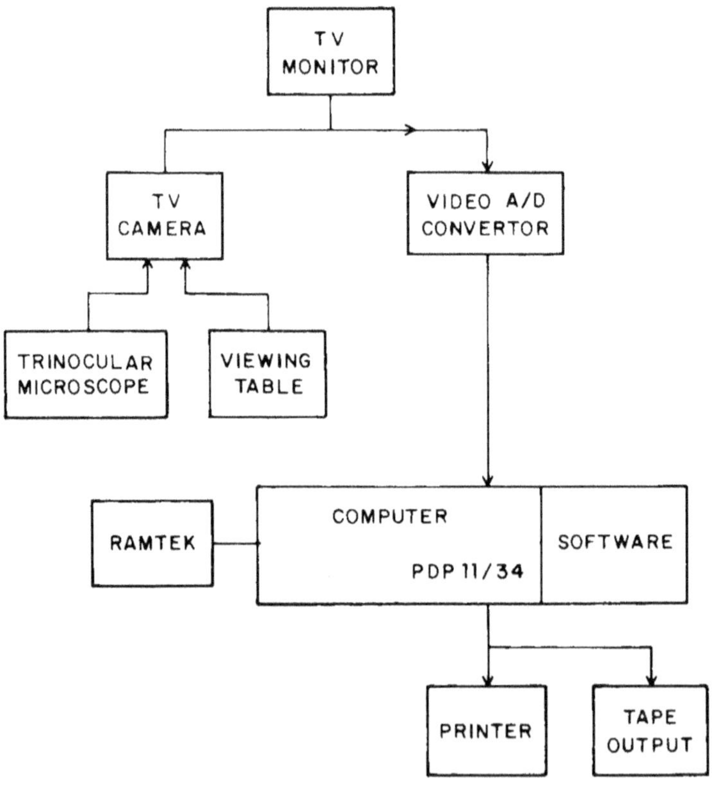


FIGURE 2. Schematic illustration of PIAS.

and so on. Note that in this case the bottom left is chosen to be (0,0), and the total effective number of scanning lines is 256.[1] Figure 3 shows this scanning mode.

## III. EDGE TRACING

Our interest concentrates on the application of this system to analyze particle shapes in two dimensions. In order to detect the edge, three programs named EDGE4, EDGE5, and EDGE5B were written in FORTRAN, based on the concept of "contour following".[2] Of the three programs, EDGE5 and EDGE5B are semiautomatic. Running EDGE5, the user will have to enter the cursor position through the cursor controller of the Ramtek to denote which particle on the Ramtek is to be traced. EDGE5B is exactly the same as EDGE5 except it provides a black background. The EDGE4 program will automatically digitize all the particles shown on the Ramtek, one by one, and kick out the unwanted particles (e.g., particles that only partially appear on the Ramtek, noises, etc.). However, because it takes time to erase the particles, the digitizing rate is considerably reduced.

Due to the limitation of a small core storage, the host computer never stores the video data; instead, all the video data are stored in the Ramtek memory. The edge-tracing routine reads the video data of a program-determined window which is fixed in size but varying in position from the Ramtek. By comparing the video data with the threshold, the edge was traced on the basis of contour-following.

A brief description of contour-following is demonstrated below:

1. Scan the picture unit from the uppermost line down until a *figure cell* is encountered
2. If you are in a figure cell, turn right and take a step in that direction
3. If you are in a background cell, turn left and take a step in that direction
4. Terminate when you meet the starting point

Note that the TV camera scans the field of view horizontally. This scan traverses the field of view vertically. The field is divided into 1024 × 1024 (or 512 × 512, or 216 × 216, depending upon the scanning mode) cells or pixels. This figure cell is one such cell within the image.

By doing this set of operations, the edge of the particle image may be traced counter-clockwise. Figure 4 illustrates the operation. Edge-tracing programs will pick up the edge points which meet the following criterion. A point is said to be on the edge if it is a figure cell, and at least one of its four immediate neighbor points is a ground cell. For example, in the following figure "0" denotes a figure cell, "X" denotes a ground cell, points 1 and 3 are edge points, while point 2 is not.

$$
\begin{array}{ccccccccc}
0 & 0 & 0 & 0 & X & X & X & X & X \\
0 & 0 & 0 & 0 & 0^1 & X & X & X & X \\
0 & 0 & 0 & 0 & 0_2 & 0_3 & X & X & X \\
0 & 0 & 0 & 0 & 0 & 0 & X & X & X
\end{array}
$$

When running the tracing program, the user will be asked to enter a threshold number which will not be changed throughout the running of the program unless the computer is asked to do so. This threshold number is the estimate of the gray level that best differentiates between the image and background. Points with gray levels lower (that is darker) than the threshold number are considered to be within the image; those with gray levels higher than the threshold number are considered as background. Based on this, the edge of the particle image may be traced. Some results and original pictures are shown in Figure 5.

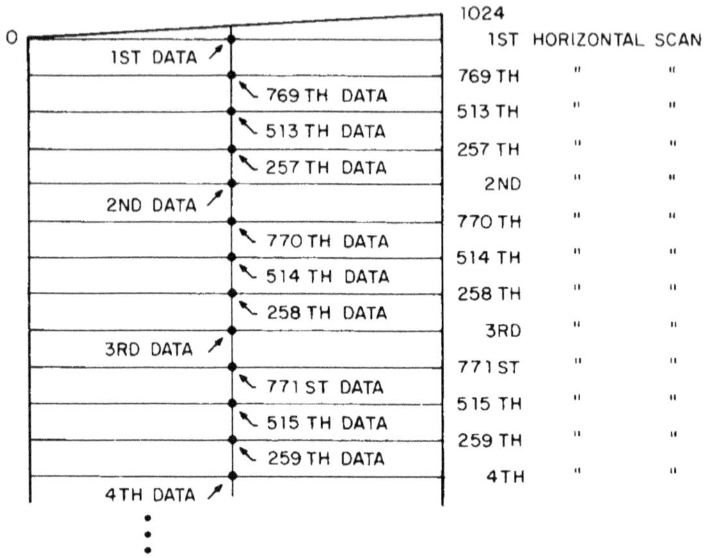


FIGURE 3. Scanning of the TV camera.

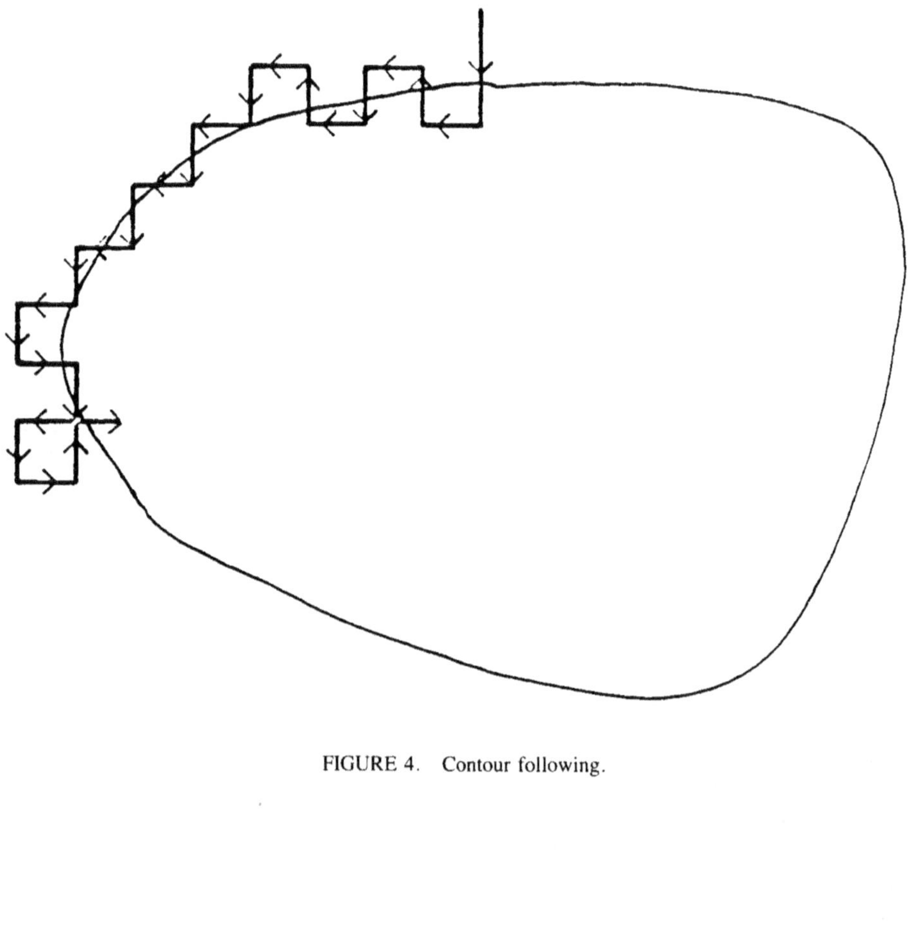

FIGURE 4. Contour following.

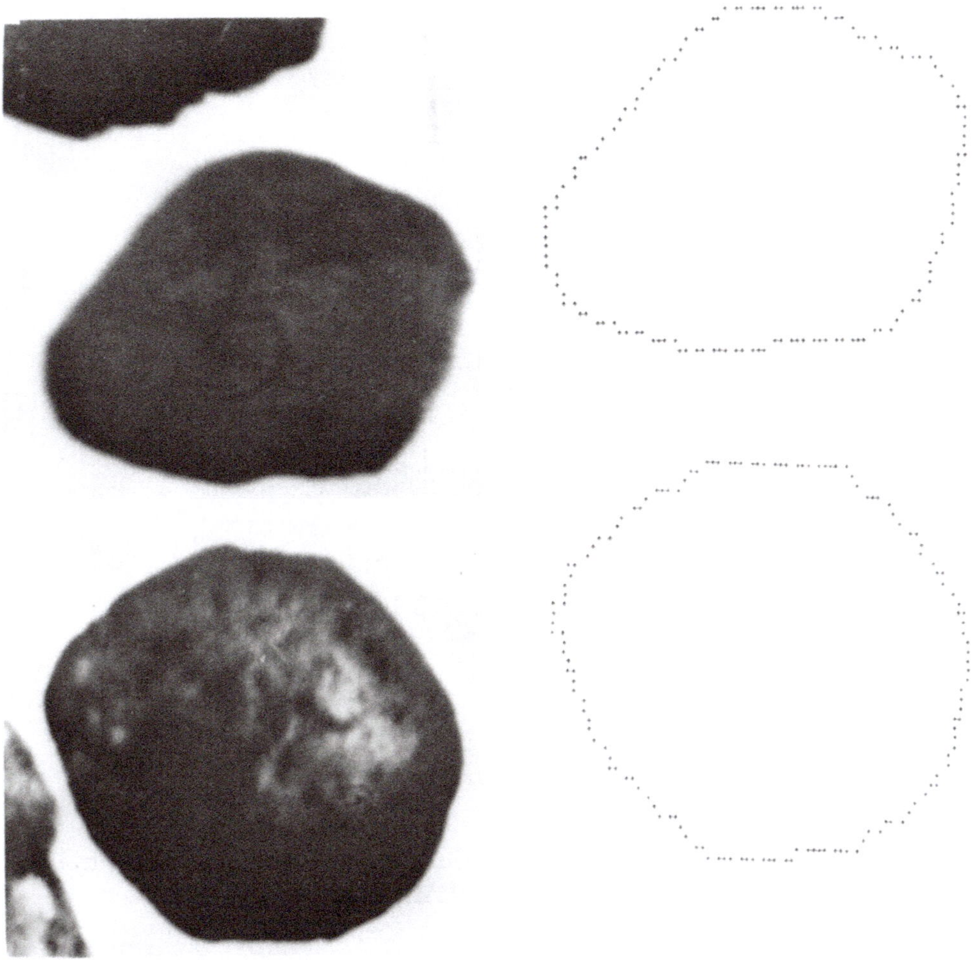

FIGURE 5. Examples of extracted edges.

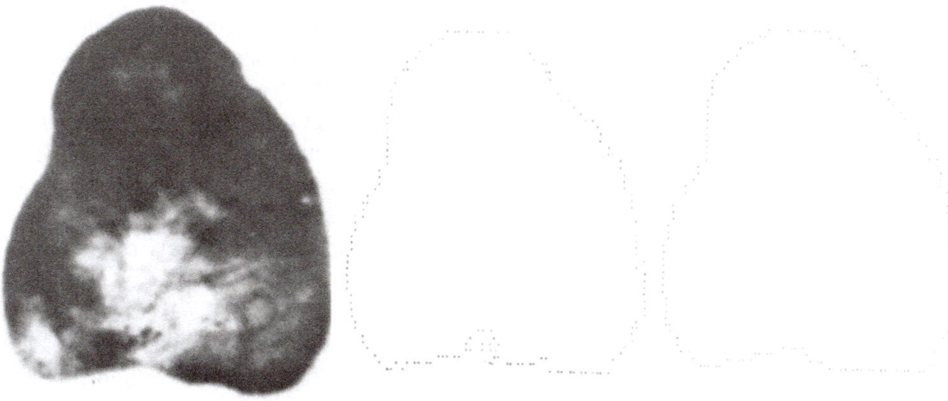

FIGURE 6. Example of misselection of threshold.

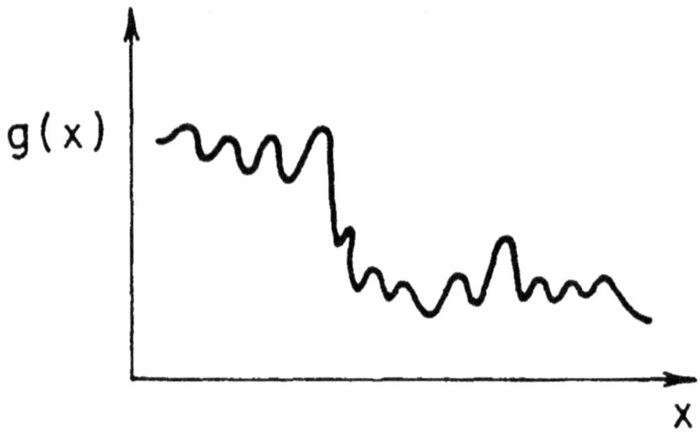


FIGURE 7. Gray levels as a function of position near an actual edge.

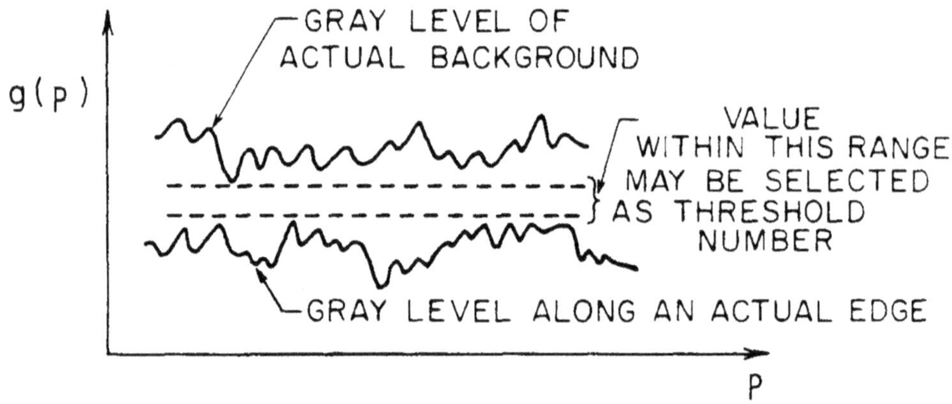


FIGURE 8. Gray level curves along both sides of an actual edge.

## IV. DISCUSSION

Edge detection errors may occur if the pictures are not good enough. For example, a picture like that in Figure 6 may cause some trouble. However, pictures with very distinct (i.e., nonfuzzy) objects and a background as in Figure 5 present little difficulty in the extraction of their edges since these image edges are nearly ideal.

For an idealized edge, there will be a sharp step-change in gray levels at the edge. However, in actual cases, the edge of the object will have some fluctuations and a plot of gray levels near an actual edge against position will be more like the one shown in Figure 7. Furthermore, the gray levels of the points along the profile are not the same throughout. Figure 8 shows the gray levels along an actual profile of an object and the gray levels of background immediately beside the edge as a function of position.

In general, the higher the magnitude of fluctuation of the background and edge gray levels, the narrower the range of usable threshold number. Under this circumstance, the selection of a threshold number becomes very important. Figure 6 illustrates by an example the error that might be caused by misselection of a threshold value. In order to examine whether a threshold value is correctly selected, the program has been written to show the extracted edge immediately superimposed on the particle image being digitized.

A limitation of this technique is that the lowest (or the highest, in case a black background is used) gray level of the background around the desired particle image must be higher (or lower) than the highest (or lowest) gray level of the object edge. In the case that the lowest gray level of the background around the desired particle image is lower than the highest gray level of the object edge, a fixed threshold value will no longer work. A possible solution for this problem is to use a varying threshold value which is determined locally by the average gray level of a small window of, say, 5 × 5 in size around the operating point.

## REFERENCES

1. Instruction manual, *Video A/D Converter M1004 Introduction Manual and PDP-11/DR11C, DRV11 Interface M999-01 Instruction*, Hamamatsu Systems, Inc., Waltham, Mass., 1982.
2. **Duda, R. O. and Hart, P. E.,** *Pattern Classification and Scene Analysis*, John Wiley & Sons, New York, 1973, 290.

Chapter 4

# FAST FOURIER TRANSFORM

**Ken-ichi Kanatani**

## TABLE OF CONTENTS

## I. INTRODUCTION

Recently, much attention has been focused on the use of the Fourier transform for particle shape analysis.[1-9] Much of this attention is due to the recent development of image processing technology and the easy access to computers. However, aside from actual methods of computation, many authors have adopted the classical Fourier series expansion. This is, in a sense, not consistent with its purpose, for the classical Fourier series expansion gives an infinite series whose finite terms reproduce only an approximation of the original shape. Additionally, the evaluation of the Fourier coefficients involves integrations. If these integrations are numerically computed, the results are again approximations. As a whole, two stages of approximation are overlapped, and it is not easy to determine the total amount of numerical errors.

Meanwhile, if we are to retain only the first $N$ terms of the expansion, there exists a formula that gives the best approximation, in a sense, within the $N$ terms. This is called the discrete Fourier transform (DFT). From the viewpoint of the particle shape analysis, the DFT is far superior to the classical Fourier series expansion in the following senses: (1) It can be computed from the values of the original function at prescribed $N$ points; (2) the values at the $N$ points are exactly reproduced; (3) the evaluation of the coefficients does not involve integrations, and hence the coefficients can be computed exactly; (4) the fast Fourier transform (FFT) is available as an efficient algorithm of computer computation;[10,11] (5) mathematical properties such as the convolution theorem and the Wiener-Khinchin theorem for the classical Fourier transform also hold for the DFT; (6) if the number $N$ is sufficiently large, the DFT coefficients coincide with the classical Fourier coefficients.

Owing to these desirable characteristics, the DFT is widely used in the fields of electric and communications engineering.[12] Some people still regard the FFT as a method of computing the classical Fourier transform. It must be emphasized that the FFT is an algorithm designed to compute the DFT. From this viewpoint, we first rewrite the existing particle shape analysis by the Fourier series expansion in terms of the DFT. This is the first aim of this chapter.

Another method of particle shape characterization is to calculate the correlation coefficients. The Wiener-Khinchin theorem, then, states that the DFT of the correlation coefficients is the power spectrum, i.e., the squared absolute values of the DFT of the original shape. To illustrate this relationship is the second aim of this chapter.

Finally, we show an algorithm of the FFT in detail. Of course, the FFT is available in the programming libraries of most computers and there are many books in which a complete computer program of the FFT can be found. However, it is sometimes quite dangerous to use a program that one does not understand well, written by other people. Moreover, one cannot make good use of it unless one is familiar with its characteristics. In many books, the derivation of the FFT is still quite complicated. This is partly because the authors followed the historical paper of Cooley and Tukey.[10] However, if the FFT is viewed from the standpoint of the mathematical theory of computer algorithms, it is given a simple explanation.[11] The actual programming is also easy and those who are familiar with, say, FORTRAN can easily write the program immediately. To show this is the third aim of this chapter.

## II. THE DFT AND THE POWER SPECTRUM

Let us consider, for simplicity, a two-dimensional projection of the particle shape in which the radius, or the modulus, $r$ is a single-valued function of the angle, or the argument, $\theta$ measured from a fixed direction around a fixed origin, say the center of gravity (Figure 1). Consider the following expansion

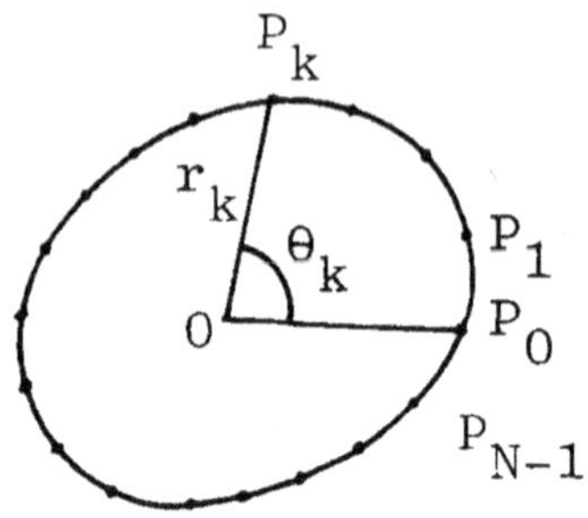


FIGURE 1. Divide the angle $2\pi$ around a fixed origin, say the center of gravity, into $N$ parts, and let $P_0, P_1, \ldots, P_N$ be the corresponding points on the contour, where $\theta_k = 2\pi k/N$.

$$r(\theta) \sim A_0 + \sum_{k=1}^{N-1} (A_k \cos k\theta + B_k \sin k\theta)$$

$$= \mathrm{Re} \sum_{k=0}^{N-1} C_k e^{ik\theta}. \quad (C_k = A_k - iB_k, \; i = \sqrt{-1}) \tag{1}$$

This is the classical Fourier series with only the first $N$ terms retained. In the limit of $N \rightarrow \infty$, it reduces to the classical Fourier series. However, we fix the number $N$ and try to obtain the coefficients $A_k$, $B_k$, and $C_k$ in such a way that Equation 1 best holds some sense.

In order to do so, partition the interval $[0, 2\pi]$ of $\theta$ into $N$ equal subintervals, and let $P_k$ be the corresponding $k$th dividing point on the contour. Let $r_k$ be the radius of the point $P_k$. The angle of $P_k$ is $\theta_k = 2\pi k/N$. Let us determine the coefficients $A_k$, $B_k$, and $C_k$ in such a manner that Equation 1 is exact at $\theta = 0, \theta_1, \theta_2, \ldots, \theta_{N-l}$, i.e., the right hand side passes the points $P_0, P_1, \ldots, P_{N-1}$. Putting $\theta = \theta_k$ $(= 2\pi k/N)$ in Equation 1, we demand

$$r_k = \sum_{l=0}^{N-1} C_l e^{i2\pi kl/N}, \quad k = 0, 1, \ldots, N-1 \tag{2}$$

The matrix form of Equation 2 is

$$\begin{bmatrix} 1 & 1 & 1 & \cdots & 1 \\ 1 & \omega_N & \omega_N^2 & \cdots & \omega_N^{N-1} \\ 1 & \omega_N^2 & \omega_N^4 & \cdots & \omega_N^{2(N-1)} \\ & \cdots & \cdots & \cdots & \cdots \\ 1 & \omega_N^{N-1} & \omega_N^{2(N-1)} & \cdots & \omega_N^{(N-1)(N-1)} \end{bmatrix} \begin{bmatrix} C_0 \\ C_1 \\ C_2 \\ \vdots \\ C_{N-1} \end{bmatrix} = \begin{bmatrix} r_0 \\ r_1 \\ r_2 \\ \vdots \\ r_{N-1} \end{bmatrix} \tag{3}$$

where we have put $\omega_N = e^{i2\pi/N}$ (the $N$th primitive root of 1). The coefficient matrix is $[\omega_N{}^{kl}]$ $(k, l = 0, 1, \ldots, N-1)$. Now, we show that its inverse is $[\omega_N{}^{-kl}/N]$. Multiplying these two matrices, applying the summation formula of a geometric progression, and noting that $\omega_N{}^N = 1$, we find that

$$\sum_{m=0}^{N-1} [\omega_N^{km}][\omega_N^{-ml}/N] = \frac{1}{N} \sum_{m=0}^{N-1} \omega_N^{m(k-l)}$$

$$\begin{cases} = 1 & \text{(if } k - l \text{ is a multiple of } N) \\ = \{1 + \omega_N^{k-l} + (\omega_N^{k-l})^2 + \cdots + (\omega_N^{k-l})^{N-1}\} & \end{cases}$$

$$= \frac{1}{N} \frac{1 - (\omega_N^{k-l})^N}{1 - \omega_N^{k-l}} = 0 \text{ (if } k - l \text{ is not a multiple of } N) \tag{4}$$

Hence, the product is the identity matrix. Multiplying both sides of Equation 2 by the inverse matrix, we can solve it for $C_k$ in the form

$$C_k = \frac{1}{N} \sum_{l=0}^{N-1} r_l e^{-i2\pi kl/N} \tag{5}$$

which is the desired formula of the DFT. For $A_k$ and $B_k$, we have

$$A_k = \frac{1}{N} \sum_{l=0}^{N-1} r_l \cos \frac{2\pi kl}{N} \tag{6}$$

$$B_k = \frac{1}{N} \sum_{l=0}^{N-1} r_l \sin \frac{2\pi kl}{N} \tag{7}$$

In particular, $C_0 = A_0 = \Sigma_{l=0}^{N-1} r_l/N$ is the average radius of the particle. We can choose

$$|C_k|^2 = A_k^2 + B_k^2, \quad k = 0, 1, \ldots, N/2 \tag{8}$$

as a measure of the particle shape characteristics, since they are invariant to rotations and reflections of the particle image. The whole set of $|C_k|^2$ may be termed the power spectrum after their use in electric engineering. If the image of the particle is enlarged $t$ times, the power spectrum is multiplied by $t^2$. Hence, if we put

$$p_k = |C_k|^2/|C_0|^2 = (A_k^2 + B_k^2)/A_0^2 \tag{9}$$

this quantity represents the relative magnitude of the component having frequency $k$ of the particle image normalized so that the average radius becomes unity. A schematic representation is given in Figure 2. The reason why the frequency $k$ is taken only up to $N/2-1$ is discussed in the next section.

## III. THE CONVERGENCE OF THE DFT

The coefficients of the expansion of Equation 1 are given by Equation 5 or Equations 6 and 7. Since they do not involve integrations, these values can be computed exactly in principle, and since Equation 1 contains only a finite number of terms, the values at the $N$ points can be exactly reproduced in principle. In compensation, Equation 1 gives rapidly fluctuating values with no physical meaning at points other than the original $N$ points. On the other hand, if the values of the coefficients in Equation 1 depend sensitively on the number $N$, then we must always specify the number $N$ for each analysis. However, there is almost no need to do so if the number $N$ is sufficiently large. As a matter of fact, these coefficients may be regarded as the classical Fourier coefficients in a sense, if $N$ is large. We now show how large it should be.

Write the classical Fourier series expansion in the form

$$\begin{aligned} r(\theta) &= \frac{a_0}{2} + \sum_{k=1}^{\infty} (a_k \cos k\theta + b_k \sin k\theta) \\ &= \sum_{k=-\infty}^{\infty} c_k e^{ik\theta}, \quad (c_k = (a_k - ib_k)/2,\ c_{-k} = \overline{c_k}) \end{aligned} \tag{10}$$

where the overbar designates the complex conjugate and

$$c_k = \frac{1}{2\pi} \int_0^{2\pi} r(\theta) e^{-ik\theta} d\theta \tag{11}$$

or

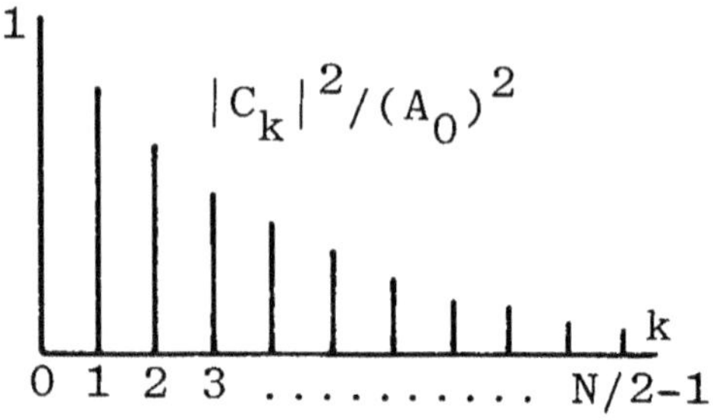


FIGURE 2. An example of the normalized power spectrum of a particle shape, indicating the relative magnitude of each frequency component.

$$a_k = \frac{1}{\pi}\int_0^{2\pi} r(\theta)\cos k\theta d\theta \tag{12}$$

$$b_k = \frac{1}{\pi}\int_0^{2\pi} r(\theta)\sin k\theta d\theta \tag{13}$$

Substitution of Equation 10 in Equation 5 yields

$$\begin{aligned} C_k &= \frac{1}{N}\sum_{l=0}^{N-1} r\left(\frac{2\pi l}{N}\right) e^{-i2\pi kl/N} \\ &= \frac{1}{N}\sum_{l=0}^{N-1}\left(\sum_{m=-\infty}^{\infty} c_m e^{i2\pi lm/N}\right) e^{-i2\pi kl/N} \\ &= \sum_{m=-\infty}^{\infty} c_m \left(\frac{1}{N}\sum_{l=0}^{N-1} e^{i2\pi(m-k)l/N}\right) \end{aligned} \tag{14}$$

Here, the expression inside the last parentheses is 1 when $m-k$ is a multiple of $N$ and is 0 otherwise due to Equation 4. Hence, we obtain

$$C_k = c_k + \sum_{l=\pm 1, \pm 2, \ldots} c_{k+lN} \tag{15}$$

or

$$A_k = \frac{1}{2}\{a_k + a_{N-k} + (a_{N+k} + a_{2N-k}) + \ldots\} \tag{16}$$

$$B_k = \frac{1}{2}\{b_k - b_{N-k} + (b_{N+k} - b_{2N-k}) + \ldots\} \tag{17}$$

From these expressions, we can conclude that if and only if the particle image does not contain components of frequencies equal to or higher than $N/2$, all the terms except the first ones on the right-hand sides vanish for $k = 0, 1, \ldots, N/2-1$, i.e., $A_k = a_k/2$, $B_k = b_k/2$ and $C_k = c_k$ for $k = 0, 1, \ldots, N/2-1$. If the original image contains components of higher frequencies, the DFT coefficients are mixtures of various classical Fourier coefficients. This phenomenon is known as aliasing.

Meanwhile, the component of frequency $N/2$ is a function oscillating with the $N$ points as its nodes, i.e., its half-wavelength is $2\pi/N$. This component is called the Nyquist component or the component of the Nyquist frequency. It is apparent that, as long as only $N$ points are used, the DFT cannot express components oscillating with frequencies above the Nyquist frequency. In actual problems, any particle has small fluctuations to such an extent that if the measurement has infinite precision, the ultimate scale of fluctuation is that of molecules. However, such a detailed description is not necessary for practical purposes. If the image is drawn with pen and ink, it cannot represent fluctuations smaller than the width of the line. In image processing, fluctuations smaller than the resolving capacity of the device must be ignored. As a result, what we analyze is not the real shape of the particle but is a shape which does not contain components of frequencies higher than some value. Hence, if the shape is well represented by taking $N$ points on the contour, this means that the components oscillating more than $N/2$ times in the interval $0 \leq \theta \leq 2\pi$ do not exist, i.e., components of frequencies above the Nyquist frequency do not exist. Thus, if the particle shape is well represented by taking $N$ points, the coefficients $C_k$ of the DFT coincide with the classical Fourier coefficients $c_k$ for $k = 0, 1, \ldots, N/2$. The remaining coefficients $C_{N/2+1}, C_{N/2+2}, \ldots, C_{N-l}$ equal $\overline{C_{N/2-1}}, \overline{C_{N/2-2}}, \ldots, \overline{C_1}$. In other words, the latter half is complex conjugates of the mirror reflection of the former half. This is a consequence of Equation 5 when $r_0, r_1, \ldots, r_N$ is a real sequence.

Thus, we have shown that the classical Fourier coefficients can be computed exactly by the DFT if the number $N$ is sufficiently large. We have also shown how large it should be. These facts are some of the consequences of the Nyquist sampling theorem.[12]

## IV. THE CORRELATION COEFFICIENTS AND THE WIENER-KHINCHIN THEOREM

Another measure of the particle shape are the following correlation coefficients.

$$R_k = \frac{1}{N}\sum_{l=0}^{N-1} r_l r_{l-k}, \quad k = 0, 1, \ldots, N-1 \tag{18}$$

Here, the index of $r$ is to be read in the sense of modulo $N$, i.e., $r_{-1} = r_{N-1}, r_{-2} = r_{N-2}$, etc. Thus, $\Sigma_{l=0}^{N-1}$ represents the cyclic summation, i.e., the summation over one cycle wherever the initial term may begin. Thus, the coefficient $R_k$ is obtained by rotating the particle image around the origin by angle $\theta_k$ and averaging the products of the corresponding radii around the origin (Figure 3). The whole set of the correlation coefficients is a measure of symmetry of the particle shape.

Let us show that all the coefficients do not exceed $R_0$. Consider the relation

$$\begin{aligned} &\frac{1}{N}\sum_{l=0}^{N-1}(r_l - tr_{l-k})^2 \\ &= \frac{1}{N}\sum_{l=0}^{N-1} r_l^2 - \frac{2t}{N}\sum_{l=0}^{N-1} r_l r_{l-k} + \frac{t^2}{N}\sum_{l=0}^{N-1} r_{l-k}^2 \\ &= R_0 - 2tR_k + t^2 R_0 \end{aligned} \tag{19}$$

The first side is nonnegative for arbitrary real $t$. The last side is a quadratic polynomial in $t$, and hence the discriminant must be zero or negative:

$$R_k^2 - R_0^2 \leq 0, \quad \therefore R_k \leq R_0 \tag{20}$$

If we divide the coefficients by $R_0$ and put

$$\rho_k = R_k/R_0 = \sum_{l=0}^{N-1} r_l r_{l-k} \Big/ \sum_{l=0}^{N-1} r_l^2, \quad k = 0, 1, \ldots N-1 \tag{21}$$

then $\rho_0 = 1$ and $\rho_k \leq 1$ for $k \neq 0$. Hence, if $\rho_k = 1$ for $k \neq 0$, the particle image rotated by angle $\theta_k$ precisely coincides with the original image at the $N$ points. If the image is a circle, then $\rho_k = 1$ for all $k$. If the image is an ellipse and if $N$ is even, then $\rho_0$ and $\rho_{N/2}$ equal 1 and the others are less than 1. Thus $\rho_k$ is a quantity measuring the degree of overlap

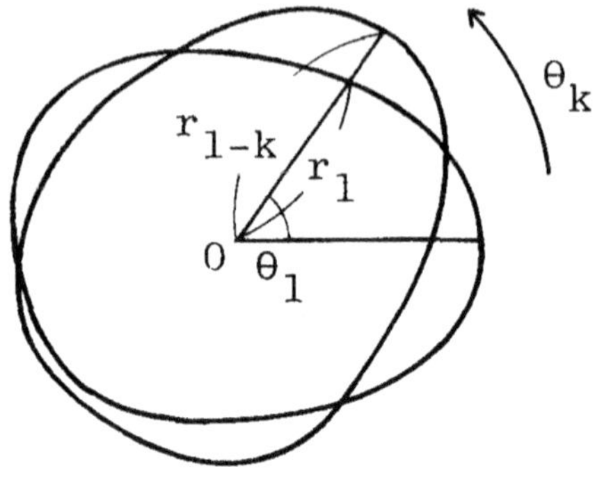


FIGURE 3. The correlation coefficient $R_k$ is obtained by rotating the particle image by angle $\theta_k$ and averaging the products of the corresponding radii.

after rotation by $\theta_k$. If these coefficients are like those in Figure 4, the particle has a 120° symmetry.

Note that $R_0$ is the mean square of the radius. Hence the variance is

$$\frac{1}{N}\sum_{k=0}^{N-1}\left(r_k - \frac{1}{N}\sum_{l=0}^{N-1} r_l\right)^2 = R_0 - A_0^2 \tag{22}$$

Next, consider the DFT of the $R_k$s. Put

$$R_k = \sum_{l=0}^{N-1} P_k e^{i2\pi kl/N} \tag{23}$$

Then, from Equation 5 we have

$$\begin{aligned} P_k &= \frac{1}{N}\sum_{l=0}^{N-1} R_l e^{-i2\pi kl/N} \\ &= \frac{1}{N}\sum_{l=0}^{N-1}\left(\frac{1}{N}\sum_{m=0}^{N-1} r_m r_{m-l}\right)e^{-i2\pi kl/N} \\ &= \frac{1}{N^2}\sum_{m=0}^{N-1}\left(\sum_{l=0}^{N-1} r_{m-l}e^{-i2\pi kl/N}\right)r_m \\ &= \frac{1}{N^2}\sum_{m=0}^{N-1}\left(\sum_{l=0}^{N-1} r_l e^{-i2\pi k(m-l)/N}\right)r_m \\ &= \frac{1}{N}\sum_{l=0}^{N-1} r_l e^{i2\pi kl/N}\,\frac{1}{N}\sum_{m=0}^{N-1} r_m e^{-i2\pi km/N} \\ &= |C_k|^2 \end{aligned} \tag{24}$$

In summary,

$$R_k = \sum_{l=0}^{N-1} |C_l|^2 e^{i2\pi kl/N} \tag{25}$$

$$|C_k|^2 = \frac{1}{N}\sum_{l=0}^{N-1} R_l e^{-i2\pi kl/N} \tag{26}$$

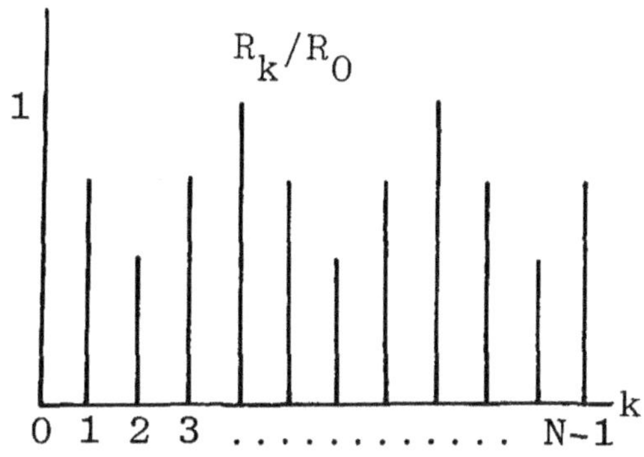


FIGURE 4. An example of the normalized correlation coefficients of a particle shape, indicating the symmetry of the shape. This example shows a 120° symmetry.

Thus, the DFT of the correlation coefficients is the power spectrum. This implies that in order to compute the correlation coefficients, one can first compute the power spectrum and then take the DFT to obtain them. This is because the FFT can, as we shall see in the following sections, compute the DFTs much faster than the direct computation of Equation 18. Equations 25 and 26 are special forms of the convolution theorem of the DFT, and the interpretation is the same as that of the Wiener-Khinchin theorem in the random signal theory.[12] Hence, it is appropriate to also term it the Wiener-Khinchin theorem. In particular, if $k = 0$ in Equation 25, we have

$$\frac{1}{N}\sum_{l=0}^{N-1} r_l^2 = \sum_{l=0}^{N-1} |C_l|^2 \tag{27}$$

which is the discrete version of the Parseval identity.

## V. WHAT IS FFT?

What we want to compute are Equations 2, 5, 25, and 26. Hence, the problem is reduced to computing

$$b_k = \sum_{l=0}^{N-1} a_l \omega_N^{kl}, \quad k = 0, 1, \ldots, N-1 \tag{28}$$

for given sequence $a_0, a_1, \ldots, a_{N-1}$ (complex in general), where

$$\omega_N = e^{\pm i2\pi/N} \tag{29}$$

is a $N$th primitive root of 1. We can choose either of the signs of the exponent according to the type of Equations 2 and 25 or the type of Equations 5 and 26. If the sign of the exponent is +, the roots of 1, $\omega_N, \omega_N^2, \ldots, \omega_N^{N-1}$ are positioned counterclockwise on the unit circle of the complex plane (Figure 5). If the sign is −, the same holds except they go about clockwise.

Now, consider a computation of Equation 28 by a computer. If the complex constants $\omega_N, \omega_N^2, \ldots, \omega_N^{N-1}$ are already stored in memory registers, the evaluation of Equation 28 requires $N$ complex multiplications and $N - 1$ complex additions/subtractions. Then, the computation of $b_0, b_1, \ldots, b_{N-1}$ requires $N^2$ multiplications and $N(N-1)$ additions/subtractions. If we note that $\omega_N^0 = 1$ and avoid multiplications by unity, we still need $(N-1)^2$ multiplications. If $N = 1000$, for example, the necessary multiplications are 810,000 times and the additions/subtractions 900,000 times. In total, about 2 million operations are

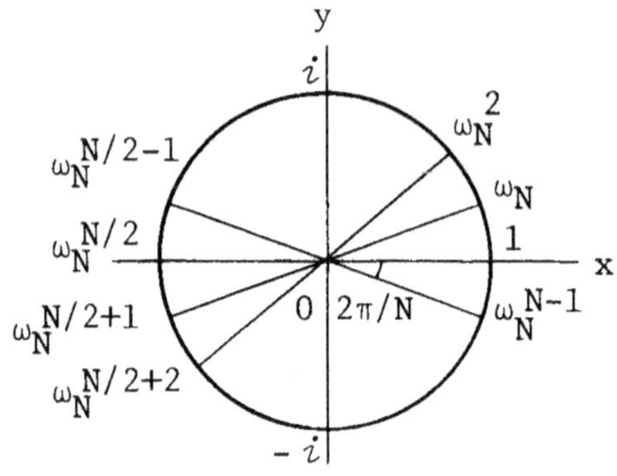


FIGURE 5. The positions of the $N$th roots of 1 in the complex plane, where $\omega_N = e^{i2\pi/N}$. If $\omega_N = e^{-i2\pi/N}$, they go around clockwise.

necessary. This amount is too much for practical purposes. However, if we use the FFT, we can compute them more than 100 times faster.

The first step in understanding the FFT is to consider the polynomial

$$f(x) = a_0 + a_1x + a_2x^2 + \ldots + a_{N-1}x^{N-1} \tag{30}$$

for a given sequence $a_0, a_1, \ldots a_{N-1}$. Comparing this with Equation 28, we can observe that

$$b_k = f(\omega_N^k), \quad k = 0, 1, \ldots, N - 1 \tag{31}$$

Thus, the computation of the FFT is reduced to evaluation of a given $(N - 1)$th degree polynomial at the $N$th roots of unity $x = 1, \omega_N, \omega_N^2, \ldots, \omega_N^{N-1}$ on the unit circle of the complex plane. An efficient method of this evaluation is the FFT about to be explained in detail.

## VI. THE FFT ALGORITHM

In general, $N$ may be an arbitrary integer, but taking $N$ to be a power of 2 is most convenient. In the following, we put $N = 2^K$. The essence of the FFT is the concept of mathematical induction, or in the terminology of programming languages, the recursive procedure. If $N = 1$, then $f(x) = a_0$ and no computation is necessary. All we have to do is construct a method of the $N$th order FFT, i.e., a way of evaluating a given $(N - 1)$th degree polynomial at the $N$th roots of 1 $x = 1, \omega_N, \omega_N^2, \ldots, \omega_N^{N-1}$ on the assumption that we already have the $N/2$th order FFT.

Now, let us rewrite Equation 30 as follows:

$$\begin{aligned} f(x) &= a_0 + a_2x^2 + a_4x^4 + \ldots + a_{N-2}x^{N-2} \\ &\quad + x(a_1 + a_3x^2 + a_5x^4 + \ldots + a_{N-1}x^{N-2}) \\ &= p(x^2) + xq(x^2) \end{aligned} \tag{32}$$

where we have put

$$p(x) = a_0 + a_2x + a_4x^2 + \dots + a_{N-2}x^{N/2-1} \quad (33)$$

$$q(x) = a_1 + a_3x + a_5x^2 + \dots + a_{N-1}x^{N/2-1} \quad (34)$$

and they are both $(N/2 - 1)$th degree polynomials whose coefficients are those of $f(x)$ with even and odd indexes, respectively.

Now, Equation 32 must be evaluated at $x = 1, \omega_N, \omega_N^2, \dots, \omega_N^{N/2-1}, \omega_N^{N/2} (= -1), \omega_N^{N/2+1} (= -\omega_N), \dots, \omega_N^{N-1} (= -\omega_N^{N/2-1})$, and the latter half is negatives of the former half (Figure 5). Thus $p(x^2)$ and $q(x^2)$ must be evaluated at $x^2 = 1, \omega_N^2, \omega_N^4, \dots, \omega_N^{N-2}, \omega_N^N (= 1), \omega_N^{N+2} (= \omega_N^2), \dots, \omega_N^{2N-2} (= \omega_N^{N-2})$, and the latter half repeats the former half. Hence, the values of $p(x^2)$ for the latter half are the same as those for the former half, and the values of $xq(x^2)$ for the latter half are the negative of those for the former half. Therefore, we only have to compute $p(x)$ and $q(x)$ at $x = 1, \omega_N^2, \omega_N^4, \dots, \omega_N^{N-2}$. But $\omega_N^2$ is the $N/2$th primitive root of 1, i.e., $\omega_N^2 = e^{\pm i2\pi/(N/2)} = \omega_{N/2}$. Hence, the problem is reduced to computing "$(N/2 - 1)$th degree polynomials $p(x)$ and $q(x)$ at the $N/2$th roots of 1 $x = 1, \omega_{N/2}, \omega_{N/2}^{2}, \dots, \omega_{N/2}^{N/2-1}$ on the unit circle of the complex plane." This can be done by the $N/2$th order FFT according to the assumption of the induction, in other words, by the recursive procedure. Then multiply $q(x^2)$ by $x (= 1, \omega_N, \omega_N^2, \dots, \omega_N^{N/2-1})$, which is referred to as the rotation factor or the twiddle factor, add it to $p(x^2)$ for $x = 1, \omega_N, \omega_N^2, \dots, \omega_N^{N/2-1}$, and subtract it from $p(x^2)$ for $x = \omega_N^{N/2}, \omega_N^{N/2+1}, \dots, \omega_N^{N-1}$. This completes the $N$th order FFT, where $N$ is an arbitrary power of 2.

Following this recursive construction of the algorithm, we can see that the $N/2$th order FFT is further decomposed into $N/4$th order FFTs, which are again decomposed into $N/8$th order FFTs and so on, until finally we reach first order FFTs.

## VII. THE EFFICIENCY OF THE FFT

In order to see how efficient the FFT is, let $M(N)$ and $A(N)$ be the number of complex multiplications and the number of complex additions/subtractions necessary for computation of the $N$th order FFT, respectively. Consider $M(N)$ first. The evaluation of $p(x^2)$ and $q(x^2)$ using the $N/2$th order FFT requires multiplications $M(N/2)$ times for each (due to the inductive hypothesis). Multiplication of $q(x^2)$ by the rotation factor $x$ is necessary only for $x = 1, \omega_N, \dots, \omega_N^{N/2-1}$ on the upper half of the unit circle of the complex plane, as we saw before. However, no multiplication is necessary for $x = 1$ and for $x = i$, because multiplication by $i$ is achieved by exchanging the real and the imaginary parts and by changing the sign of the real part. Thus, the number of complex multiplications actually necessary is $N/2 - 2$. Similarly, additions/subtractions are necessary $A(N/2)$ times both for $p(x^2)$ and for $q(x^2)$, and $N$ times for $p(x^2) + xq(x^2)$. In conclusion,

$$M(N) = 2M(N/2) + N/2 - 2 \quad (35)$$

$$A(N) = 2A(N/2) + N \quad (36)$$

For $N = 2$ (i.e., $f(x) = a_0 + a_1x$), only $x = \pm 1$ are substituted. Hence,

$$M(2) = 0$$

$$A(2) = 2 \quad (37)$$

The solutions of the recursive Equations 35 and 36, respectively, are

$$M(N) = \frac{N}{2}\log N - \frac{3}{2}N + 2 \tag{38}$$

$$A(N) = N\log N \tag{39}$$

where the base of the logarithm is 2. Table 1 lists these numbers together with the numbers of operations for the direct computation of the DFT. In most of the existing computers, a multiplication takes longer than an addition/subtraction. Table 1 shows that the number of the time-consuming multiplications for the FFT is drastically less than for the direct computation, especially when $N$ is large. Note that these figures are the numbers of operations necessary to compute the whole set of the DFT coefficients. If only a part of them is necessary, as is usual for the power spectrum, one may think that the direct computation is more efficient. However, this is not true for most cases. The figures in the square brackets are the maximum numbers of coefficients for which the direct computation is more efficient. As for the multiplication, for example, if $N = 128$, the computation of all the 128 coefficients by the FFT is faster than the direct computation of more than two coefficients. If $N = 1024$, the computation of all the 1024 coefficients by the FFT is faster than the direct computation of more than three coefficients. This demonstrates how fast the FFT actually is.

## VIII. THE FFT PROGRAMMING

The construction of a FFT program is most easily shown by the flow diagram representation (Figure 6). What we want to do is construct a box (or a subroutine) named FFT($N$) (Figure 7). The flow diagram representation of the recursive construction of the FFT described in Section VI is shown in Figure 8, where FFT($N$/2) means exactly the same procedure except that $N$ is halved. Since recursive calls are possible in such computer languages as ALGOL and PASCAL, it is possible to write this diagram directly in such languages. However, the recursive call is usually inefficient for most computers and they are forbidden in many computer languages such as FORTRAN, i.e., a subroutine cannot call out itself.

Writing inside the boxes until we reach FFT(1)s, we obtain Figure 9, $N = 8$, for example. This diagram is rearranged into the form of Figure 10, where B($n$) is the operation shown in Figure 11. From this, we can deduce that the $N$th order FFT goes as follows: (0) rearrangement of the coefficients, which is designated by R($N$); (1) pairwise operation of B(2); (3) operation of B(4) for each fourtuple; . . . ; ($k$) operation of B($2^k$) for each $k$tuple; . . . ; ($K$) operation of B($N$) for the $N$tuple once for all. Thus, B($n$) is operated log$N$ times and hence the total number of arithmetic operations is of order $N$log$N$.

A close examination of the operation B($n$) of Figure 11 reveals an important fact: the pair of $X_k$ and $X_{n/2+k}$ computes the pair of $Y_k$ and $Y_{n/2+k}$. This is explicitly shown in Figure 12. This diagram is referred to as the butterfly for its resemblance of the shape. After all, the operation B($n$) consists of $n/2$ parallel butterflies with rotation factors 1, $\omega_n$, $\omega_n^2$, . . . , $\omega_n^{n/2-1}$, respectively.

Meanwhile, an algorithm of the rearrangement of the coefficients, R($N$), is shown in Figure 13 for $N = 8$, where P($n$) is the operation described in Figure 14, i.e., the operation to separate the even- and odd-indexed inputs. The operation B(2) is actually not necessary.

## IX. THE BIT REVERSAL

We have completed a general flow diagram of the FFT. It can easily be written in, say, FORTRAN. However, it is sometimes helpful to know the final positions of the coefficients after the operation R($N$). To see them, write the indexes in binary numbers. The results are shown in Figure 15 for $N = 8$. We can observe that the binary representation of the index

**Table 1**
**THE NUMBER (*N*) OF ADDITIONS/ SUBTRACTIONS NECESSARY TO COMPUTE THE *N*th ORDER FFT AND THE NUMBER *M(N)* FOR MULTIPLICATIONS**

| *N* | A(*N*) | | M(*N*) | |
|---|---|---|---|---|
| 2 | 2<br>(2) | [2] | 0<br>(1) | [0] |
| 4 | 8<br>(12) | [2] | 0<br>(9) | [0] |
| 8 | 24<br>(56) | [3] | 2<br>(49) | [0] |
| 16 | 64<br>(240) | [4] | 10<br>(225) | [0] |
| 32 | 160<br>(992) | [5] | 34<br>(961) | [1] |
| 64 | 384<br>(4,032) | [6] | 98<br>(3,968) | [1] |
| 128 | 896<br>(16,256) | [7] | 258<br>(16,129) | [2] |
| 256 | 2,048<br>(65,280) | [8] | 642<br>(65,025) | [2] |
| 512 | 4,608<br>(261,632) | [9] | 1,538<br>(261,121) | [3] |
| 1,024 | 10,240<br>(1,047,552) | [10] | 3,586<br>(1,046,529) | [3] |
| 2,048 | 22,528<br>(4,192,256) | [11] | 8,194<br>(4,190,209) | [4] |
| 4,096 | 49,152<br>(16,773,120) | [12] | 18,434<br>(16,769,025) | [4] |
| 8,192 | 106,496<br>(67,100,672) | [13] | 40,962<br>(67,092,481) | [5] |
| 16,384 | 229,376<br>(268,419,072) | [14] | 90,114<br>(268,402,689) | [5] |

*Note:* The figure in parentheses is the number of operations for the direct computation ($N[N - 1]$ for additions/subtractions and $[N - 1]^2$ for multiplications). The FFT is faster than the direct computation if the number of desired outputs is larger than the one in the brackets. This shows how effective the FFT really is.

of each output is the reverse of the digits of that of the corresponding input. Thus, the operation R($n$) does nothing but convert the input indexes to binary numbers and move each one to the position of its reversed bits. The output is said to be in the order of bit reversal. How it happens is shown in Figure 16. By definition, even-indexed inputs are carried upward and odd-indexed downward. In the binary representation, this means that those inputs whose last digits of their indexes are 0 are carried upward, otherwise they move downward. This process is carried out recursively. In the second stage, those inputs whose second to the last digits of their indexes are 0 are carried upward and the others downward, and so on. After all, we are rearranging the inputs, reading the binary digits of the indexes from right to left, and hence, the order of bit reversal results.

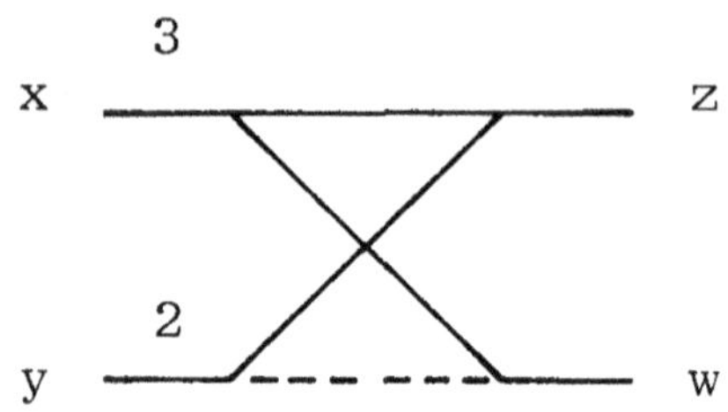


FIGURE 6. A flow diagram representation of computing $z = 3x + 2y$ and $w = 3x - 2y$. Follow the branches from left to right and pass the variables along them, multiplying them by the values attached to them. At a branching point, deliver the same value to both. Branches with no attached values designate passing the variables intact (i.e., multiplication by 1) and dashed branches designate changing the sign (i.e., multiplication by −1). Sum is taken at a meeting point.

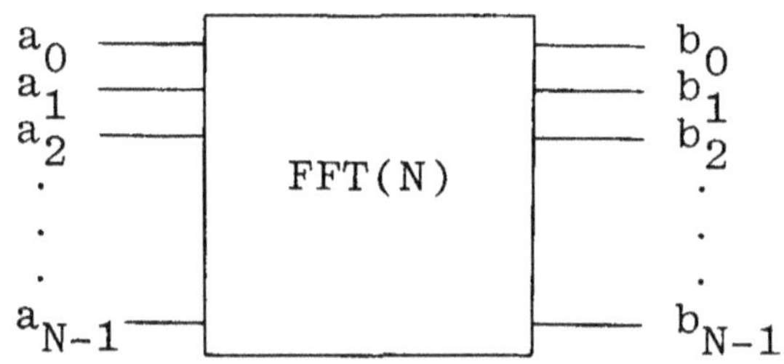


FIGURE 7. The flow diagram of the *N*th order FFT to be constructed. This box may be regarded as a subroutine. The inside of this box is shown in Figure 8.

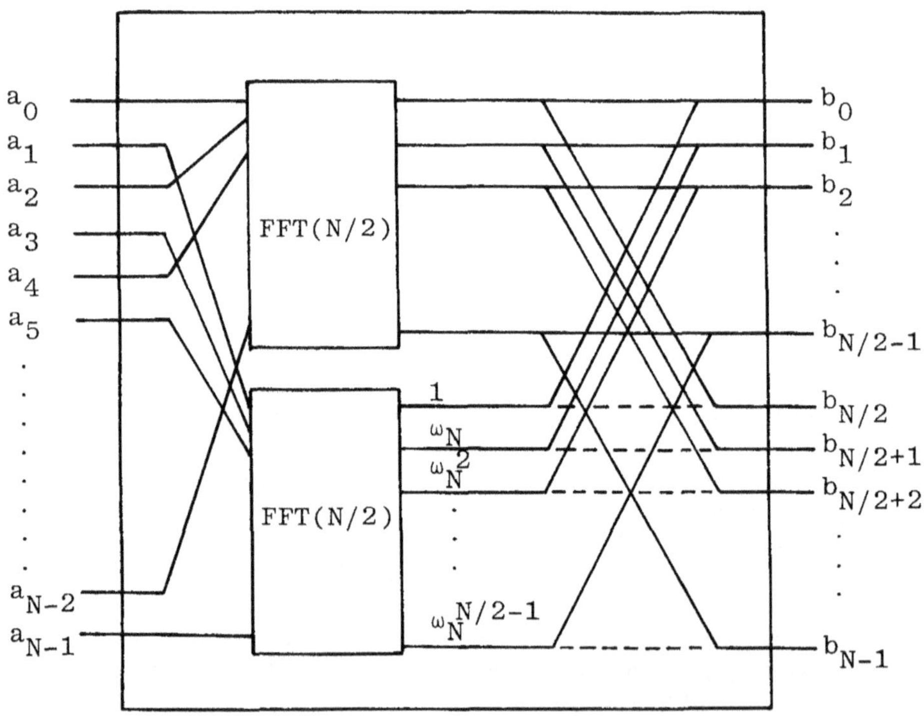


FIGURE 8. The recursive flow diagram of the inside of the FFT(*N*) of Figure 7. It contains two FFT (*N*/2)s. Thus, it is finally reduced to FFT(1)s.

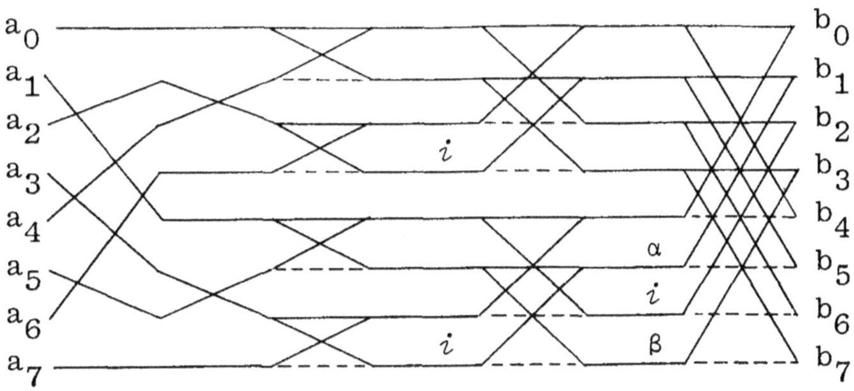


FIGURE 9. An explicit flow diagram of the FFT for $N = 8$, where $\alpha = (1 \pm i)/\sqrt{2}$ and $\beta = (-1 \pm i)/\sqrt{2}$. Only two complex multiplications are involved.

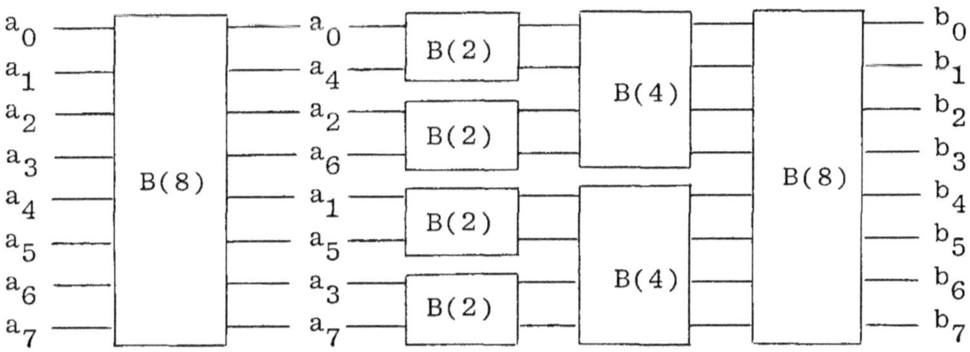


FIGURE 10. The scheme of FFT(8): (0) rearrangement R(8) of the input; (1) pairwise operation of B(2); (2) operation of B(4) for each fourtuple; and (3) operation of B(8) for all the variables. Here, B($n$) is the operation described in Figure 11.

The bit reversal can be computed directly. For $N = 2^K$, all the indexes are expressed in $K$ bit binary numbers. Let us compute the bit reversal $J$ of $I$. Let $l_K l_{K-1} \ldots l_2 l_1$ be the binary representation of $I$, where each $l_i$ is 0 or 1:

$$I = l_k \cdot 2^{K-1} + l_{K-l} \cdot 2^{K-2} + \ldots + l_2 \cdot 2 + l_1 \tag{40}$$

Hence, $I$ divided by 2 makes quotient $l_K l_{K-1} \ldots l_3$ and residue $l_1$. The quotient $l_{KIK-1} \ldots l_2$ divided by 2 makes quotient $l_K l_{K-1} \ldots l_3$ and residue $l_2$, and so on. Thus, successive division of $I$ by 2 gives $l_1, l_2, \ldots, l_K$ as residues in this order. The bit reverse of $I$ is

$$\begin{aligned} J &= l_1 \cdot 2^{K-1} + l_2 \cdot 2^{K-2} + \ldots + l_{K-1} \cdot 2 + l_K \\ &= (\ldots(((l_1 \cdot 2 + l_2) \cdot 2 + l_3) \cdot 2 + l_4) \ldots) \cdot 2 + l_K \end{aligned} \tag{41}$$

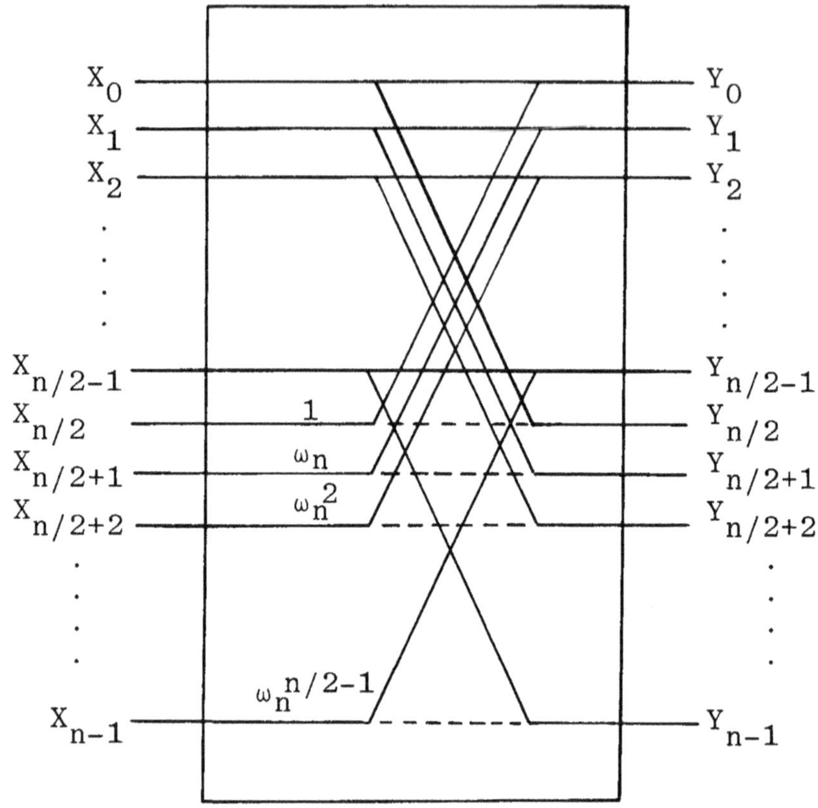


FIGURE 11. The inside of the box of B($n$), where $\omega_n = e^{-i2\pi/N}$. It consists of $n/2$ parallel butterflies depicted in Figure 12.

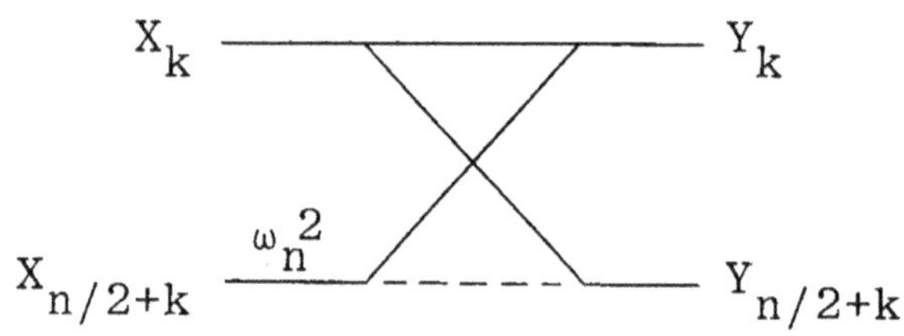


FIGURE 12. The butterfly. It is the most basic operation of the FFT, where $\omega_n^k = e^{-i2\pi k/n}$ is called the rotation factor or the twiddle factor.

In FORTRAN, this may be written as follows:

```
J=0
DO 10 M=1,K
L=MOD(I,2)
I=I/2
J=J*2+L
10 CONTINUE
```

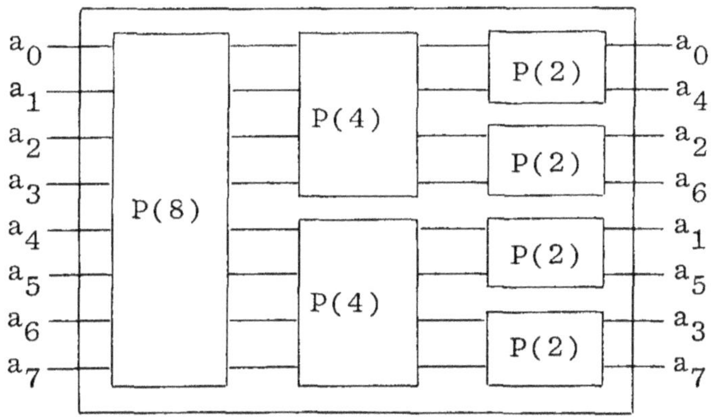


FIGURE 13. The inside of the box of R(8). This rearranges the inputs, where P($n$) is the operation described in Figure 14. The operation P(2) is not necessary in practice.

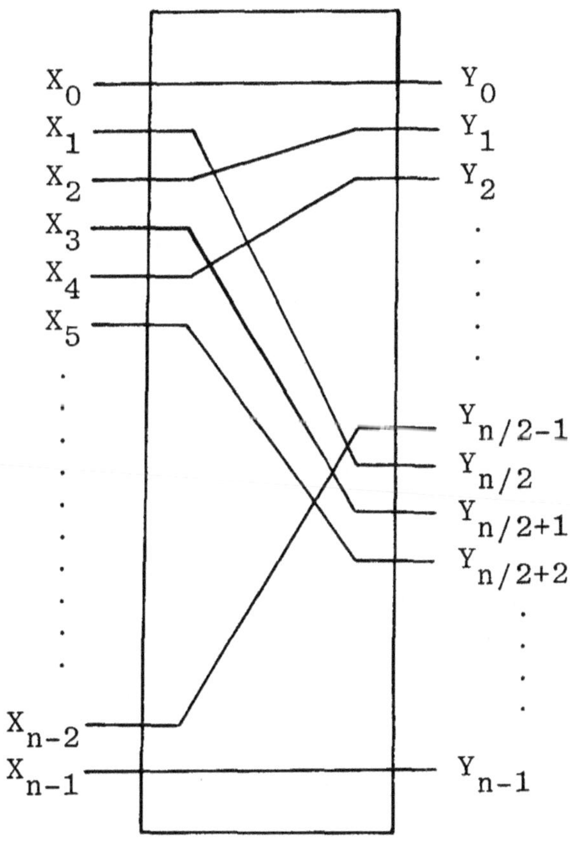


FIGURE 14. The inside of the box of P($n$). This separates the even- and the odd-indexed inputs. If $n$ = 2, no operation is actually necessary.

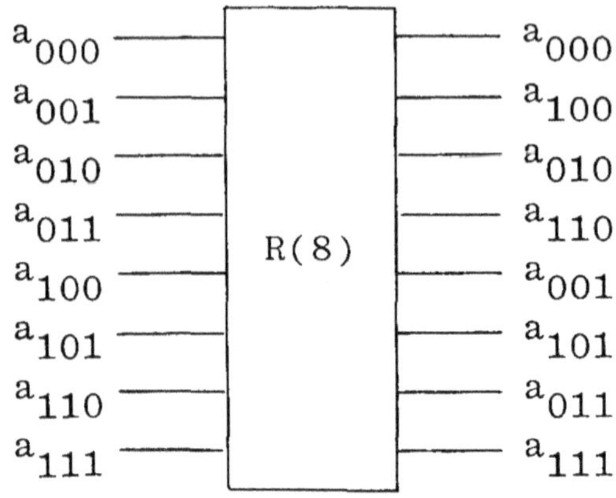


FIGURE 15. The binary representation of the input and output indexes. The operation R($n$) rearranges the inputs into the bit-reversed order.

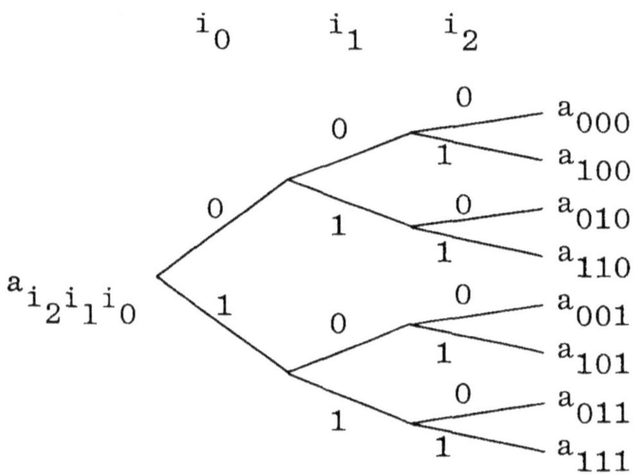


FIGURE 16. An illustration of how the bit reversal occurs. The binary representation of even integers has a 0 bit at the rightmost position, and they are carried upward. The others are carried downward. Repeating this process yields the bit-reversed order.

By this computation, we can know the final positions of the inputs to be rearranged. Hence, we do not have to rearrange the data themselves. Instead, we can compute the indexes and directly address the input data. This helps a great deal when the word length of the data is long and it takes time to move the data between the registers. If we are to compute the FFT many times with fixed $N$ in particular, we can preprocess the data, i.e., rearrange the data in the bit-reversed order or assign them the bit-reversed addresses beforehand. This kind of consideration is especially important for processing by microcomputers.

## XI. MISCELLANEOUS NOTES AND TECHNIQUES

### A. Numerical Errors

As we have seen, the DFT does not involve integrations and consists of a finite number of terms, unlike the classical Fourier series. Hence, it can be computed exactly, in principle. Of course, the use of a floating-point finite precision computer yields accumulated round-off errors. However, it has been noticed that the FFT drastically increases not only the speed of computation but also the accuracy. Roughly speaking, each output value undergoes only log $N$ butterfly operations, and accordingly, the accumulation of error is small.

### B. Saving Memory Space

The capacity of the memory space in small computers, especially microcomputers, is limited. Even a large-scale computer has a maximum memory size available for arrays. The FFT is suitable for processing from the standpoint of the memory space. In fact, it requires an initial array of size $N$ to input the data plus only a few extra registers for computation. This is easily seen if we look into the operation B($n$) of Figure 11. It consists of the butterflies of Figure 12. It is observed that once $Y_k$ and $Y_{n/2+k}$ are computed from $X_k$ and $X_{n/2+k}$, the values of $X_k$ and $X_{n/2+k}$ are no longer necessary. Therefore, the values of $Y_k$ and $Y_{n/2+k}$ can be stored back in the places that $X_k$ and $X_{n/2+k}$ have occupied. Repeating this process, we do not have to prepare an auxiliary array of size $N$.

As we stated before, we do not have to move the input data into the bit-reversed positions if we compute the bit reversals of the indexes directly. If we are to move the data, however, there is a more suitable way than the successive operations of P($n$) of Figure 14 from the viewpoint of saving memory space. Note that if $a_j$ is in the bit-reversed position of $a_i$, then $a_i$ is in the bit-reversed position of $a_j$. Hence, all we have to do is interchange the inputs with their bit reversals: leaving out $a_0$, exchange $a_1$ and its bit reversal, $a_2$ and its bit reversal, and so on. If $a_j$ is the bit reversal of $a_i$ and $j < i$, skip the exchange $a_j$, because they have already been exchanged. This implies that we do not have to prepare another auxiliary array of size $N$ to check whether the exchange has already taken place or not.

If $a_0, a_1, \ldots, a_{N-1}$ are stored in an array $A(1), A(2), \ldots, A(N)$, and the exchange is done successively for $I = 1, 2, \ldots, N$, the above procedure goes as follows in FORTRAN:

```
J= 1
N1 =N-1
N2=N/2
DO 10 I=1, N1
IF(I.GE.J)GO TO 1
P=A(J)
A(J)=A(I)
A(I)=P
1 K=N2
2 IF(K.GE.J)GO TO 3
J=J-K
K=K/2
GO TO 2
J=J+K
10 CONTINUE
```

This procedure corresponds to successive decimal computations of the sequence of binary numbers 00 . . . 0, 10 . . . 0, 01 . . . 0, 11 . . . 0, . . . in parallel to the sequence 0 . . . 00, 0 . . . 01, 0 . . . 10, 0 . . . 11, . . . It also takes into account the fact that the index begins with 1, not 0.

Thus, the capability of *in situ* computation using only an initial array of size $N$ is one of the striking characteristics of the FFT.

### C. Generation of Complex Constants

Necessary complex constants are $\omega_N^k = \cos 2\pi k/N + i\sin 2\pi k/N$ for $k = 1, 2, \ldots, N/2 - 1$. The remaining powers of $\omega_N$ are just negatives of these, as we have seen from Figure 5. The use of sin and cos functions increase the accuracy compared with the computation of powers, although it takes so much time for function generation. Since the same constant is used a number of times in the FFT computation, it saves time to compute all of them first and to store them in memory registers for constants, instead of computing them whenever they occur though so much memory space becomes necessary. In order to save further time or memory space, we have only to compute $\omega_N^k$ for $k = 1, 2, \ldots, N/4 - 1$, since those on the upper left quarter of the unit circle of the complex plane are obtained by changing the sign of the real part. As a matter of fact, the computation of $\omega_N^k$ for $k = 1, 2, \ldots, N/8$ is sufficient, since those for $k = N/8 + 1, \ldots, N/4 - 1$ are obtained by exchanging the real and the imaginary parts. However, the greedier we are, the more complicated the program becomes.

### D. The DFT and Its Inverse

As we stated before, equations of the type represented by Equations 2 and 25 and the type represented by Equations 5 and 26 can be computed by the FFT if we choose the appropriate sign of the exponent of $\omega_N$. However, the same program can also compute both types. Suppose, for example, the program is written for $\omega_N = e^{i2\pi/N}$ and we are to compute the inverse type of transformation of $a_0, a_1, \ldots, a_{N-1}$. First, prepare $\overline{a_0}, \overline{a_1}, \ldots, \overline{a_{N-1}}$ by taking complex conjugates (i.e., by changing the signs of the imaginary parts). Next, feed them to the FFT. If $b_0, b_1, \ldots, b_{N-1}$ are the outputs, then the complex conjugates $\overline{b_0}, \overline{b_1}, \ldots, \overline{b_{N-1}}$ (divided by $N$ if necessary) are the desired values. This is readily verified from Equation 28.

### E. Parallel Computation of Real Sequences

In general, the FFT inputs a complex sequence and outputs a complex sequence. All internal operations are complex arithmetics. Hence, if the input is a real sequence, the FFT is only working up to half the capacity. Therefore, it can process two real sequences simultaneously. Suppose $a_0, a_1, \ldots, a_{N-1}$ and $a_0', a_1', \ldots, a_{N-1}'$ are two sequences of which we are to compute the DFT. View them as a complex sequence $c_k = a_k + ia_k'$, $k = 0, 1, \ldots, N - 1$, and feed them to the FFT. Let $f_0, f_1, \ldots, f_{N-1}$ be the outputs. Compute

$$b_k = (f_k + \overline{f_{N-k}})/2$$

$$b_k' = i(-f_k + \overline{f_{N-k}})/2$$

for $k = 0, 1, \ldots, N/2$ (the former half). The desired DFTs are $b_0, b_1, \ldots, b_{N/2}, \overline{b_{N/2-1}}, \ldots, \overline{b_1}$ and $b_0', b_1', \ldots, b_{N/2}', \overline{b_{N/2-1}'}, \ldots, \overline{b_1'}$, respectively. (The latter half of the DFT of a real sequence is complex conjugates of the former half in reversed order.)

### F. The Use of External Memories

If the number $N$ of the data exceeds the internal capacity of the computer, the computation of the FFT can be done in two stages. First, make two sequences of the data, separating the odd- and even-indexed terms. Then, feed them to the FFT separately and store the outputs in external devices such as tapes and disks. Finally, combine them into one sequence, following the scheme of Figure 8.

### G. The Best Program of the FFT

There exist many variants of the FFT other than the one we described in the text, though they are essentially the same. Various improvements are possible, depending on the purposes and the aims of the analysis, and the types and the characteristics of the computer used. Often a number of conceivable criteria of a good program such as accuracy, speed, memory space, and simplicity of the program are in conflict with each other. Therefore, there exists no best program in all aspects. It is necessary to design a suitable program for a specific application by taking into account the principle and the characteristics of the FFT.

## LIST OF SYMBOLS

$A_0$, $A_k$, $B_k$ = Fourier coefficients
$r(\theta)$ = Radius at angle $\theta$
$(c_k)^2$ = Power spectrum
t = A magnification
$R_0$ = Mean square of radius
1 = An integer

## REFERENCES


1. **Gotoh, K. and Finney, J. L.,** Representation of the size and shape of a single particle, *Powder Technol.*, 12, 125, 1975.
2. **Meloy, T. P.,** A hypothesis for morphological characterization of particle shape and physiochemical properties, *Powder Technol.*, 16, 233, 1977.
3. **Meloy, T. P.,** Fast Fourier transforms applied to shape analysis of particle silhouettes to obtain morphological data, *Powder Technol.*, 17, 27, 1977.
4. **Beddow, J. K., Philip, G. C., and Vetter, A. F.,** On relating some particle profile characteristics to the profile Fourier coefficients, *Powder Technol.*, 18, 19, 1977.
5. **Fong, S. T., Beddow, J. K., and Vetter, A. F.,** A refined method of particle shape representation, *Powder Technol.*, 22, 17, 1979.
6. **Tsubaki, J. and Jimbo, G.,** A proposed new characterization of particle shape and its application, *Powder Technol.*, 22, 16, 1979.
7. **Tsubaki, J. and Jimbo, G.,** The identification of particles using diagrams and distributions of shape indices, *Powder Technol.*, 22, 171, 1979.
8. **Gotoh, K.,** Shape characterization of two-dimensional forms, *Powder Technol.*, 23, 131, 1979.
9. **Fong, S. T., Beddow, J. K., and Vetter, A. F.,** Acid-zinc reaction predictably affects particle morphology, *Powder Technol.*, 23, 187, 1979.
10. **Cooley, J. W. and Tukey, J.,** An algorithm for the machine calculation of complex Fourier series, *Math. Comput.*, 19, 297, 1965.
11. **Borodin, A. and Munro, I.,** *The Computational Complexity of Algebraic and Numeric Problems*, Elsevier, New York, 1975.
12. **Oppenheim, A. V. and Schafer, R. W.,** *Digital Signal Processing*, Prentice-Hall, Englewood Cliffs, N.J., 1975.

Chapter 5

# FORTH — ITS POTENTIAL AS THE COMPUTER LANGUAGE FOR SHAPE ANALYSIS

**Toshio Inoue**

## TABLE OF CONTENTS

## I. INTRODUCTION

For morphological studies and the development of their applications, one of the most important decisions to be made is on what sort of system should the work be based. Improper choice of the system will eventually confine the activity of the investigators in an unduly narrow domain. The commercial image analyzers using the TV camera, dedicated data acquisition and control circuits, and microprocessors are useful tools for predicated areas of image analysis, but extension of their applications to the flexible methods of usage which may be called for according to the progress of specific research schemes is quite limited. This situation is mainly caused by the rigid configuration of the software packages offered by the manufacturers. The dedicated software system may be well organized for the convenience of users who can command the system with a series of simple instructions, but only within the predicted limit.

By purchasing options, the user will regain some extent of freedom and will be able to command the system for data analysis based on the primary output of the software package, however it still remains a blackbox. Thus, the user's freedom is limited to the straightforward extension of the basic system, and can hardly reconstruct a program meeting his requirements from the simple data acquisition available.

Without good programming tools it is impossible to develop satisfactory software for the advanced image analyses expected in the future. The software, in this context, whether it is supplied by the machine manufacturers or tailored by the users, has to be efficient in running time, easy and efficient to develop, easy to include in the more sophisticated tasks which will be required in the future, and easy to maintain even if the hardware system might be upgraded later on. Considering the labor-intensive nature of programming, the software products must never be consumable items.

Suppose you have obtained a computer program written by someone else for image analysis on a different system and you want to use it on your own system. If the program is written in machine code or assembler language, and both systems are based on different types of CPUs, it is very hard to modify the program to fit your system since the programs are strongly dependent on the hardware structure of the computer. If the major part of the program is written in FORTRAN or BASIC, the situation may be better than in the case considered above. High-level languages, compilers (translators), and interpreters have been developed in pursuit of more efficient and machine-dependent tools for programming. In spite of this design target, it is widely recognized that the programs written in FORTRAN, for example, are strongly dependent, though not so much as the assembler programs, on the system environment. You might have bitter experiences in reading ill-documented, lengthy FORTRAN programs.

The situation is even worse for BASIC. BASIC was designed as a computer language (interpreter) for beginners in the time-sharing environment. Because of the easy usage and the relatively compact size of the interpreter, it has been popularized in the field of microcomputers. With the rapid development in the hardware technology of the microcomputers, which has resulted in increased capability and reduced hardware cost, the functions of BASIC have been extended in various ways without any control. This has resulted in a variety of BASIC dialects and has reduced the portability of programs. Apart from this, large programs written in BASIC tend to be entangled agglomerates, and one will easily get lost in tracing the line numbers used for GOTO (jump) or GOSUB (call procedure) designations. Reading a large BASIC program is just like trying to find both ends of a spaghetti noodle in a bowl. All the variables are common to the entire program, since local definitions within each subprogram are impossible in BASIC.

The modern high-level computer languages like C and PASCAL are much better organized in logical structures. The programs can be written in small, self-contained modules. Various types of data appearing in the program are declared explicitly in the well-defined domains, local or global. Thus the machine-dependent parts can be clearly limited to separate and normally very small modules. As a result the programs written in the so-called structured programming techniques using these languages are easier to extend and easier to transport to other environments.

Suppose you are supplied by someone with a program package written with proper care, skill, and documentation in C or PASCAL and you want to implement this program into your own system. You will be able to do this with much less effort than you would have to expend in developing the program from the beginning for yourself. The enhanced data structure offered by PASCAL is very attractive for programming for shape analysis, too.

The programming language FORTH allows for an extremely discretized style of program. It is somewhat similar to APL in appearance, but is enormously different from other languages such as FORTRAN, BASIC, ALGOL, PL/I, COBOL, C, PASCAL, LISP, etc., and even from APL in many respects. In a word, it is a very unique language. Like APL, the strange appearance of the programs written in FORTH is apt to cause disfavor. However, once familiarized through endurance or enjoyment, one will be strongly impressed with the unbelievable capability of this language and will be convinced that the effort to master it will certainly be repaid with much more profit. Most of the modern high-level languages are written in the later versions in themselves. In a sense, it is firm proof that the languages are portable to other environments, and the compilers written in themselves are good examples how complex programs can be written, as well. Thus, the PASCAL compiler written mostly in PASCAL appears beautiful in its modular structure, the largest modules being only as big as about 100 lines in the source (PASCAL) codes.

In the modularized small-program unit it is easy to write, maintain, and analyze because the references can be confined to a narrow range within easy reach of the normal human power of thinking. The FORTH system can mostly be written in FORTH and the system is an assemblage of very small modules, as large as 10 lines in source codes at the most. The profound capability and the extremely flexible and extensible function of FORTH facilitates the development and maintenance of programs. Good FORTH programs are not only executable by the computers, but also provide human beings with the thinking logic as to how the problem is to be solved.

It is especially emphasized that, among other reasons why FORTH is recommended as a programming tool for shape analysis, it is of great advantage to image analysis with dedicated small computers since the total program size is unbelievably small and the time-critical part of the task can be written in FORTH assembler*, the modules developed thereby being able to nicely fit in the program. No doubt, people capable of developing the logic of image analysis are inherently capable of using FORTH, too. FORTH will be of much help to the logical evolution in morphological analysis.

## II. ENVIRONMENT AROUND FORTH

Before going "forth" let us briefly review the historical aspects of, and the present environment around, FORTH. The concept of FORTH was conceived by Charles H. Moore at the end of the 1960s when he was hired by a carpet manufacturer as a programmer. Discontented with the conventional programming languages like FORTRAN, ALGOL, etc., which were running on the third-generation computer systems at that time, he developed his own ideas and the resultant product was named FORTH, implying it to be the fourth-generation language. By 1975 this new language was nearly complete and Moore applied it to the development of programs to control the radiotelescope in the National Astronomic Radio Observatory (NARO) at Kitt Peak, Arizona. Since then, FORTH has become standard in this specific scientific field and Moore has established a company, FORTH Inc., to distribute the FORTH system on a commercial basis.

* Assembler language written in FORTH.

FORTH has been intensively used in some isolated fields where people happened to recognize the extraordinary power of this somewhat strange-looking language. The environment changed for FORTH toward the end of the 1970s when microcomputers rapidly became popular and the FORTH Interest Group (FIG) was organized in San Carlos, California. FIG is a group of enthusiastic volunteers whose aim is the popularization of FORTH. The group publishes a bimonthly journal called *Forth-Dimensions* (FD), whose articles amply prove how intensive a language FORTH is. FIG has developed a unifed system of FORTH and implemented it on various types of mini- and microcomputers such as PDP11, NOVA, 8080, 8086/8088, 9900, 6800, 6809, 6502, 68,000, etc. It organizes regular meetings for discussion about new developments in FORTH and also has published FORTH standards.

## III. FUNDAMENTALS OF FORTH

The objective of this section is to introduce the programming language FORTH, to give a basic idea of how the FORTH system is organized, and to present concepts of how FORTH can be useful for shape analysis. For the detailed programming techniques the readers are requested to refer to books and/or manuals.[2,4] Unless otherwise stated the following descriptions are compatible with the FORTH '79 standard by FIG.[11]

Let us get started with some simple examples of programs in FORTH. Now, suppose a FORTH system is up and running in front of you (see Figure 1). The VDU shows "OK" prompting next input from the keyboard. Key in 2 5 + and . in sequence, with one or more spaces between each entry, then press the RETURN key and the computer answers with "7" and OK. This example shows some important features of FORTH. In this sequence the computer received the stream of input characters into the text input buffer and after RETURN key was pressed the input stream was processed item by item within the text interpreter (also called the external interpreter).

In FORTH there is no hierarchical difference among commands, functions, operators and even variables and constants. They are equally called "words". Every "word" in the input stream is delimited by a blank or blanks.

The FORTH system has a set of inherent "words" for fundamental operations. The set of those primitive "words" is called the FORTH nucleus and is normally implemented by a range of minute sets of machine codes. The entity of these machine code sets configure a "virtual FORTH machine" (such as the P-code interpreter in PASCAL) and it is called the internal interpreter.

Besides the text interpreter, the FORTH system is endowed with the compiling functions. The function of the FORTH compiler is to build up in the computer memory the list of already-defined words and assign this entity a name. Thus a new "word" can be added to the current vocabulary. The definition of a new "word" is normally started with a word ":" and ends in another word ";". Thus, to compile a "word" with the name "SUM" to sum up 2 and 5 and print out the result, one demands:

```
: SUM 2 5 + . ;
```

After the RETURN key is pressed the FORTH system compiles this input stream and registers the sequence of operations under the word name SUM in the memory. The "word", once defined, can be used at any time and any number of times by the text interpreter or the compiler. In fact, by keying SUM followed by RETURN the computer answers with "7" (see Figure 1).

The FORTH system uses two stacks to execute a given sequence of "words": the parameter (or data) stack is used mainly for passing and processing various types of data while the

```
OK
2 5 + .  7 OK
: SUM   2 5 + . ;  OK
SUM  7 OK
: QUAD   ( x --- y )   DUP 3 * 12 - * 20 + ;  OK
4 QUAD .    -4 QUAD .   20  116  OK
: PLOT   ( n --- )   CR 60 MIN 0 DO 42 ( * ) EMIT
                     LOOP :  OK
10 PLOT
**********  OK
200 PLOT
************************************************************  OK
: PICTURE   ( n1 n2 --- )
  1+ SWAP   DO  I QUAD PLOT  LOOP  ;  OK
-5 5 PICTURE
************************************************************
************************************************************
************************************************************
********************************************************
***********************************
********************
***********
********
***********
********************
***********************************  OK
```

FIGURE 1. VDU screen run by the interactive FORTH system. The underlined part represents the printout by the computer. Other part is input on keyboard.

return stack is used for nesting the addresses for sequential execution of "words", stacking DO-LOOP parameters and temporarily sheltering data.

In programming in FORTH, one has to attend to the state of the parameter stack in each step of the execution of the "words", since it is used intensively for parameter passing between "words". Principally, the functions in FORTH are commanded and executed in the so-called inverse Polish notation (IPN). The principle of IPN is to put the operands prior to the operation to be performed. In favor of the description of formulas, the simplicity of computer architecture, and the efficiency in execution this notation has been adopted by Hewlett-Packard for their programmable calculators. The calculus

$$(9-2)*(5+3)$$

can be written in IPN as:

$$9\ 2\ -\ 5\ 3\ +\ *$$

with no parentheses required. The above calculus is straightforward if one observes the change in the stack in the sequence (Figure 2). Thus, input of numbers 9, 2, etc. induces the interpreter to push down these data onto the stack. The binary operators +, −, *, and / pops up two items on the top and the second-top of the stack and pushes down the resultant

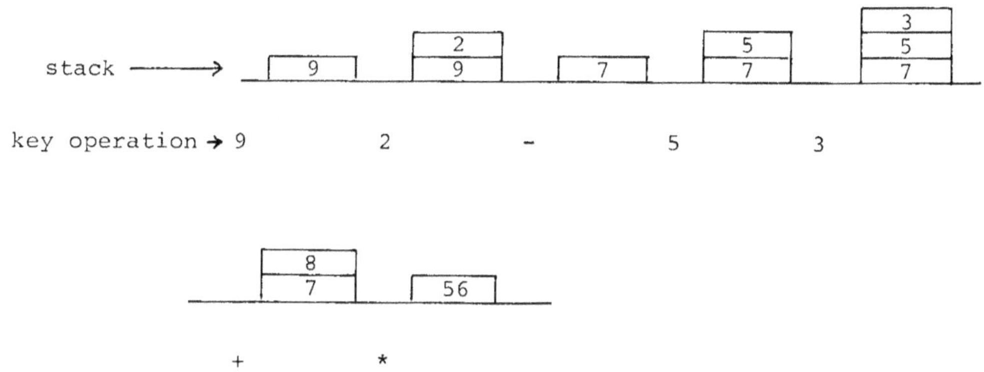


FIGURE 2. Stack operation by inverse Polish notation (IPN).

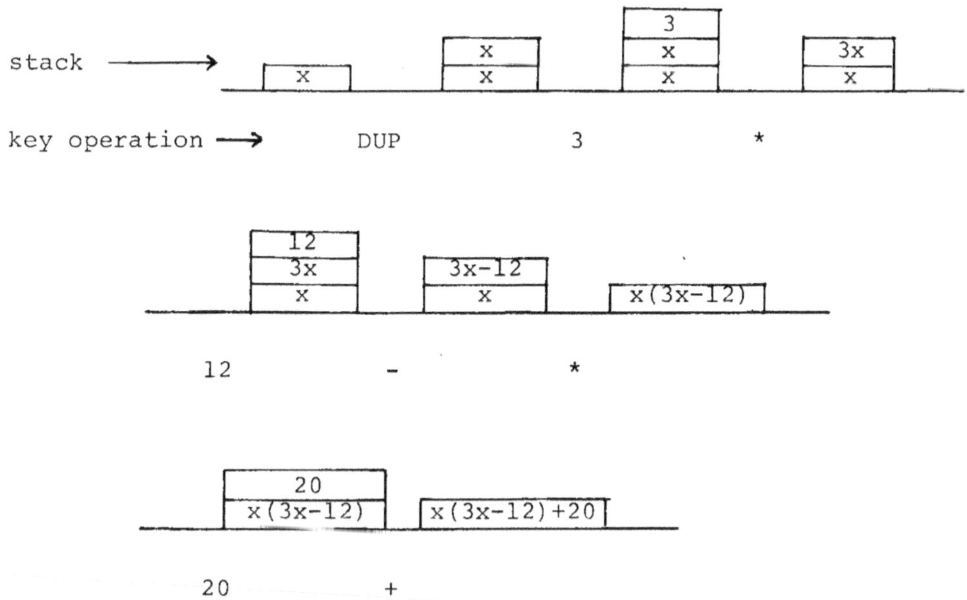


FIGURE 3. Execution of the "word" QUAD.

data onto the stack. Likewise the word "." (dot) in Figure 1 pops up the data on the top of the stack, converts it to a series of digits, and prints them out.

Let us see the example of another procedure named QUAD in Figure 1. Suppose a value $x$ is on the stack before the word QUAD is activated. Following the sequence shown in Figure 3, the word DUP, one of the FORTH primitives, duplicates the top data on the stack. As obvious from the illustration in Figure 3, QUAD is a definition to convert the value of $x$ on the stack to the function y:

$$y = 3x^2 - 12x + 20$$

In the definitions of QUAD in Figure 1, "(" is a "word" to specify that the following text, until ")" on the same line, is a comment having no effect on the program. In the comment ($x$---$y$), $x$ refers to the content of the stack before operation and $y$ refers to that after operation. The concise comment such as this to show the transition in the stack will be very useful for later reference of the word. The evocation of the word QUAD with its operand in advance and the word "." for printing after the operation will give the results as shown on line 7 in Figure 1.

Let us define another "word" under the name PLOT to print out a given number of asterisk marks on the VDU or the printer (Figure 1). At this point we introduce some other primitive FORTH words. The word EMIT is used to print out the letter whose ASCII code value is on the top of the stack. The ASCII code for "*" is 42 in decimal numbers. The word "CR" emits a carriage return and a line feed code in succession to the VDU and/or the printer. The word "MIN" compares the two data on the stack (top and second-top) and leaves their minimum value. The word "DO" picks up two data on the parameter stack as the limit value and the initial index value and pushes them onto the return stack. These values are referenced by the word "LOOP" which increments the index value and compares it with the limit value to discriminate whether or not the execution of the "words" between DO and LOOP should be repeated.

Thus the activation of PLOT with the number on the stack in advance will print out the specified number (if larger than 60 it is abbreviated to 60) of asterisk marks. Now we combine the words QUAD and PLOT to draw a graph of the quadratic function. See Figure 1 again. The word "I" pushes the current index value onto the stack. The value n2 is incremented by 1 and then SWAPped with n1 on the stack. Then these values are picked up by DO as the limiter and the index, respectively.

The length of a FORTH word is normally between 2 and 50 words. Splitting overlong words into functional groups as separate words, the program becomes surprisingly more readable. Higher levels of functions can be realized very easily by structuring the lower levels of words in stages, as will be seen from the example of PICTURE in Figure 1.

Before proceeding further let us summarize. The FORTH system has a text interpreter for immediate execution of words and a compiler for defining words for later execution in addition to an internal interpreter for realization of the virtual FORTH machine. The system observes the inverse Polish notation, so that each operation is preceded by its operands.

The system is operated either in the interpret mode or the compiler mode. Switching between these two different modes can be normally made by the words, ":" and ";". In the interpret mode, or when the compiling DICTIONARY has to be searched for the given names of the words appearing in sequence, this process is not so fast. In a sense compilation is also a process to interpret words within ":" and ";" for immediate execution in the later time. Once the word is compiled, however, the execution is commissioned solely by the internal interpreter, which picks up the address of the words in sequence and this process is very fast. The hybrid compiler-interpreter system of FORTH is extremely beneficial for efficient production of programs. On top of this, there is no conceptual difference among commands, main programs, subprograms, functions, or even variables and constants. All these things can be executed immediately after definition (compilation) for testing without any external support whatsoever. This unique feature* is a great advantage of FORTH, in view of the high productivity in program development that it permits.

## IV. STRUCTURE OF FORTH DICTIONARY

The FORTH system, when supplied by the vendor or having been implemented in any other way, has a dictionary — an assemblage of "words". In the case of fig-FORTH**, this basic FORTH system contains about 200 different words excluding the FORTH editor and the FORTH assembler. These "words" occupy about 7 kB of memory, and about 40 out of them are the most primitive FORTH words which are written in machine codes. The

* APL and LISP share this feature.

** A FORTH system supplied by FIG.

latter words were called the FORTH nucleus in the previous context. The residual part is coded essentially in the FORTH language. In effect, up to approximately 80% of the basic FORTH words are written in FORTH, and further extensions including the editor, the assembler, and user's programs can mostly be written in FORTH. The dictionary can be divided into different groups called vocabularies. Normally, the editor and the assembler are grouped in separate vocabularies.

The user can switch the CURRENT vocabulary and the CONTEXT vocabulary from time to time. New definitions of "words", by input via keyboard or disk, are entered into the CURRENT vocabulary, whereas the dictionary search is made in the CONTEXT vocabulary and then, if not found, in the "FORTH" vocabulary. This provides us with a way to distinguish the words with the same name and different functions in different vocabularies.

Figure 4 illustrates how the "word" is configured in the fig-FORTH dictionary. (There can be minor differences among other FORTH systems.) The first byte of the word is the header of the word. The header marks the top of the word definition and contains the number of letters in the name of the word. Only the first 31 letters and the total number of letters of the name are identifiable in fig-FORTH, although longer names are permitted. The header also includes a function to identify the attribute of the word. The header, followed by the ASCII codes to represents the name, is called the name field (NF). Next to the name field is located the link field (LF). The link field is used to link the words in the form of a linear list. Next to the link field is the code field (CF). The code field points to the first machine code which should be executed when this word is invoked. In the fig-FORTH definitions, this part corresponds to the primitive word ":". The code field is followed by a series of the parameter fields (PF) until it is ended by the word for ";". Each of the parameter fields contains the address of the code field of the word which should be executed. The word "LIT", in the parameter fields 4 and 8 in this example, are inserted automatically by the compiler. The function of LIT is to push down the following data in the parameter field as a 16-bit-long piece of data onto the parameter stack instead of being executed. LIT is absent ahead of "3" (PF2), because "3" is already defined (compiled) as a constant in the FORTH dictionary. The link field, the code field, and the parameter fields occupy 2-byte (16-bit) areas each. Basically, all the data and addresses are also represented by the 2-byte-long units, which are called "cells".

The higher-level words in FORTH are composed of lower-level words in multiple stages merely by the addresses stored in the parameter fields. As a result the FORTH program, or the highest level word, looks like a chain of threaded codes, where any higher-level words are connected directly or indirectly in (multiple) steps down to the primitive virtual FORTH machine "words", which are composed of executable native machine codes. This particular program structure is called the indirect threaded code (ITC). FORTH is a unique language realized in ITC structure. Since threading is fulfilled by only a few steps in the native machine codes, the overhead on the CPU time can compare favorably with the inefficiency inherent in other high-level languages. The program size is minimized by the use of the ITC structure. The ITC structure is illustrated schematically in Figure 5.

## V. PROGRAMMING IN FORTH

Because of its well-organized structure and interactive capability, it is easy to get started with FORTH if you sit down in front of the FORTH system terminal. In line with words which are easy to learn, however, there are small numbers of words whose functions are fairly difficult to comprehend in spite of their simple appearance in definitions. These difficult words play an essential part in evaluating FORTH, however. The present section is intended only as a general outline of FORTH. Readers are requested to refer to the dedicated publications for tutorial descriptions.[2-4]

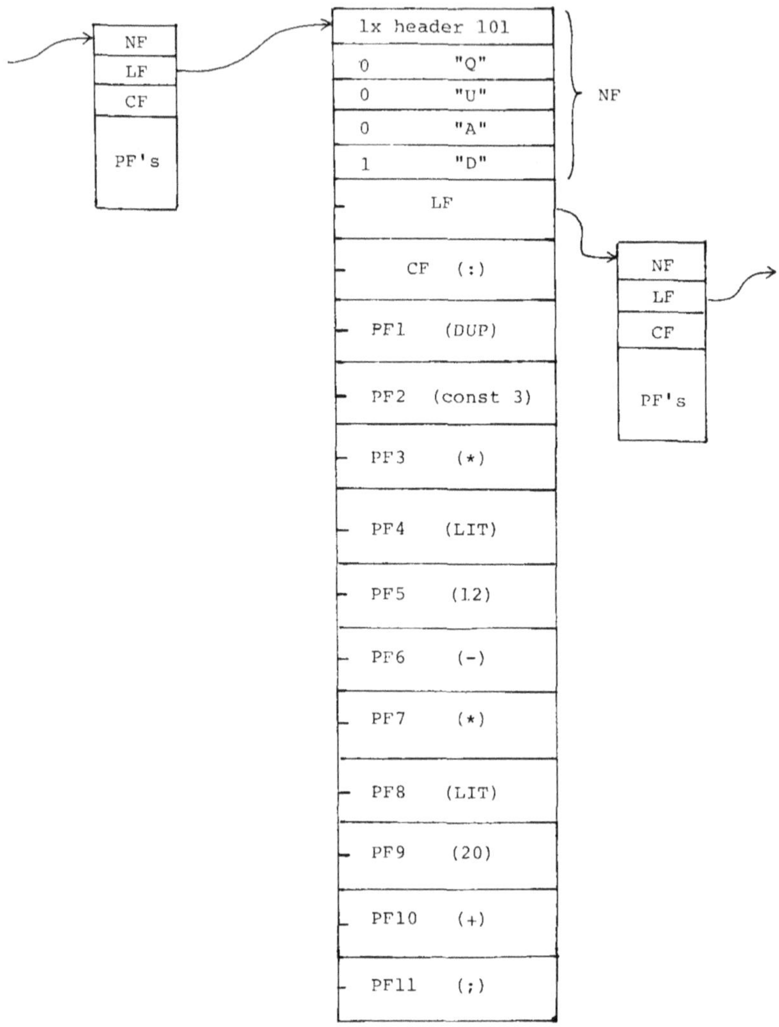


FIGURE 4. The "word" QUAD in the FORTH dictionary.

The typical FORTH system consists of a small computer, micro or mini, with a VDU terminal, a printer, and dual diskette drive units. Most systems have their own assembler and editor, both written mainly in FORTH. The FORTH assembler is designed to assemble "words" which can be used just like other words defined in FORTH, and are used only when time-critical and/or machine-coherent processes have to be described. Many FORTH systems can generate ROMable codes. The program baked in ROM (read-only memory) may or may not include the FORTH compiler, external interpreter, and/or the name fields of the words. Thus, the daughter systems can be designed in minimum size for the microprocessors built in various machines for dedicated uses, such as game machines, machine controllers, and analyzers.

It is also interesting to note that by using such systems one can develop higher-level "hyper-FORTH" systems. Some commercial systems are provided with multitasking capability, while some others, written by earnest FORTH lovers, are subordinate to the large-scale computer systems.

As a rule of thumb, FORTH is equivalent to an operating system including a FORTH language interpreter/compiler, a FORTH assembler, and an editor. Access to the disk (or diskette) is performed by the logical block numbers or "screen" numbers using simple virtual memory management functions provided within the FORTH system. The concept

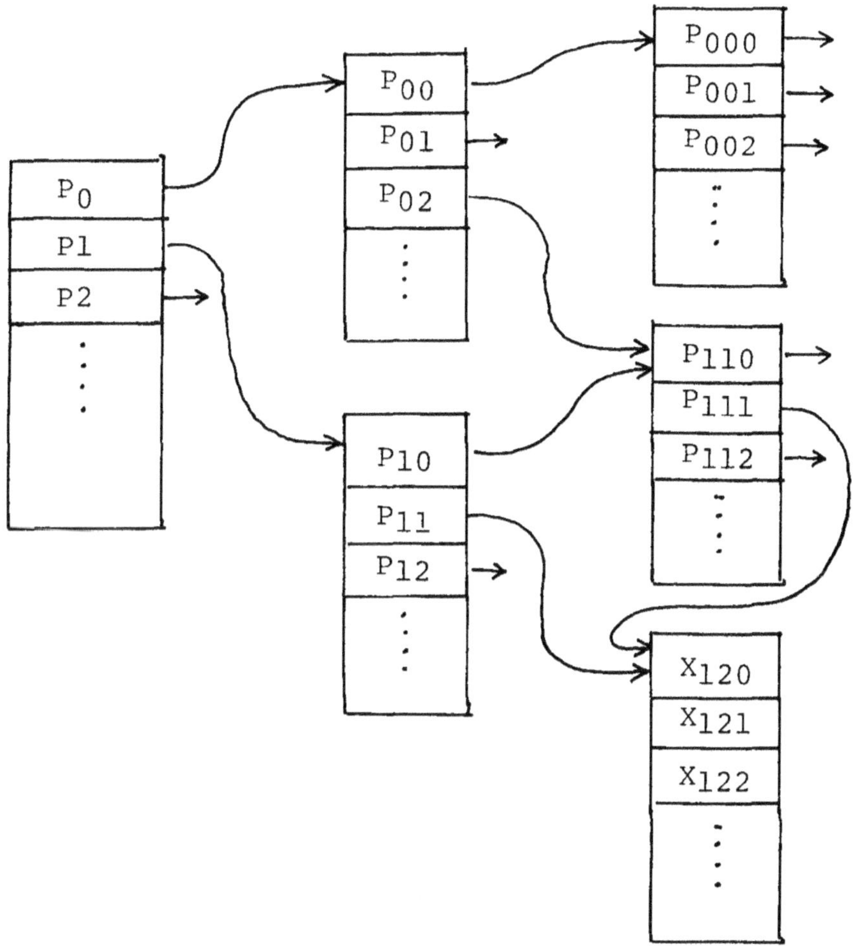


FIGURE 5. Indirect threaded code (ITC) structure. P. . .: FORTH codes, X . . .: machine codes.

"screen" is represented by 64 characters by 16 lines each of regions in the disk by which program loading (execution and/or compilation) and edition is performed on the VDU screen (see Figure 6).

As stated before, FORTH is based on a virtual machine using a parameter stack for various operations. In fact, the parameter stack is useful for passing various types of parameters between or across "words". The following is a convention to specify the change in the stack before and after execution of a word:

( x1 x2 x3 --- x4 x5 )

In the above example x1, x2, and x3 on the left-hand side of --- indicate the data before execution with x3 at the top of the stack, while x4 and x5 indicate the data after execution with x5 at the top of the stack. The x can be various, such that:

| | | | |
|---|---|---|---|
| n: | 16-bit data (for integers) | f: | flag |
| a: | 16-bit address | tf: | true flag (non-zero) |
| c: | 7-bit data for ASCII character code | ff: | false flag (zero) |
| b: | 8-bit data | d: | 32-bit data |

```
SCR # 87
( Extension                                              TSI 82/04/20)
BASE @ HEX     3 CC16 !
: 0!     ( a ---- )     0 SWAP ! ;
: D=     ( d1 d2 ---- f )     ROT = >R = R> AND ;
: 2DUP    2DUP >R  R ROT DUP >R ROT ROT R> R> R> ;
: 3DROP   DROP DROP DROP ;
: FLEX      9 F193 C!    CD03 CC16 !     DOS ;
: ?EMPTY     ( ---- f )     S0 @ SP@ 2+ = ;
: DEFAULT    ( n ---- )     >R ?EMPTY IF R> ELSE R> DROP THEN ;
: THRU     ( n1 n2 ---- )     CR 1+ SWAP DO ." Loading SCR #"
   I 4 .R CR I LOAD LOOP ;
: DUMP     ( a n ---- )     OVER + SWAP DO I 10 MOD 0= IF CR
   I 4 U.R ."   :" THEN I C@ 3 U.R I 0 10 U/ DROP 0F - IF
   ."  : " -1 0F DO J I -  C@ 7F AND DUP BL < OVER 7F = OR IF
   DROP 2E THEN EMIT -1 +LOOP ."   :" THEN LOOP CR ;
: PICK     ( n1 ---- n2 )     DUP + SP@ + @ ;          BASE !       ;S

SCR # 89
( Copy from original tFORTH+ disks                        TSI 82/04/20)
BASE @    HEX
: ORIGCPY     ( offset n-blocks ---- )
   0 DO 0A 0 DO DR0 20 J * I + BLOCK
                    DR1 OVER 0A J * I + + BLOCK
                    100 CMOVE UPDATE
                 LOOP
     LOOP DROP FLUSH ;
BASE !      ;S

       m n ORIGCPY

       copies n blocks of contents in Drive-0 which is written
       in 10 sectors by 74 tracks to disk in Drive-1 in the
       format: 32 sectors by 40 cylinders.   m = offset on
       Drive-1.

SCR # 91
( Aristotelian sieve                                      TSI 82/04/20)
: PRIMES     ( n ---- )
   DUP 2/ 1+ SWAP CR ." START" CR CR
   0 OUT ! 1 DO DUP R 1 ROT 2
      DO DROP DUP R /MOD DUP 0= IF DDROP 1 LEAVE ELSE
      1 = IF DROP 1 ELSE DUP 0> IF DROP 1 ELSE 0= IF
      0 LEAVE THEN THEN THEN THEN LOOP IF OUT @ 59 > IF CR
      0 OUT ! THEN 6 .R ELSE DROP THEN LOOP
   DROP CR ." END      " ;
;S
```

FIGURE 6. Example of FORTH screens.

Attributes of various data can be neatly described by the notation given above.

It is also possible to use named variables and constants after definitions, e.g.,

```
      VARIABLE  x              VARIABLE p/\x
31416 CONSTANT PI
```

where x, p/\x, and PI are named variables and a constant, respectively. Due to the discipline of the original inventor, floating-point numbers have not been commonly used in FORTH. In effect, this tradition seems to have restricted the use of FORTH to limited fields of applications. Thanks to the high productivity of programming in FORTH it is not difficult to implement the floating point tools individually, but there are some FORTH systems which feature the floating-point capability. It will be a natural course in the future to provide FORTH with the floating-point operation as standard or option.

FORTH is a homogeneous language in the sense that all the "words" in the system or those defined additionally by the users in the FORTH vocabulary or other named vocabularies behave entirely alike. Variables and constants are also qualified as "words". They are defined by using the words "VARIABLE" and "CONSTANT", respectively, as shown above. VARIABLE and CONSTANT are already defined in the basic FORTH system. They can be redefined in FORTH as follows, however:

```
: VARIABLE      CREATE 0, DOES> ;
: CONSTANT      CREATE , DOES> @ ;
```

Here we can see a remarkable function in FORTH. That is, in the above definitions the "words" CREATE and DOES> are used to define the "words" VARIABLE and CONSTANT, which in turn are useful for defining a family of variables and constants having different names but similar attributes. Likewise, we can define various types of words which are capable of defining a family of words. Let us examine this important feature of FORTH in more detail.

As we have seen before ":" is a word with which the definition of a new FORTH word (VARIABLE or CONSTANT in the above cases) should be started. Basically, the name field and the link field of the word are created by the sequence ": name" (see Figure 4). Similarly, the word CREATE creates the name field and the link field WHEN IT IS EXECUTED. When VARIABLE is executed the first word invoked by VARIABLE is CREATE. Thus, the word CREATE reads the character string which follows immediately and creates a name field of the new word in the dictionary. The following "0" is also a word, already defined as a constant 0. Execution of "0" effects in pushing the data 0 onto the top of the parameter stack. The next word "," pops up the data (0 in this case) on the top of the stack and appends it to the next position in the dictionary. DOES> is a special word used in pair with CREATE in most cases. Its function is to direct the code field of the word being defined by CREATE to the position next to DOES> by pushing its address down to the parameter stack. Finally ";" is the word to conclude the definitions and its behavior in execution is to secure linkage between words in coordination with ":".

In spite of the simple appearance, and, in fact, simple definitions (not shown here) of the words: , ; CREATE and DOES>, their behaviors are considerably difficult to comprehend. As one advances in FORTH programming, he will become capable of solving the magic, if he reads the definitions of these words. But, for beginners, it is not necessary to know all the details of their internal mechanisms. It is sufficient to write:

```
: name-of-new-word CREATE the-functions-to-be-
  performed-in-the-stage-of-compilation DOES>
  the-functions-to-be-performed-in-the-stage-of-execution ;
```

In this way one can define a family of "words" having similar behavior in execution.

Let us examine the simplest cases in which the defining words, VARIABLE and CONSTANT, are used. When defining a new variable, VARIABLE reserves a 16-bit space in the parameter field and fills it with 0. When the new variable is "executed" it pushes the address of the parameter field, whose content is 0 when compiled, onto the top of the stack. The function of CONSTANT is also obvious. In the compilation stage it places the data on the top of the stack (31416 in the above example of PI) into the parameter field of the word. In the execution stage, evocation of the word (PI for the present example) pushes the address of the parameter field on to the stack and then replaces it with its value (31416) by the function of the word "@" following DOES>. In the interactive FORTH system the execution of a constant PI will be as follows:

```
31416 CONSTANT PI OK        .... compilation
PI . 31416 OK               ...... execution
```

(The underlined part specifies the computer answer-back.)

Generation of various data types is very easy using the word-defining capability of FORTH. Take the example of a structured data type, or "record", containing an 8-byte-long character string, a 2-byte-long integer, and a 4-byte-long floating-point number. Thus:

```
: RECORD    CREATE    ALLOT ;
```

could be a word to generate arbitrary length of data given on the top of the stack with a name pointing at the top address when quoted. Thus, for this example:

```
2 8 2 4 +  +  + RECORD REC
```

will generate a region for the storage of a data with 16 bytes in total.

By the definition of another word-defining word "MEMBER", which follows, and the repeated use thereof, one can define a 2-byte-long data LINK as a pointer, an 8-byte-long data STRING, a 2-byte-long data INTEG, and a 4-byte-long data FLOT, respectively. The FORTH sequence

```
3 INTEG !
```

will store the data 3 into the position INTEG in REC.

```
: MEMBER CREATE , DOES> @ REC + ;

 0  MEMBER LINK     2  MEMBER STRING
10  MEMBER INTEG   12  MEMBER FLOT
```

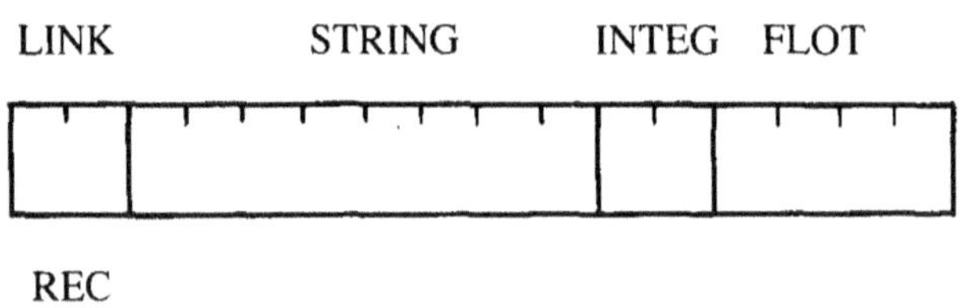


A simple design of the data base system is presented [6] to manipulate a family of structured data records in disk files.

In the normal processing of source programs and data the space code is reckoned as the only delimiter to separate words. This allows the programmer to use any special characters except blank(s) as a part of the name of the word. Thus, for example,

```
: ** (n1 — n2) DUP * ;
```

can be defined as a word to calculate the square of data on the parameter stack.

```
-7 ** . 49 OK
```

It was mentioned before that the FORTH words are homogeneous in the sense that no difference can be detected among the concepts as job control commands, main programs, subprograms (functions, subroutines, procedures, etc.), and variables and constants except for their position in the hierarchical and modular structure of the dictionary. The above examples are implicit of the vigorous proof of this fact. That is, it was shown that new data type could be generated in the example of REC, while a new operation type was appended to the system by "**". It is also possible to provide new program structures, e.g., CASE in addition to the structures such as ....IF....ELSE....THEN, BEGIN....WHILE....REPEAT, BEGIN....UNTIL, etc. which are available in the basic FORTH system.[8] It should also be noted that these things are (or can be) all written in FORTH. At this point we discover that "homogeneity" is synonymous to "extensibility".

There are even more sparkling evolutions in FORTH such as the definitions of tools to enable recursive use of words[9] and sealing of local variables, local procedures, etc., within separate modules,[10] nevertheless these techniques had not been anticipated during the original design of FORTH.

Some people may be disappointed to find out that in FORTH the arithmetic operations +, − ,*, /, etc. are not common to various data types as they are in other high-level languages. Some others may well complain that

```
C = A + B
```

in FORTRAN, has to be written as

```
A @ B @ + C !
```

in FORTH, where A, B, and C are integer variables. In FORTH, however, it is possible to define a word under the name X*, for instance, to calculate the inner product of matrices A and B and put the result in C:

```
A B C X*
```

where X* checks the consistency of these matrices for types and sizes automatically.

FORTH "words" are written normally in 5 to 50 words of lower levels. In most cases, a FORTH word can be executed for immediate debugging after its definition. The ease in testing each word by virtue of the homogeneity of words, interactive system organization, and the compactness of each word is incomparable with other languages. FORTH is akin to APL.

Figure 7 shows the set of FORTH words included in '79 fig-FORTH Standard.[11]

## VI. SOME HINTS FOR SHAPE ANALYSIS

The outstanding characteristics of FORTH have been depicted broadly. At this point let us speculate how programming in FORTH would be achieved for shape analysis.

A dedicated computer system will be used for data acquisition and processing in shape analysis, in most cases. The tasks assigned to the computer will be categorized as follows:

1. Control of the scanning system
2. Acquisition of signals generated by scanning the particulate objects
3. Modification of primary data to cancel out the errors caused in data acquisition
4. Entry of data into a table, each item of which corresponds to the object identified in the scanned area
5. Processing data in the table to characterize each item according to specific criteria
6. Statistics, report preparation, and storage of results for possible later reference
7. Output of data to the host computer system for more extensive data processing requiring very large storage and very fast calculation speed

```
Nucleus Words

!  *  */  */MOD  +  +!  +loop  -  /
/MOD  0<  0=  0>  1+  1-  2+  2-  <
=  >  >R  ?DUP  @  ABS  AND  begin  C!
C@  colon  CMOVE  constant  create  D+
D<  DEPTH  DNEGATE  do  does>
DROP  DUP  else  EXECUTE  EXIT  FILL  I
if  J  LEAVE  literal  loop  MAX  MIN
MOD  MOVE  NEGATE  NOT  OR  OVER  PICK
R>  R@  repeat  ROLL  ROT  semicolon
SWAP  then  U*  U/  U<  until  variable
while  XOR

(note that the lower case entries refer to just the run-time code corresponding
to a compiling word.)


Interpreter Words

#  #>  #S  '  (  -TRAILING  .
79-STANDARD  <#  >IN  ?  ABORT  BASE  BLK
CONTEXT  CONVERT  COUNT  CR  CURRENT
DECIMAL  EMIT  EXPECT  FIND  FORTH  HERE
HOLD  KEY  PAD  QUERY  QUIT  SIGN  SPACE
SPACES  TYPE  U.  WORD


Compiler Words

+LOOP  ,  ."  :  ;  ALLOT  BEGIN
COMPILE  CONSTANT  CREATE  DEFINITIONS  DO
DOES>  ELSE  FORGET  IF  IMMEDIATE
LITERAL  LOOP  REPEAT  STATE  THEN  UNTIL
VARIABLE  VOCABULARY  WHILE  [  [COMPILE]  ]

Device Words

BLOCK  BUFFER  EMPTY-BUFFERS  LIST
LOAD  SAVE-BUFFERS  SCR  UPDATE
```

FIGURE 7. '79 FORTH standard words.

The stages of (1) and (2) and possibly also (3) will, in most cases, be performed by a slave microprocessor and will be categorized as a firmware of an intelligent digital TV camera. Except for special cases where the standard scanning method has to be modified due to the necessity of the logics of data analysis which follows, it will be rare that this part has to be designed in the framework of shape analysis.

Programming in stage (4) will call for techniques such as interrupt processing and reentrant module structure. The former is closely related with the hardware structure of the data acquisition and computer system. Writing a program involving interrupts poses no difficulty in programming in FORTH.

It is also easy to write reentrant programs, since FORTH is a stack-based language and the content in the stacks is retrieved automatically when the program happens to be revised or interrupted. In order that higher-level words using variables besides stack should be reentrant, special data structure will have to be introduced, however, this is easy in FORTH.

In stage (4), flexible data types will have to be provided for use in the memory as well as in the disk space. As has been delineated before, the original data types are very poor in FORTH, but unlimited extension is possible, instead. It should be noted that the concept of virtual memory management in the basic FORTH system makes the design of transparent and compatible usage of both the memory and the disk space an easy task.

Stages (4) and (5) will pose the programmer no difficulty in the solution of complicated logical problems. FORTH is also a good thinking tool to work on these types of problems. The programmer can depict his concept in simple logical sequences in FORTH words. If he prints out any words which have to be described more precisely, he can insert the definitions of these words ahead of the words in which they are used, so that they are recompiled in order. By the repetition of these steps, he can finally complete his target program, where every word is defined and executable. If he finds any part inconsistent with the entity or the logical description of the problem, he can go back to that point and do it again.

FORTH does not permit a FORWARD type of definitions, because of the interactive and interpretive nature of the language. In other words, FORTH programs have to be developed in the "bottom-up" direction. However, the "top-down" design of programs is more compatible with the way of thinking of human beings, and this principle should be preserved in so much as the logical structure of the problem can be clear. One of the methods to cope with this situation is the temporary use of substitutions for the modules which must eventually be developed. These temporary modules, or simulators, are called "stubs". Stubs are also useful for developing software concurrently with the development of the hardware. Stubs, in the latter cases, could be made to communicate with the operator instead of the machine for input/output commands and data. We may be motivated to append a structure to admit forward referencing capability to FORTH, and it can be done simply.

The author has developed a software tool for handling complex data structure in the disk. The data is structured in the form of binary trees in a multilevel hierarchy in which each member can consist of different types of data with variable lengths. The program was written in FORTH in about 120 lines in the source codes. The biggest module size was four lines long, involving only about 40 words. This data structure will be useful for data base as well as for data processing in shape analysis.

## VII. CONCLUSION

So far we have delineated the outstanding characteristics of FORTH with particular attention to its application to shape analysis. The reader will have noticed that FORTH is not only a compiler-interpreter language in the conventional sense but also a unique concept or a philosophy in software development from which basis all the virtues of FORTH evolved. The popularity of FORTH is increasing rapidly in parallel with the availability of microcomputers. This is also a proof that using a language is the best way to learn it. The internal FORTH interpreter is very small and fast. In spite of the overhead due to the multilevel calls of subordinate words, the execution speed of FORTH compares favorably with other compiler languages. The FORTH stack machine can also be realized in special firmware design instead of using the conventional computer architectures. In designing a dedicated firmware, it is reported that about 100 times as high throughput as the normal microcomputers has been achieved.[7]

To conclude the chapter let us summarize the features of FORTH:

1. FORTH is based on the virtual stack machine. Normally the function of the virtual machine is controlled by a small, high-speed interpreter, called the inner interpreter.
2. In principle, FORTH uses the IPN for arithmetic as well as other types of operations.
3. Besides the inner interpreter, FORTH has an external interpreter/compiler. The external interpreter takes in the data stream from the keyboard or the disk to be executed in succession. The compiler also processes data and commands (defined already) and compiles them to create the new definitions of words in the dictionary. The word thus generated can now be used by the interpreter/compiler.

4. FORTH words are homogeneous in the sense that there is no intrinsic difference between the words depending on whether they are inherent in the basic system or created by the users. Likewise, there is no difference in the qualification of words. The difference, if at all, is their attribute to different vocabularies and their difference in functions.
5. Hence, it follows that an important feature of FORTH is that users can develop various program structures and even various programming concepts which had not been considered at the original designing stage of FORTH itself. FORTH can also be configured to enable it to regenerate itself for the same as well as different computer systems. This feature qualifies FORTH as an extensible or self-breeding language.
6. It has been proved again that the productivity of programming in FORTH is amazingly high in virtue of the extreme modularity of the program structure and the interactive capability of the FORTH system.
7. In addition to the unusually compact size of the basic part of the FORTH system, the program size in FORTH, after compilation, is normally much smaller than the corresponding object codes generated by any other compilers.
8. The definition of ''words'' is possible by using the built-in assemblers, as well. Thus, one can write time-critical processes in short assembler codes. FORTH is also easily compatible with machine-dependent operations such as interrupts. These features present a great advantage, especially in developing programs for device control.
9. Depending on the system products, FORTH enables the generation of codes which can be baked on ROM with or without the labels and links together with the minimum part of the FORTH nucleus. This feature is useful for developing programs for dedicated microprocessors for device control.

## REFERENCES

1. **Hicks, S. M.,** FORTH's forte is tighter programming, *Electronics,* March 15, 1979.
2. **Brodie, L.,** *Starting FORTH,* Prentice-Hall, Englewood Cliffs, N. J., 1981.
3. Special issue, the FORTH Language, *Byte,* 5, 8, 1980.
4. Manual, *Using FORTH,* FORTH Inc., Hermosa Beach, Calif., 1982.
5. **Loeliger, R. G.,** *Threaded Interpretive Language,* Byte Publications, Peterborough, N.H., 1981.
6. **Haydon, G. B.,** Elements of a FORTH data base design, *FORTH Dimens.,* 3(2), 45, 1981.
7. **Winkel, D.,** The FORTH engine, *FORTH Dimens.,* 3(3), 78, 1981.
8. **Anon.,** CASE contest, *FORTH Dimens.,* 2(3), 36, 1980.
9. **Peterson, J. V.,** Recursion and the Ackermann Function, *FORTH Dimens.,* 3(3), 89, 1981.
10. **Schorre, D. V.,** Structured programming by adding modules to FORTH, *FORTH Dimens.,* 2(5), 132, 1980.
11. FORTH Standard Team, *FORTH '79,* FORTH Interest Group, San Carlos, Calif., 1979.

Chapter 6

# COLOR AS A MORPHIC FEATURE

**Poh-Leng Leong, Chi-Ren Chang, John Keith Beddow, Arthur F. Vetter, S. Collins, and C. Seemuth***

## TABLE OF CONTENTS



* The authors are grateful to National Science Foundation Particles Multiphase Processing Program for partial support under grant CPE-80-23868.

## I. INTRODUCTION

One of the latest developments in the morphological analysis of particulate materials has been the formulation of the set of invariant descriptors for the size and shape of particles.[1] An extension of this work is to incorporate color analysis into the set of morphic characteristics. This paper details the first attempts to extend morphological analysis to include the color of the object.

A brief introduction to the Particle Image Analyzing System (PIAS) will be given, followed by the theory which guides this research. The first part of this paper discusses an experiment to reproduce a colored image from a colored object by using PIAS, which analyzes an object in a black and white mode; the second section also using PIAS, discusses a series of experiments which have been conducted to investigate the possibility of separating and identifying sets of particles on the basis of color.

## II. PARTICLE IMAGE ANALYZING SYSTEM

PIAS is composed of a TV camera, a camera controller, a Ramtek, a slave computer, a master computer, a microscope, and three primary filters: Kodak® Wratten filter #29 (red), #47 (blue), and #61 (green) (see Figure 1).

The PIAS is a black and white system, however, the addition of the Ramtek to the circuit makes PIAS capable of being used for color analysis. The Ramtek is a CRT device which can transform inputted electronic signals into a colored image on the screen according to a predetermined (yet arbitrary) procedure.

## III. COLOR

In photography, different combinations of red, blue, and green light give the whole spectrum of colored light. Based on this principle, three primary color filters, i.e., red, blue, and green filters, were used in the analysis of the color of a particle in order to extract all the pertinent color information. For example, consider a yellow color. One should expect the amount of red and green information to be relatively higher than the blue information that we obtain when we analyze the light passing through the respective filters.

## IV. SECTION 1

In this section, the use of PIAS for colored image reproduction will be demonstrated. Look up table definitions, effects of illumination, methods of combining gray levels, and image normalizing techniques will be discussed.

### A. Experimental Method

A colored object is presented to the TV camera, the image and its color is transferred via the computer to the Ramtek unit. In the Ramtek, the color is stored as gray levels. A computer program (COL216) is then utilized to display the colored image onto the screen of the Ramtek. This image is reconstituted from the gray levels contained in the three files: RED, BLUE, and GREEN, which have been previously set up in the computer memory. It is very important to note that in using this electronic procedure the color of the final image is quite arbitrarily determined. The general direction of the experiment is to try to ensure that the color of the Ramtek image is comparable to the color of the original. Because there is a limitation within the Ramtek (a maximum of only 216 color levels are available), it is not possible to exactly match color for color. Yet another limitation is related to the color filters. For example, the "red" filter actually transmits yellow and magenta as well as red itself. This impurity in the transmission characteristics makes for inadequate color repro-

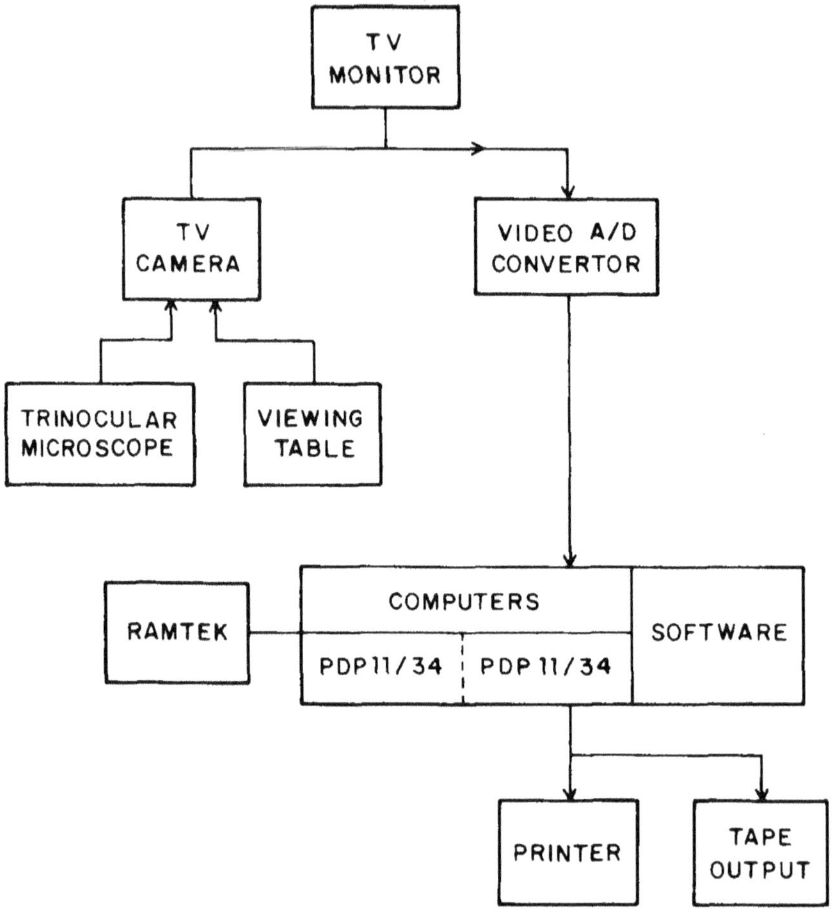


FIGURE 1. Particle image analyzing system.

duction, especially when coupled with the limitations of the Ramtek. In any case, faithful reproduction of color is not the basic requirement, rather we are looking to be able to consistently measure a given color. This is the basic requirement for the analysis of color as a morphic feature.

### B. Constituting the Image

Figures 2, 3, and 4 are Ramtek reproductions of the logo shown in Figure 5. Figure 2 was obtained using a red filter, Figure 3 was made using a blue filter, and Figure 4 a green one. All of these were made using 256 color levels (i.e., gray levels). It should be noted that the images in Figures 2, 3, and 4 are not especially red, blue, or green, respectively. This is directly related to the arbitrariness of the color reproduction (referred to later).

### C. Look-Up Table (LUT)

The Look-up table (LUT) defines the correspondence between the gray level in the Ramtek and the color of the object. Figures 6 and 7 illustrate the ranges of colors available in the case of two LUTs. In Figure 6, 64 colors are available and in Figure 7, there are 216 of them. For example, going along the rows in Figure 7 from top left to bottom right we move from a red level of 0 to a red level of 5. Within each matrix the vertical direction is the green scale (again 0 to 5 from top to bottom) and the horizontal direction is the blue scale going from 0 to 5. This combination gives a total of 216 colors.

### D. Results

The combination of the three images shown in Figures 2, 3, and 4 is shown in Figure 8. This image is reproduced from a LUT with an inventory of 64 colors. As might be expected, the larger the inventory of colors used to reproduce the measured image, the more satisfactory the reproduction appears. This is shown in Figure 9 where 216 colors were used to regenerate the figure. Compare Figures 8 and 9.

### E. The Arbitrary Aspect of Color Reproduction

Figures 2 and 10 are Ramtek reproductions of the logo using the same red filter in each case, but with different LUTs in each case. Clearly, the choice of correspondence between the color and gray level determines the appearance of the final Ramtek image. This clearly shows that the color of the final image is quite arbitrary.

### F. Effect of Illumination

In any imaging process it is always possible to either underexpose or overexpose. This is true in the case of chemical photography and it is also true in electronic image processing. Figures 11 and 12 illustrate this point. In Figure 11, too much light was focused on the central region of the image. This resulted in a central, bright, fuzzy region of poorer definition. Figure 12 was more evenly illuminated. The result is much improved. The quality of illumination affects the reproduced image. However, the image is more often than not determined by subjective judgment. We have found that the measured gray levels are very sensitive to the mode of illumination. This last point is important for morphological analysis purposes.

### G. Normalizing the Image

The quality of the reproduced image may be further manipulated by using different ways of combining the various gray levels. For example, Figure 13 was obtained in the following way: the TV camera was used in a mode with 256 gray levels. These were reduced (compressed) to six gray levels for each of the three files corresponding to the three filters, red, blue, and green, respectively. This gives $6 \times 6 \times 6 = 216$ colors in the final image reproduction. The picture appears to have a pinkish background. This was modified as shown in Figure 14 in the following way: the averages of the gray levels in the three files were normalized using the light from a white object as a normalizing source. In this procedure, the mean gray levels for red, blue, and green from the white paper were used as weightings with the gray levels of the logo picture. The result in Figure 14 is less pink, but somewhat more fuzzy. If the weightings are used in the numerator, rather than in the denominator (as was done for Figure 14, for example), the image quality can be further enhanced as shown in Figure 9.

### H. Discussion

It is possible to produce a colored image of a colored object using PIAS. This opens up a new area in the morphological analysis of fine particles. It should be possible to provide a set of three color descriptors for each particle analyzed that would uniquely identify the particle. Various approaches are possible. At the simplest level, the average gray level of the particle can be found for each of the three filters used. In more complex analysis, contours of gray levels can be digitized on PIAS and the unique contour shape and contour level can then be used to identify any particle. Correlations may be sought between these data and particle chemical properties.

The next section of this paper will discuss the steps to produce color descriptors that can be used to identify, differentiate, and classify sets of particles. In this work, accurate measurement of gray levels will be important rather than the quality of image reproduction.

# V. SECTION 2

This section discusses a series of experiments which have been conducted to investigate the possibility of separating and identifying sets of particles on the basis of color, using the PIAS. Experiments using model and actual colored particles are reported. Future work is also discussed.

## A. Purpose

The purpose of the experiments was to try to differentiate and identify sets of particles on the basis of color, using the PIAS. A systematic approach was planned. The first sets of experiments were conducted using *model particles* (a 24-color chart); next, experiments with *real particles* (crystalline chemicals) were carried out. These experiments, the results, and conclusions are discussed below.

## B. Experimental Method

The model and real particles were observed in the PIAS using selected filters, and in some experiments no filters were used. In some other experiments, colored light sources were used. In all experiments, the average, $\bar{x}$, and standard deviation, $\sigma$, of the gray level values of the particle were calculated. Particular methods of analyses for each experiment are discussed more fully under the appropriate section.

## C. Model Particle Experiments

The purpose of these experiments was to achieve a satisfactory demonstration of the use of the PIAS to differentiate quantitatively between particles of different colors. Model particles were used as they were uniform in color. In these experiments, a 24-color chart was used to represent homogeneously colored "ideal" particles (see Figure 15). A computer program calculated the average, $\bar{x}$, and standard deviation, $\sigma$, of the gray-level values within a square window of size N by N, the position and size of which is specified by the user (see Figure 16). This method is adequate for use with the "ideal" particles as the color grids are uniformly colored, with no variation in the pattern distribution of color.

The PIAS is used in the black and white mode, with a 256-level gray scale. The scale ranges from 0 (black) to 255 (white). The particles are observed using the TV camera.

### *1. Experiment 1*

Gray-level data (i.e., $\bar{x}$ and $\sigma$) for the various color grids of the color chart were obtained using the black and white mode of the PIAS, without any filters. Table 1 shows the numbering system for different color grids. Figure 17 shows the position of the different colors on the 256-level gray scale.* Darker colors give a lower $\bar{x}$ value than the lighter or brighter colors.

Figure 18 shows the position of the six gray shades of the color chart, with black giving the lowest $\bar{x}$ value, and white the highest. In between the black and white colors, progressively brighter shades of gray give progressively higher $\bar{x}$ values. These data are in line with our expectations.

### *2. Experiment 2*

In experiment 1, the ability to differentiate colored particles was limited. In this experiment, red, blue, and green filters were used when the $\bar{x}$ and $\sigma$ readings were taken. Thus, three separate readings were obtained for each color. This increased the dimensionality of the analysis to facilitate differentiation. These readings were plotted onto a three-dimensional

* Note that the standard deviation deviation, $\sigma$, around the average value, $\bar{x}$, is not shown in all the plots. A certain amount of overlap is present.

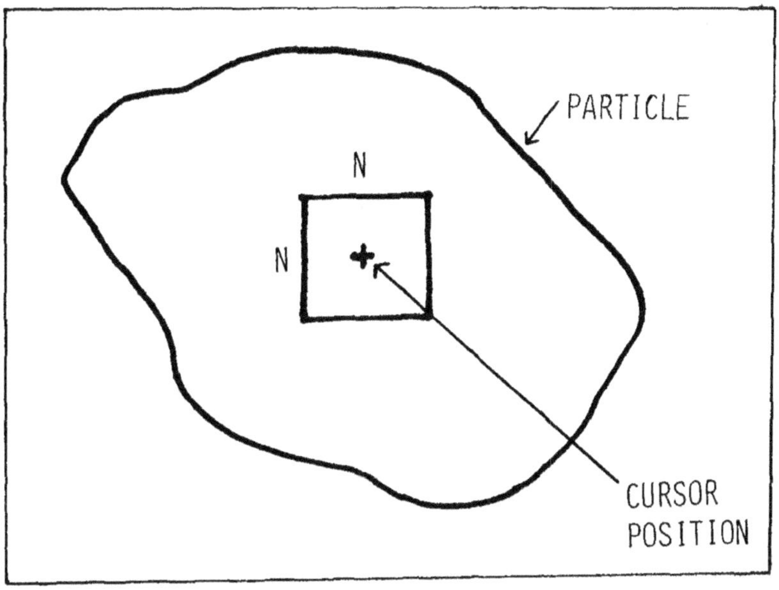


FIGURE 16. Gray-level data are sampled within the window of size N by N to compute the average, x, and standard deviation, σ, of the gray-level values.

**Table 1**
**THE NUMBERING SYSTEM FOR THE COLOR GRIDS**

| Number | Color |
|---|---|
| 1 | Brown |
| 2 | Light brown |
| 3 | Blue |
| 4 | Murky green |
| 5 | Light purple |
| 6 | Light blue-green |
| 7 | Dark orange |
| 8 | Blue |
| 9 | Orange-red |
| 10 | Dark purple |
| 11 | Apple green |
| 12 | Light orange |
| 13 | Dark blue |
| 14 | Green |
| 15 | Dark red |
| 16 | Bright yellow |
| 17 | Deep pink |
| 18 | Turquoise |

(3-D) graph with the x-axis representing the blue scale, the y-axis the green scale, and the z-axis the red scale. Figure 19 shows the results of this experiment.

The plot does not indicate much spread in the blue and green direction. This is due to the low transmission of light through these filters. The red filter gave the most spread, with high values indicating that the particular grid has more red, and low values indicating little red.

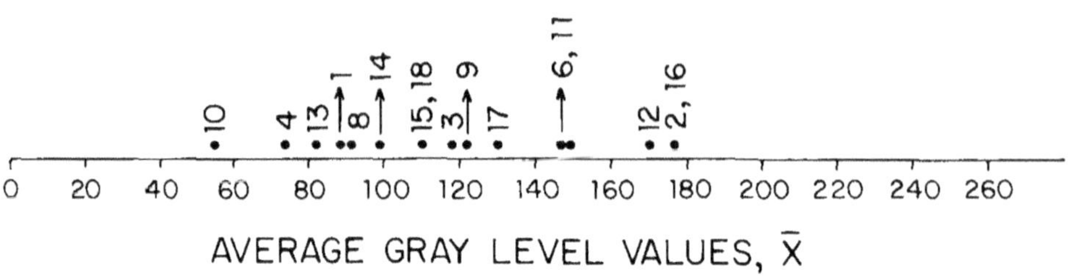


FIGURE 17. Ability to differentiate between colors is limited when using a one-dimensional analysis.

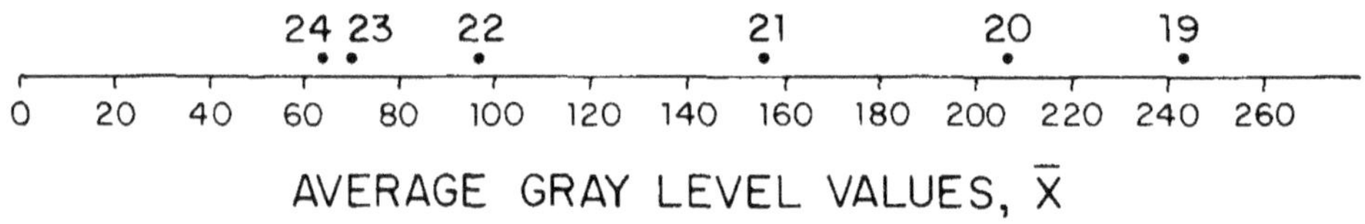


FIGURE 18. Number 19 represents a white color and number 24, black. Numbers 20 to 23 are progressively darker gray colors (see Figure 15).

*3. Experiment 3*

The low transmission of the blue and green filters reduced the differentiation capability in experiment 2. In experiment 3 a thinner yellow filter with a higher transmittance was used in place of the thicker green filter. Results are shown in Figure 20. Note the greater spread in the yellow direction. The colors are separated a little better now, thus improving the differentiation capability of the system.

*4. Experiment 4*

In this experiment, a different technique was used. Three red, blue, and green light sources were used in place of the red, blue, and green filters. The light sources have different intensities, with the blue showing the lowest intensity and red the highest. This difference is accounted for by adjusting the camera aperture until almost similar $\bar{x}$ values are obtained when the light reflected off a white piece of paper is analyzed on the PIAS.

Figure 21 shows the results of the experiment. Numbers 2, 7, 9, 12, 15, 16, and 17 are red-rich and found high up in the color space. Blue-green colors are found in the lower part.

*5. Discussion of Experiments 1 to 4*

Experiments 1 and 2 show that particles can be differentiated on the basis of color when we increase the dimensionality of color analysis from one to three dimensions. For example, color grids number 15 and 18, 6 and 11, and 2 and 16 all lie on the same position on the line chart (Figure 17), however, they do not lie on the same spot on the 3-D graph (Figure 19). Grids 6 and 11 and 2 and 16 are not too greatly separated on the 3-D graph, but grids 15 and 18 are indeed well separated. This may suggest that, perhaps, if the analysis is increased to n dimensions, the power of differentiation may be increased further.

Experiments 2 and 3 illustrate the need for thin filters that will transmit a high percentage of light. This high resolution will allow for a wider spread of the $\bar{x}$ values needed for differentiation purposes. An alternative solution is to use colored light sources and account for the different intensities of the light sources by adjusting the aperture of the camera. Compare graphs in Figures 19, 20, and 21. Another alternative is to adjust the aperture of the camera to account for the different transmittance of the filters. This will be discussed later in Experiment 8.

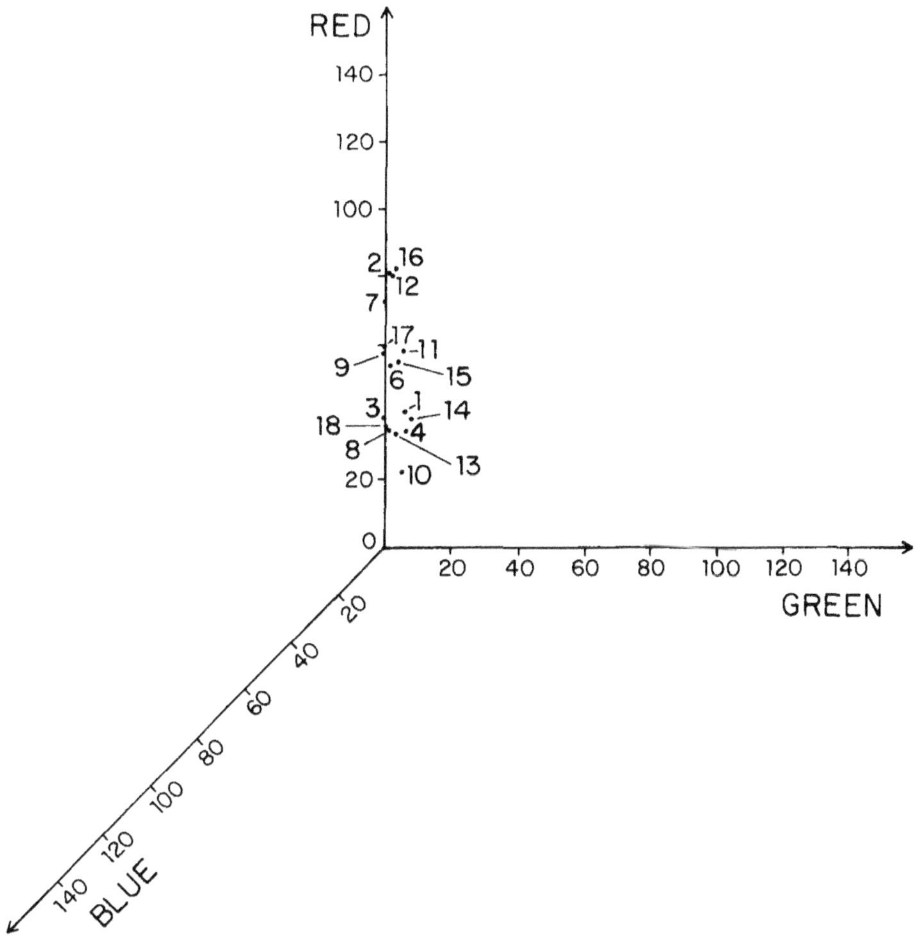


FIGURE 19. Location of the various color grids on the 3-D blue-green-red space.

*6. Conclusion*

Comparing Figures 19, 20, and 21, the method of using colored lights gives the best separation of the particles.

**D. Real Particle Experiments**

Results from the model particle experiments were encouraging. Real particle experiments were conducted next. The particles used were colored crystalline chemicals. The PIAS is used in its black and white mode and sets of particles are observed under a microscope.

*1. Experiment 5*

In this experiment, what we learned in the model particle studies is applied to a set of real particles. This experiment utilizes a computer program that traces the boundary of the crystal, and the $\bar{x}$ and $\sigma$ values are then calculated within the enclosed particle. This method was used because the particles exhibit nonuniform pattern distributions of color. Results are shown in Table 2. Within each filter group, the numbers were very similar.

It was perplexing to find little differentiation between the particles, even though they were distinctly different in color. On subsequent analysis, the reason for this was clear. For example, when we look at a green nickel sulfate crystal or a reddish potassium ferricyanide crystal with the naked eye, we are seeing the reflected color of the object. When we observe

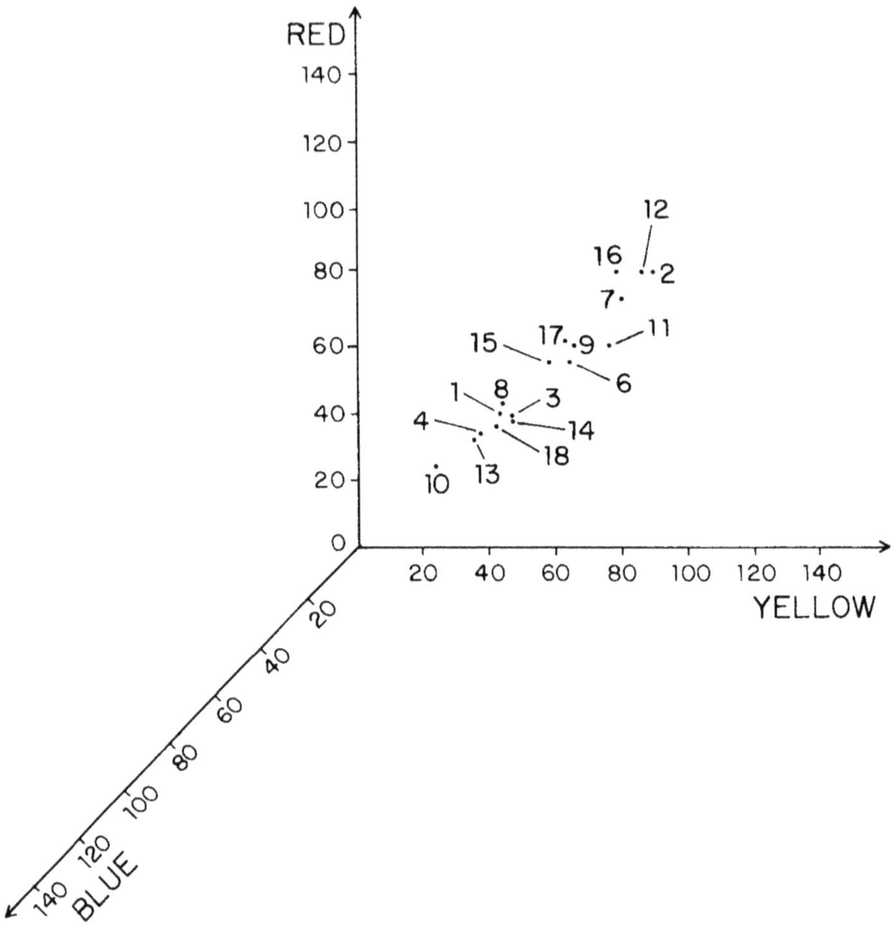


FIGURE 20. Location of the various color grids on the 3-D blue-yellow-red space.

the same crystal under the transmission microscope, the light that passes through the crystal into our eyes has been reflected multiple times inside the crystal and refracted through the crystal. For example, the nickel sulfate crystal shows bright center spots and dark edges under the transmission microscope while it appears uniformly green to the naked eye. Most of the crystals used in this experiment showed the same dark-edge and bright-center effects, thus obtaining the $\bar{x}$ values for the whole particle which gave almost bland results and full weight was not given to all the color features of the image.

*2. Experiment 6 — Measuring Color Variations Within a Crystal*

In this experiment, the gray level distribution within a nickel sulfate crystal, when observed under a transmission microscope, was obtained. Figure 22 shows the gray-level distribution of the crystal at certain areas within the crystal. These gray-level data (i.e., $\bar{x}$ and $\sigma$) were obtained without using any filters.

High values of $\bar{x}$ correspond to the bright spot while low values correspond to the dark edges. High values of $\sigma$ were obtained when the window enclosed both bright and dark areas. Thus, different readings were obtained at different areas of the crystal. Since the crystals do not appear uniformly colored when observed under the transmission microscope, a reflecting microscope was used in the next experiment.

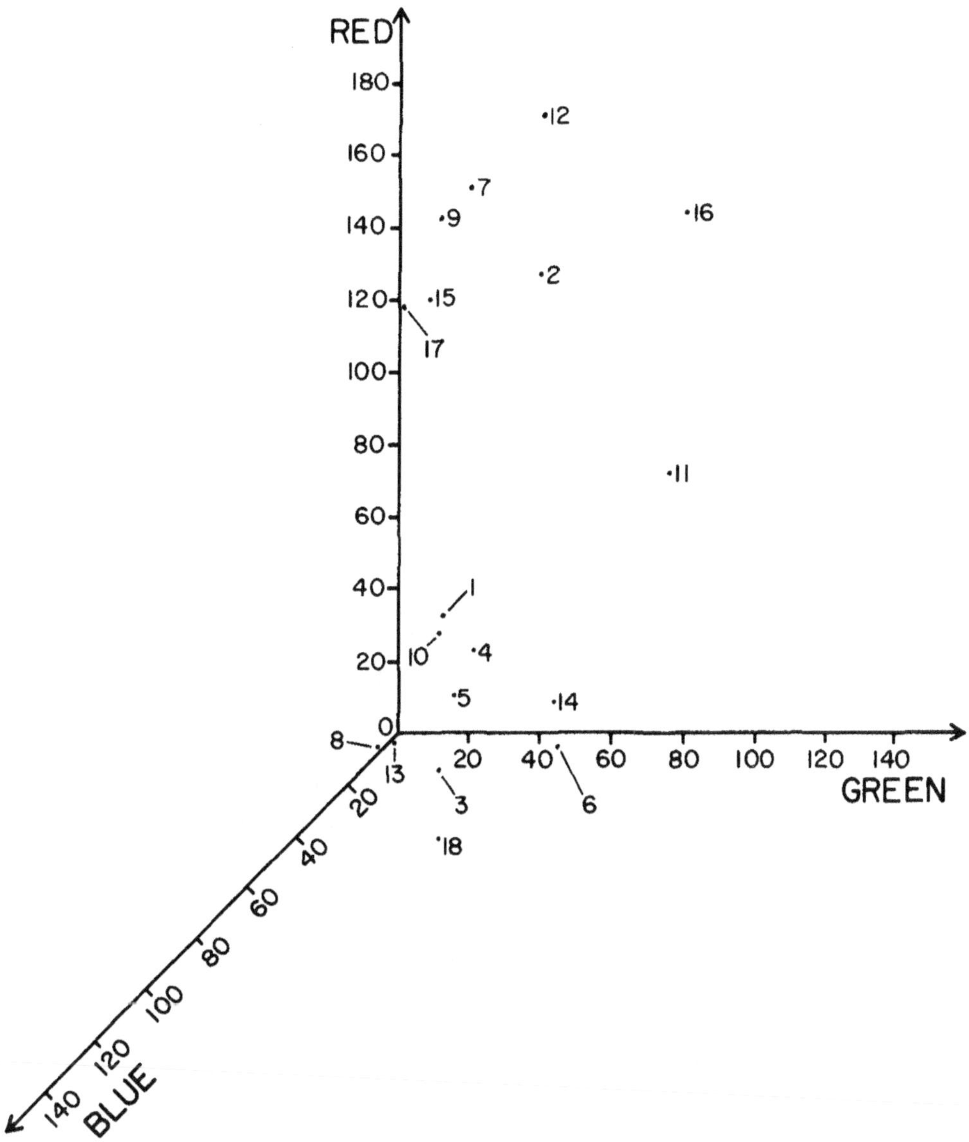


FIGURE 21. Location of the various color grids on the 3-D color space. Colored light sources are used in place of the filters.

*3. Experiment 7*

In an effort to reduce the variation of color within a field of view, a metallurgical microscope was used. The particle appeared more uniformly colored now, but the microscope was not suitable for the type of experiment performed here. The usual method of placing the filters between the particle and the objective lens was not applicable because reflections now took place off the filter, and so light that is transmitted through the filter interfered with the light reflected from the filter. When filters designed specifically for the metallurgical microscope were used, no consistent readings were obtained because the slot did not provide a tight fit for the filter. Furthermore, the filters may not be optically suitable.

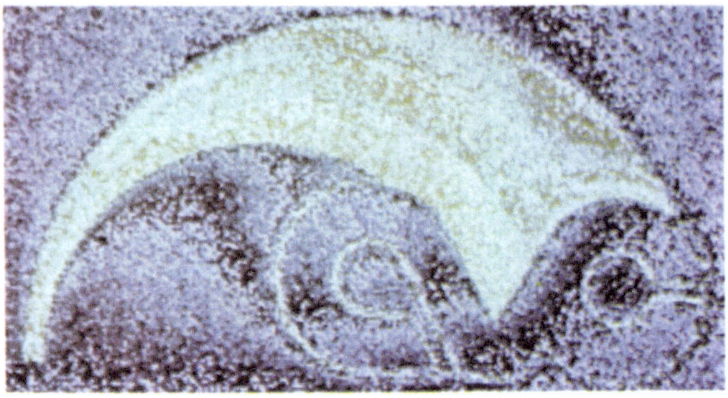

FIGURE 2. Reproduction of the logo using a red filter.

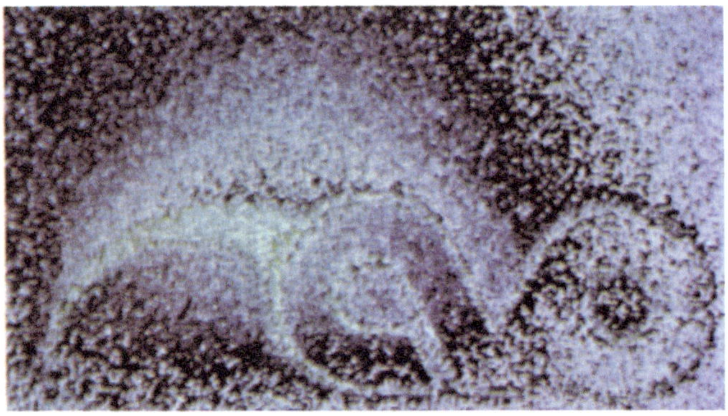

FIGURE 3. Reproduction of the logo using a blue filter.

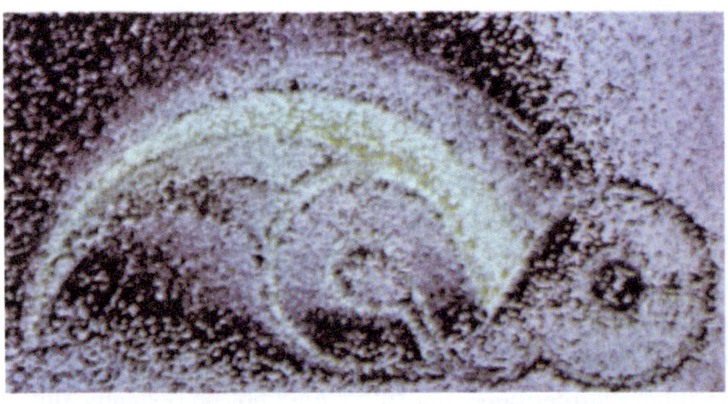

FIGURE 4. Reproduction of the logo using a green filter.

FIGURE 5. Original image.

FIGURE 6. Look-up table — 64 colors.

FIGURE 7. Look-up table — 216 colors.

FIGURE 8. Reproduced image with an inventory of 64 colors in the LUT.

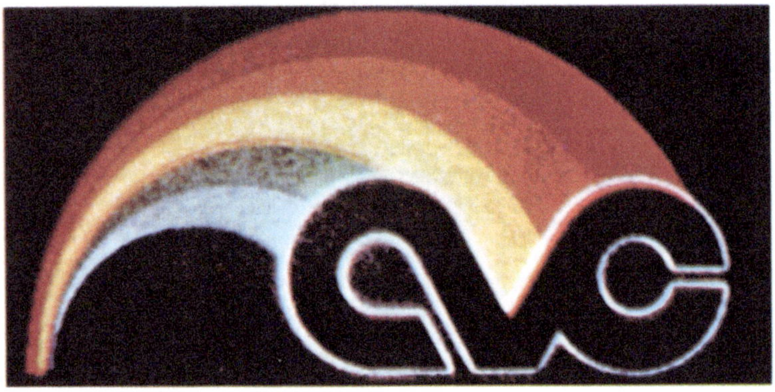

FIGURE 9. Reproduced image with an inventory of 216 colors in the LUT (normalized in the numerator).

FIGURE 10. Reproduction of the logo using a red filter (using a differently defined LUT; compare with Figure 2).

FIGURE 11. Effects of illumination — overexposed.

FIGURE 12. Effects of illumination — less overexposed.

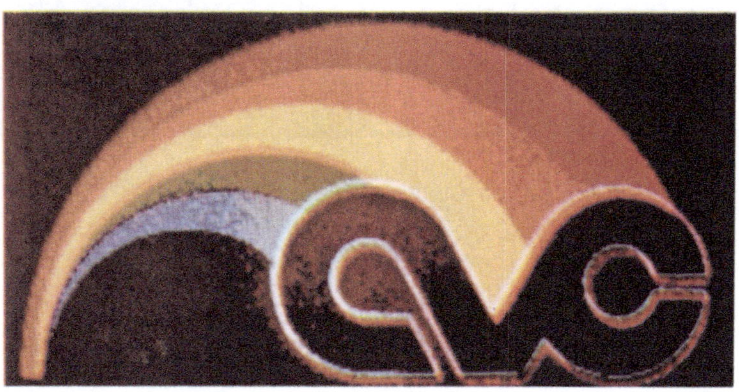

FIGURE 13. Reproduced image — unnormalized.

FIGURE 14. Reproduced image — normalized in the denominator.

FIGURE 15. 24-Color chart.

**Table 2**
**MEASUREMENT OF GRAY-LEVEL DATA USING THE TRANSMISSION MICROSCOPE**

| | | Filter | | |
|---|---|---|---|---|
| **Particle** | **Color** | **Red** | **Blue** | **Green** |
| Nickel sulfate | Green | $\bar{X}$ = 136 | 124 | 127 |
| | | σ = 9.5 | 6.9 | 8.4 |
| Potassium permanganate | Dark purple | 131 | 120 | 121 |
| | | 10.6 | 6.9 | 7.1 |
| Copper sulfate | Blue | 131 | 121 | 121 |
| | | 9.3 | 6.4 | 6.4 |
| Potassium ferricyanide | Red | 140 | 117 | 118 |
| | | 8.9 | 6.1 | 6.3 |

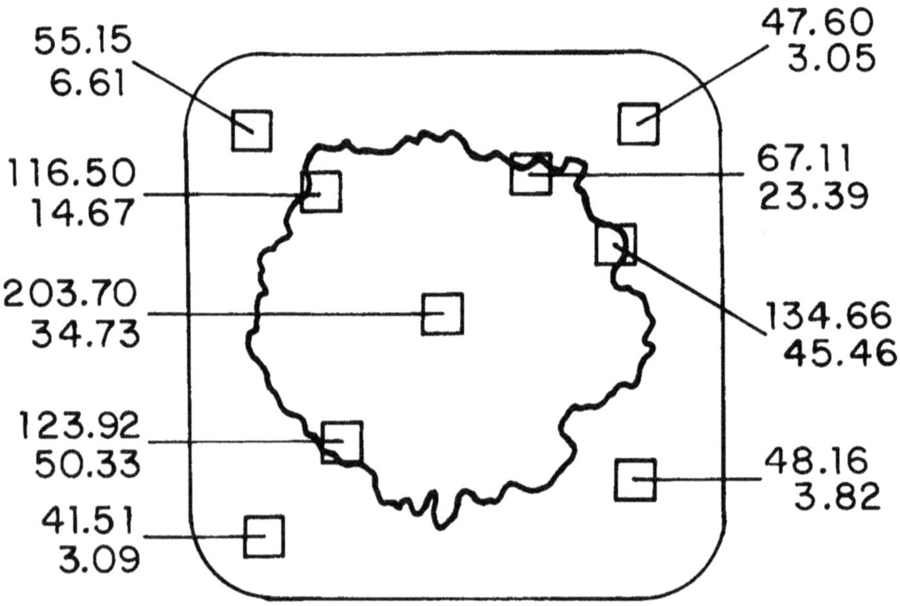


FIGURE 22. Gray-level distribution at certain areas within the crystal.

*4. Experiment 8*

An improved reflecting microscope was required. For this purpose, the BHC Olympus transmission microscope was converted into a reflecting microscope. A piece of white paper was put across the stage of the microscope. Four stage lights (the intensity of which was controlled by a transformer) were shone onto the stage. When the crystal was viewed under this "reflecting" microscope, it appeared more uniformly colored, and more like it is seen with the naked eye.

In this experiment, various crystalline chemicals were used. Table 3 shows the numbering system and color of the particles used in the experiment.

Figures 23, 24, and 25 show the results obtained under different experimental conditions. Results in Figures 23 and 24 were obtained when the particles were observed using the "reflecting" microscope, while the results in Figure 25 were obtained when the particles were observed using the TV camera.

**Table 3**
**NUMBERING SYSTEM AND COLOR OF PARTICLES**

| Number | Particle | Color |
|---|---|---|
| 1 | Potassium ferricyanide | Orange-red |
| 2 | Potassium ferricyanide | Orange-red |
| 3 | Potassium ferricyanide | Orange-red |
| 4 | Nickel sulfate | Green |
| 5 | Nickel sulfate | Green |
| 6 | Cupric acetate | Dark green |
| 7 | Copper sulfate | Blue |
| 8 | Mercuric iodide | Red |
| 9 | Potassium dichromate | Orange |
| 10 | Potassium chromate | Yellow |
| 11 | Ferric chloride | Yellow |
| 12 | Potassium permanganate | Dark purple |
| 13 | Chromium fluoride | Green |
| 14 | Potassium ferrocyanide | Light yellow |
| 15 | Tungstic acid | Yellow |
| 16 | Cupric carbonate | Darker green |
| 17 | Cupric carbonate | Lighter green |
| 18 | Vanadium sulfate | Blue |

The conclusions drawn were similar to those from the model particle experiments conducted earlier. Thinner filters have a higher transmittance and the resolution is thus higher. The ability to separate the particles, on the basis of color, is enhanced. Compare graphs in Figures 23 and 24.

Experimental difficulties were encountered when experiments with colored light sources were conducted. The intensity of the light sources could not be adjusted in the microscope, or be externally controlled, as there were no aperture controls above the microscope stage and no transformer was connected to the light sources at the time the experiment was carried out. Instead, the camera was used, and the differences in the transmittance of the filters were accounted for by adjusting the aperture of the camera. The results show that this method is superior to the other methods used in this experiment. Compare Figures 23, 24, and 25.

*5. Experiment 9*

This experiment deals with factors that produce variations in readings. The variation of the $\bar{x}$ and $\sigma$ values with different experimental conditions was found. Tables 4, 5, and 6 show the variation of $\bar{x}$ and $\sigma$ under different illumination conditions, magnification, and the effect of focus, respectively.

Illumination effects are always important. The gray-level values are very sensitive to the light intensity. Standard or constant conditions should always be provided for in all experiments. Similarly, the same magnification should always be used for a particular particle. Although focusing does not appear to be a factor of overwhelming importance, the particle should always be in focus.

*6. Consistency of Measurements*

Table 7 shows the variation of $\bar{x}$ and $\sigma$ values of four particles when the values are taken during different times of an experimental session. The amount of variation is slight.

Figure 26 shows the variation when the values of particular particles were taken during experiments performed on different days. The variation is no longer trivial. The particles numbered 14 and 15 are easily misidentified as particle 10. This apparent inconsistency in the measurement of the $\bar{x}$ and $\sigma$ values is explained by the fact that the experiments were

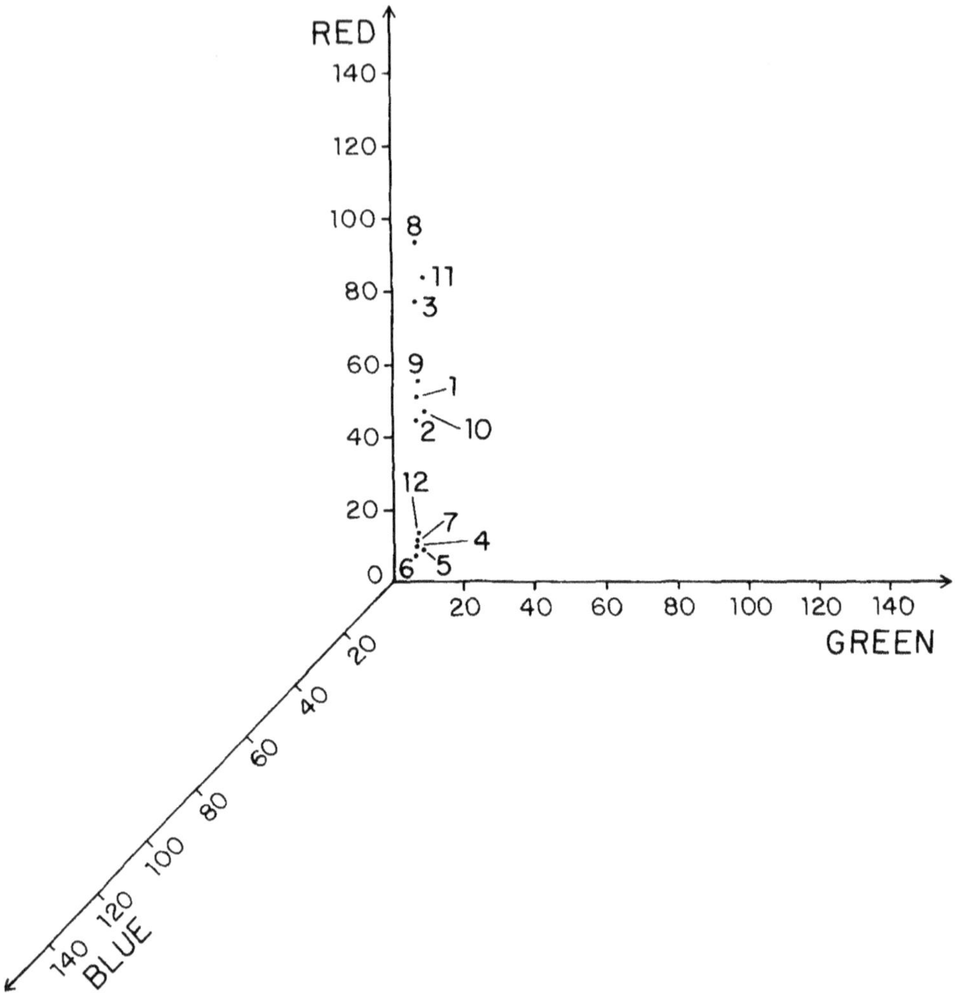


FIGURE 23. Location of differently colored crystals in the 3-D blue-green-red space. The blue and green filters transmit much less light than the red filter.

conducted on two different days. The aperture opening and magnification and other relevant factors were kept the same, but the amount of daylight entering the laboratory was beyond the control of the experimenter. Both experiments were conducted at about the same time each day. However, there must have been a difference in the daylight illumination in the laboratory. The gray-level values are very sensitive to the intensity of light. Looking at the position of the particles on the 3-D color space, all six positions are located relatively higher in the second experiment, but the general distribution of the points is the same. For example, the set of particles numbered 15 is to the left of number 14 and below number 10 in both experiments. This illustrates the need for precise illumination control in the laboratory when future experiments are carried out to characterize the color of the particles.

*7. Conclusion*

It is possible to differentiate and identify sets of particles on the basis of color on the PIAS. In order to do this, a set of high-transmittance filters are required. Alternatively, colored light sources or filters could be used when the differences in the intensity of the

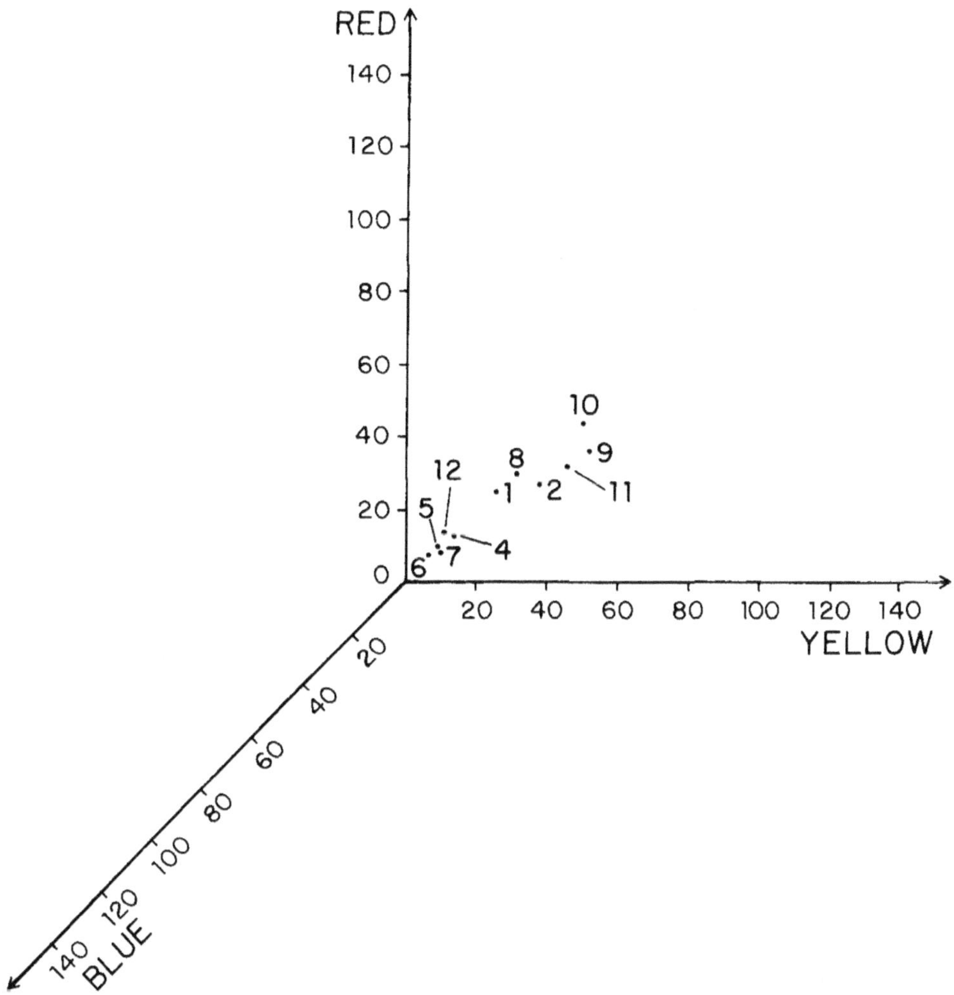


FIGURE 24. Location of the differently colored crystals in the 3-D blue-yellow-red space. Thinner filters are used here.

colored light or transmittance of the filters are accounted for by adjusting the aperture of the TV camera or microscope.

In this study, only uniformly colored particles were observed. Future work would entail characterizing the pattern distribution of color for all particles: uniformly, multicolored, or nonuniformly colored.

## VI. FUTURE WORK

The analysis of color on the PIAS is extremely sensitive to the illumination in the laboratory. Some recommendations on future experiments are given below:

1. The size of the particle influences the color. This has to be examined in detail.
2. The reflection off the surface of the particle will influence the reading. This reflectivity is a characteristic of the particle. Potential information regarding the reflectivity can be gathered from the analysis.
3. The illumination in the room must be controlled precisely so as to obtain consistent, and hence, reproducible results.

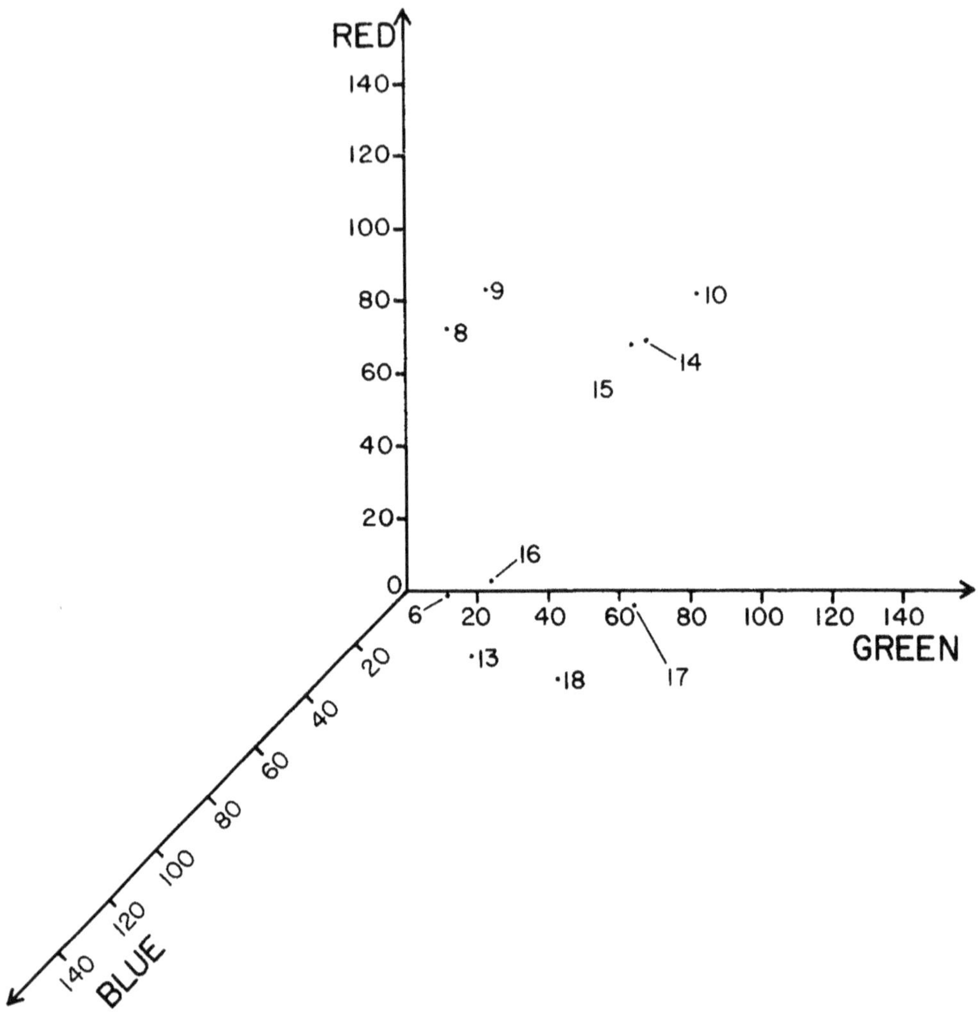


FIGURE 25. Location of the differently colored crystals in the blue-green-red space. The differences in the transmittance of the filters are accounted for by adjustments in the camera aperture.

4. Further increase in the dimensionality of the analysis with the use of complementary color filters, which are yellow, cyan, and magenta; or a set of 12 interference filters, each covering 25 mm of the 400- to 700-mm visible wavelength range, should be considered.

## VII. POTENTIAL APPLICATIONS OF THE COLOR ANALYSIS OF THE PARTICLES

This area of research has potential applications in manufacturing processes involving the processing of particulate materials. Of course, in the broader context, in conjunction with the size and shape characteristics of the particle, it appears to offer a useful analytical tool for object identification and classification. Other areas of application may include robotics, the food industry, the paint industry, etc.

**Table 4**
**EFFECT OF ILLUMINATION**

| Voltage of transformer | Aperture of camera at 16 | Aperture of camera at 11 |
|---|---|---|
| 80 V | $\overline{X}$ = 240 | $\overline{X}$ = 211[a] |
| | σ = 5.9 | σ = 4.0 |
| 70 V | 167 | 239 |
| | 6.3 | 6.9 |
| 60 V | 93 | 185 |
| | 4.7 | 6.0 |

[a] Reading is overexposed.

**Table 5**
**EFFECT OF MAGNIFICATION**

| Length of camera extension used | Gray-level measurement |
|---|---|
| 5 mm | $\overline{X}$ = 234 |
| | σ = 7.6 |
| 20 mm | 229 |
| | 5.9 |
| 50 mm | 210 |
| | 3.9 |

**Table 6**
**EFFECT OF FOCUS**

| Focus at | Gray-level measurement |
|---|---|
| 0.6 m | $\overline{X}$ = 218 |
| | σ = 4.6 |
| 0.7 m | 220 |
| | 4.8 |
| 1.0 m | 224 |
| | 5.3 |
| 1.2 m | 226 |
| | 5.6 |
| 1.5 m | 227 |
| | 5.8 |
| 2.0 m | 229 |
| | 6.1 |
| ∞ | 233 |
| | 6.3 |

**Table 7**
**CONSISTENCY OF MEASUREMENTS**

| Color | Filter | Measurement of gray-level data | | |
|---|---|---|---|---|
| Blue | Red | $\bar{X} = 87$ | $\bar{X} = 87$ | $\bar{X} = 88$ |
| | | $\sigma = 7.1$ | $\sigma = 7.1$ | $\sigma = 7.2$ |
| Murky green | None | 74 | 76 | 74 |
| | | 6.9 | 7.1 | 7.0 |
| Light brown | Blue | 96 | 97 | 97 |
| | | 8.7 | 7.9 | 8.0 |
| Light orange | Red | 226 | 221 | 221 |
| | | 12.8 | 5.9 | 6.7 |

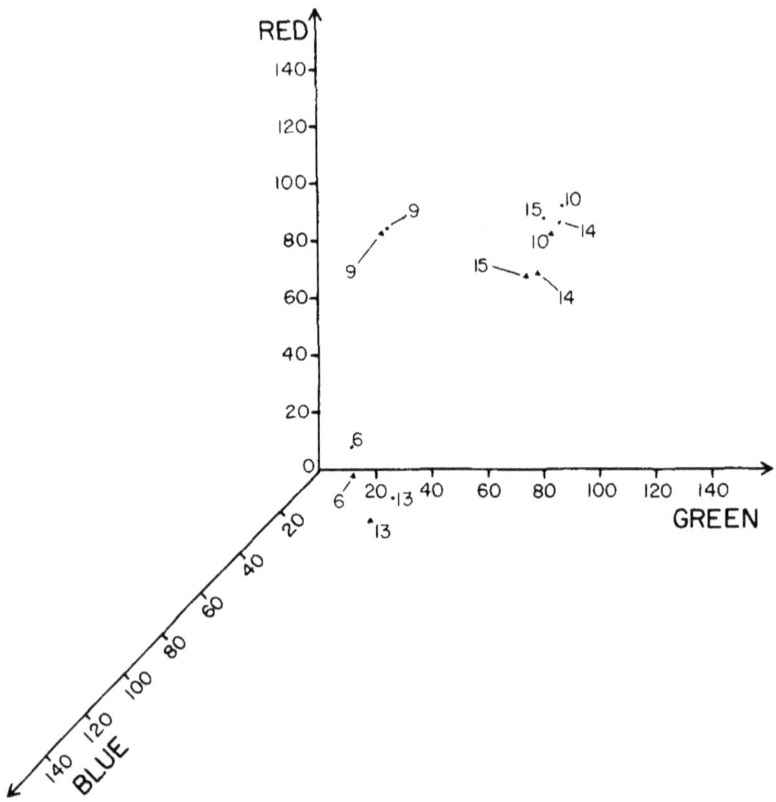


FIGURE 26. Test of consistency of measurements on different experimental sessions.

## REFERENCES

1. **Beddow, J. K. et al.,** Principles and applications of morphological analysis in particulate technology, Proc. Int. Symp. Powder Tech., Society of Powder Technology, Kyoto, Japan, 1981, 10.
2. **Leong, P.-L. et al.,** Analysis of color as a morphic feature, Proc. 13th Annu. Meet. Fine Particle Soc., 1982.
3. **Leong, P.-L. et al.,** Differentiating particles using color as a morphic feature, Proc. Powder Bulk Solids Conf., AMF O'Hare, Chicago, 1982, 191.

# *Part II*
# *Data Analysis*

Chapter 7

# FOURIER SHAPE ANALYSIS — A MULTIVARIATE PATTERN RECOGNITION APPROACH

**Robert Ehrlich and William E. Full**

## TABLE OF CONTENTS

## I. INTRODUCTION

One of the general objectives of sedimentology is to determine the origin ("provenance") of sediments. Commonly, attempts to determine provenance are predicated on the assumption that sediment sources yield a characteristic suite of minerals. Hence, once mineralogy is determined, the presence of diagnostic minerals is an unequivocal indication of the contribution of a specific sediment source. A problem arises, however, if there is little compositional contrast between sources. This may arise because either the sources have common mineral assemblages, or the sources in question are not primary but consist of quartz-rich sediments which have lost most nonresistant minerals to dissolution or abrasion.

The most resistant, common, rock-forming mineral is quartz. In general quartz abundance increases in the sand-silt size range while other, less resistant, minerals are selectively removed. Because of its resistance to chemical and physical attack, the shapes of quartz particles change very slowly under the influence of processes associated with transport and deposition. Not surprisingly, the shapes of quartz sand and silt particles from different source terranes (or with different transport histories) are generally different.[7]

Most sedimentary deposits are of mixed provenance and often the number and nature of sources are not known a priori. Thus, any attempt at using quartz shape for provenance purposes must determine three things:

1. The number of sources
2. The shape signature of each source
3. The relative contribution of each source to each sample

In addition, shape data must be acquired economically, allowing analysis of hundreds of particles per sample and tens or hundreds of samples in any particular investigation. To this end, a simple image analysis procedure generating edge points of particles and a set of pattern recognition algorithms accomplishing end member (source) characterization and mixing proportions has been designed.

## II. IMAGE ANALYSIS

The perimeters of two-dimensional projections of quartz particles are digitized using a microprocessor-controlled, microscope-mounted video scanning system (Figure 1). The associated video digitizer accepts the analog signal from the video scanner and generates a spatial array of points (pixels), each of which has an associated shade of gray. Because the digitizing unit has an on-board memory acting as a buffer, it is synchronized with both the scan rate of the video scanner and the acquisition rate of the microprocessor. This permits digitization of a scene in 1/60 sec.

The microprocessor uses simple thresholding (pixel value equal to a dark shade of gray) to find the first edge point. Then, using a search pattern involving small excursions, first away from the edge and then inward, the particle outline is traced pixel by pixel until closure is obtained.[8] Depending on the magnification, 200 to 1000 edge points are generated for each particle. These edge points are used as raw data for the next step in the analysis.

## III. SHAPE MEASUREMENTS

Amplitude spectra of a finite Fourier series in closed form are used as shape descriptors of each particle. A Fourier series in closed form can be used to describe the outline of a maximum projection silhouette of a grain to any desired precision.[6] Each term in this series is of the form $R_n \cos(n\,\theta - \phi_n)$ where $R_n$ is the amplitude of the nth term or harmonic and $\phi_n$ is the phase angle associated with this specific harmonic. Each harmonic in the series represents the contribution of a specific shape component to the total shape of the grain.

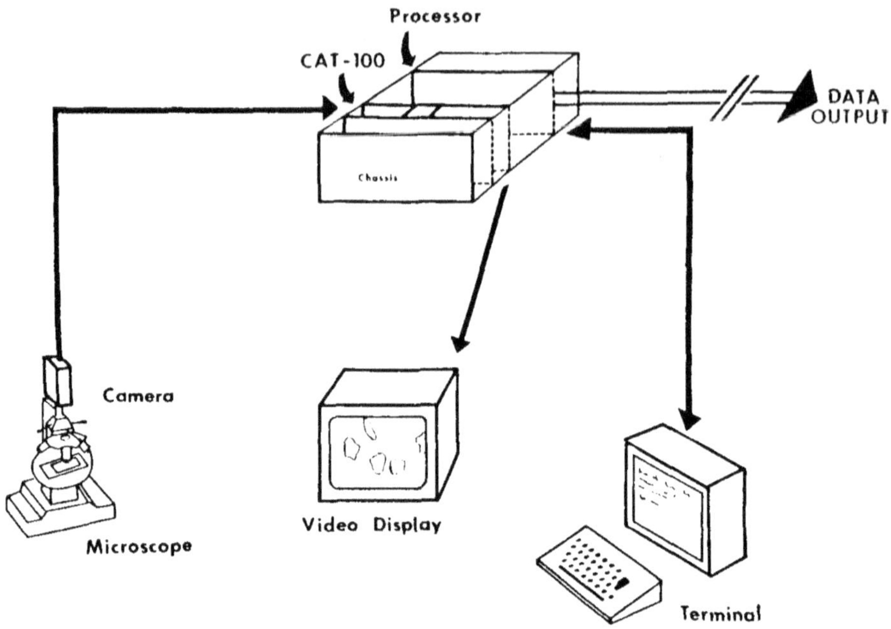


FIGURE 1. System hardware configuration.

For example, the second harmonic represents the contribution of a "figure-eight", the third represents a trefoil, the fourth a quadrefoil, and so on. In general, the nth harmonic represents the contribution of an n-leaved clover to the empiric shape of the grain (Figure 2). Thus, the lower harmonics are a measure of gross shape while the higher harmonics measure increasingly fine-scaled surface features. Because of the limited resolution of the measuring device, only the first 24 harmonics are calculated and used in the shape analysis.

The coordinates of the maximum projection outline derived from the video digitizing system are used to calculate the Fourier series. To produce homology between grains, the centroid of a subset of 48 points extracted from the original grain outline is always used as an origin for the Fourier series.[10] Although any origin will produce a series that converges to the same shape, the use of this centroid ensures that two grains having exactly the same shape will have the same harmonic values for their respective Fourier series. This allows the Fourier harmonic amplitudes between grains to be compared. The magnitude of the first harmonic amplitude (an offset circle) is a measure of accuracy in finding this centroid and is not used in any shape comparisons.

For each sediment sample, 200 or more randomly selected quartz grains (generally within a narrow size range) are analyzed. Each sample is thus represented by 200 Fourier series. At any specific harmonic of those series, a sample is represented by 200 amplitude values (one for each particle). These values can be displayed in a "shape-frequency distribution" at each harmonic wherein each harmonic amplitude is plotted against frequency of occurrence.[7] Thus, each sample is represented by 23 shape-frequency distributions — one for each harmonic from 2 to 24.

Aspects of information theory[21] can be used to cast the above harmonic amplitude distributions in an optimal configuration.[12] The algorithm that produces such an optimal configuration is based on information entropy. The information entropy (as defined by Shannon[21]) of any frequency plot is a measure of contrast between intervals. A low entropy value indicates that the amplitude value of any grain has a much greater probability of falling in certain intervals than in others. Thus, a frequency plot with low entropy is a distribution

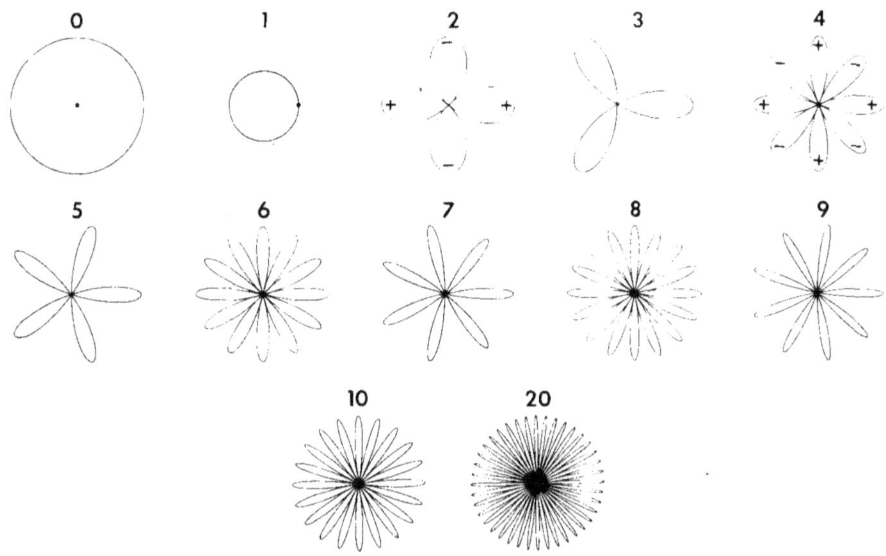


FIGURE 2. Plots of circular harmonics; amplitude = 1.00; phase angle = 0°.

with well-developed modes and antimodes exhibiting large contrast between intervals. A high entropy value indicates relatively small contrast between intervals. "Maximum entropy" is attained when the amplitude value of any grain has equal probability of occurring in any interval,[21] i.e., when the frequencies in each interval are equal.

Using one given harmonic, if amplitude values from all samples are pooled together into a single shape-frequency distribution, then the widths of the intervals can be adjusted such that the frequency plot displays maximum entropy. Thus the width of the intervals can be unequal and depends only on the shape of the distribution of the entire pooled data set. Individual sample frequency plots cast in the form of variable interval width spectra have been shown to provide maximum information upon any subsequent analysis.[12]

The problem exists in which the harmonic(s) (2 to 24) have the greatest potential for a clear, unambiguous solution upon subsequent shape analysis. We need to choose the harmonic(s) that exhibits the greatest contrast between individual sample frequency plots. The "relative entropy" of the entire data set can be used to measure this contrast.[12] The relative entropy is the average entropy of the individual sample plots divided by the maximum possible entropy. The lower the relative entropy of a data set, the more contrast exists between samples contained in that set. Thus, the relative entropy can be used as a feature extractor.[12] Choosing the harmonic with the lowest relative entropy will produce optimal results from the unmixing algorithms described below.

## IV. PATTERN RECOGNITION

### A. CABFAC/EXTENDED CABFAC

Once each sample is cast as a frequency distribution, with class intervals defined by the maximum entropy argument, the array is in an optimal configuration for subsequent analysis. Proportions of particles in the class intervals at a given harmonic comprise a constant-sum variable set (they sum to 100% in this case). Each sample can be viewed as a vector defined on a set of orthogonal axes, one for each class interval of the frequency distribution. Because the number of class intervals is generally large (say, >10) this "measurement space" is of high dimensionality and so is difficult for most normal persons to visualize. The fact that the values of each sample-vector sum to 100% immediately confines the data to a hyperplane of dimensionality one less than the number of intervals.[10] In addition, if we assume that two or more sets of intervals wax and wane together from sample to sample (maintain constant relative proportions), then the number of independent sets of intervals doing so is equal to the number of "end members". For reasons exactly analogous to the aforementioned constant-sum restriction, this further reduces the number of directions in which the data array can spread in measurement space. In fact, the data must be confined to a hyperplane of one dimension lower than the number of end members. For instance, if all samples were mixtures of only two end members (each may consist of many characteristic modes), then, in the absence of random perturbations, the vertexes of the sample vectors will be arrayed along a line embedded in a measurement sufficient to contain the data array, called "mixture space."[12]

A method to determine the dimensionality of mixture space has evolved over the past decade.[15-17,19] This has resulted in the algorithm EXTENDED CABFAC.[17] The algorithm CABFAC[16] evolved from conventional factor analysis.[13] Conventional factor analysis seeks, through analysis of correlated variables (via either a correlation or a variance-covariance matrix), to produce a set of new variables (factors) which are, themselves, uncorrelated. Often, most of the variance associated with the old variable set is associated with fewer variables in the new set. Imbrie and students[14,15] utilizing the general mathematical-analytical framework of conventional factor analysis (eigenvector-eigenvalue evaluation of a scatter matrix), built in analogous analytical system for determining the relationships between multivariate sample vectors. Sample vector pairs can be related via their vector cosines (termed similarity coefficients), which are exactly analogous to correlation coefficients between variable pairs. However, when the number of samples is large, calculations of a complete similarity matrix overwhelms the capacity of many computers. To alleviate this, Klovan and Imbrie[16] derived the necessary information by diagonalizing the cross-product matrix formed by pairwise comparison of variables. Klovan and Imbrie[16] thus demonstrated the functional relationship between conventional (variable-based) factor analysis (termed "R-mode" factor analysis) and sample-vector analysis (sometimes termed "Q-mode factor analysis").

Program CABFAC determines a new set of orthogonal reference axes in place of those defined by the original class-intervals. Commonly, in order to simplify interpretation, this new set is rotated using the varimax criterion.[15]

Program EXTENDED CABFAC[17] contains, as its basis, program CABFAC of Klovan and Imbrie.[16] However, two important features have been added: (1) the results (scores of reference axes) are reported in the same units as the input data vectors, and (2) a new criterion is developed to determine the dimensionality of mixture space.[19] Once the true dimensionality of the data is known, the ends of the sample vectors are projected into this "reduced space" within which individual points (each representing a multivariate vector) can be defined that, when connected by straight lines, will form a geometric figure (termed a "polytope") enclosing the data cloud. The minimum number of polytope vertexes needed is one more than the dimensionality of this reduced space. For instance, the vertexes of a triangle in a plane are sufficient to classify all enclosed samples in terms of positive linear combinations of end members (vertex compositions).

### B. QMODEL

Program EXTENDED CABFAC does not, itself, identify the end members. Accordingly, Klovan and Miesch developed program QMODEL whose object is to determine end members with more realistic compositions (i.e., positive class interval proportioning in data space). Program QMODEL defines end members in one of three ways. The first way is to use the principal axes after the varimax rotation as end members. Although this method generally assures positive composition loadings (proportions of each end member in mixing space), positive composition scores (actual interval proportions in data space) are not guaranteed.

The second method, optional in QMODEL, is to manually locate the end members. To aid in this, Miesch[20] has devised program EXSPIN. But unless the user has extremely tight control on the model being developed, he will have difficulty in finding a solution to the problem, especially when four or more end members are involved.

The third method available in QMODEL is an oblique solution which, by searching either the varimax space or data space, locates extremal points (vectors) as end members. These points, representing real samples, become reference axes. After locating these points within the varimax data set, the coordinates of points in this space are normalized by dividing the squared varimax factor matrix loadings by the communalities as derived by program EXTENDED CABFAC.[19] Relative distances of all sample points about these new reference axes can be interpreted as mixing proportions. These proportions are called oblique composition loadings and the space defined by the above reference axis is referred to as oblique space.

With this method, the remaining configuration of data (i.e., the nonextreme samples) in varimax space is not involved in end members determination. Thus, the QMODEL algorithm may be said to depend on a relatively small number of "outliers" to establish a metric for the entire data set. If the data envelope has bulges or invaginations, lines connecting the QMODEL end members may penetrate that envelope. Points falling outside the mixing polytope must contain negative proportions of one or more end members. The points with negative proportions are not easily interpretable in cases where mixing is considered an additive process. The presence of such negative mixing proportions indicates that one or more true end members have not been captured in the original data array. Therefore, new reference axes (end members), external to the data, must be located such that the envelope defined by these axes contains the data.

## C. EXTENDED QMODEL

In a recent paper, Full et al.[9] introduced a modification of the Q-mode unmixing algorithm QMODEL,[17] wherein an iterative procedure was developed to locate feasible end members not captured within a data set. This new algorithm, EXTENDED QMODEL, finds end members whose compositions are positive (or nearly so) and can be linearly proportioned to regenerate all of the data without resorting to negative proportions. Additionally, the sum of the proportions for each data point is constrained to be unity. The subroutine that locates these end members (herein called the DENEG procedure) does so by making an "edge-adjustment" of the mixing polytope if negative mixing proportions occur in QMODEL, followed by a "vertex adjustment" if the new polytope vertexes (extremal end members) contain negative proportions. The DENEG end members represent vectors more extreme than any sample in the data set.

The general steps in the DENEG algorithm are as follows:

1. Each column of the oblique composition loadings matrix of QMODEL is scanned for negative composition loadings. Negative loadings are divided into three categories: negligible (0.00 to $T_1$), adjustable ($T_1$ to $T_2$), and nonadjusted (less than $T_2$). Nonadjusted loadings are ones that correspond to "outliers" which will not effect definition of reference axes.
2. If there are no adjustable loadings, stop. Otherwise proceed to (3).
3. For each column, identify the most negative adjustable loading, say of value $d_i$. Add the absolute value of $d_i$ to every entry of the column being adjusted. Geometrically, this translates the edge represented by this column "out" a distance equal to the absolute value of $d_i$ in the initial oblique space, parallel to the original edge defined by the initial oblique end members (Figure 3A).
4. Change the coordinates of all sample vectors based on the amount of "outward" movement of the edge of the reference polytope. That is, if $X_i - x_{i1}, \ldots, x_{in}$

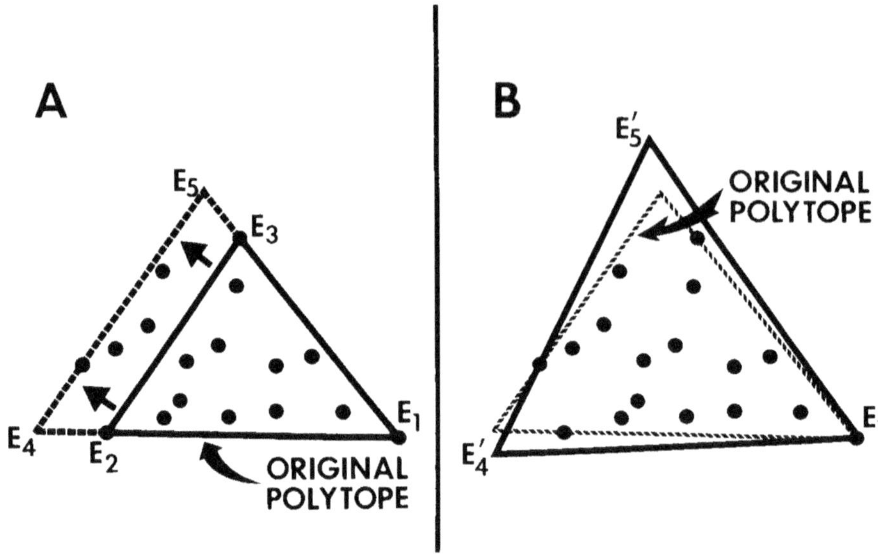


FIGURE 3. Pictorial representation of the edge adjustment (A) and following vertex adjustment (B) steps in the DENEG procedure.

represents the coordinates of each vector in the initial oblique space and $d_1, \ldots, d_n$ is defined as in (3) above, then $x_i^* =$ new coordinate $= (x_{i1} + d_1)(Z), \ldots, (x_{in} + d_n)(Z)$; where $Z = 1/(1 + d_1 + \ldots + d_n)$.

5. Locate the new vertexes of the reference polytope in QMODEL oblique space. The coordinates of the vertex will be the corresponding distance each side was translated added to the old vertex coordinate corresponding to the translated direction, which is determined by recalling that the sum of coordinates at each vertex is unity. That is, the oblique coordinates (loading) of each new end member $E_i = (d_1, \ldots, d_{i-1}, C, d_{i+1}, \ldots, d_n)$; where $d_1, \ldots d_{i-1}, d_{i+1}, \ldots, d_n$ is the distance in oblique space each respective side was translated, and $C = 1 - (d_1 + \ldots + d_{i-1} + d_{i+1} + \ldots + d_n)$.
6. Find compositions of each vertex using the QMODEL oblique transformation matrix.
7. Test the composition scores (corresponding end member composition in the original variable space) of each vertex for positivity. If all composition scores are greater than cutoff $T_3$, stop. Otherwise, for each unacceptable end member, zero all negative composition scores and reproportion this vertex to a constant sum. A result of this procedure is that edges of the new polytope need not be parallel to those of the initial QMODEL polytope (Figure 3B).
8. Submit the new vertex compositions to QMODEL. Generate the corresponding varimax loadings.
9. If the varimax loadings of the new end members fall within a distance $T_4$ of the previous iteration, stop. Otherwise, continue.
10. Generate a new oblique solution.
11. If the maximum number of iterations $T_5$ have been obtained, stop. Otherwise, return to (1).

Two variable parameters in the algorithm add flexibilty to the modeling process. One parameter ($T_2$) defines the maximum distance a polytope side can potentially move. Points beyond this maximum specified distance are considered outliers or impurities and will not affect the overall solution. "Distance" in this sense represents the degree of movement of a side measured relative to the distance from the opposite vertex. A distance of 0.25, for instance, means that a polytope side had moved a distance equal to 25% of the distance of the opposite vertex to the original polytope side in QMODEL oblique space. Appropriate values for this parameter must b determined from the nature of the problem to be solved. In many cases where QMODEL cannot converge to a unique set of end members via the extreme normalized composition subroutine (EXNORC) or the extreme raw composition subroutine (EXRAWC), modest movement of polytope sides provides a satisfactory solution. Of course, lack of convergence of these subroutines might also indicate a wrong choice in the number of end members.

The other variable parameter ($T_4$) defines the termination point of this algorithm in terms of closeness of the varimax loadings to the end members from iteration to iteration. Failure to converge may arise from either a poor end member fit, or oscillation in a narrow region between realistic varimax space and unrealistic varimax space.

Three cutoffs are used to control the amount of computer time needed to find a proper solution. One cutoff ($T_5$) determines the maximum number of iterations. Another cutoff permits solutions with slightly negative composition loadings ($T_1$), whereas the last cutoff permits slightly negative composition scores ($T_3$). The latter two cutoffs are necessary because, in some cases, accepting such a factor model may better reproduce the original data set in terms of geological modeling; the small negative values may be "close enough" to zero to be considered zero when compared to the original precision of the empirical data.

**D. FUZZY QMODEL**

In the absence of *a priori* knowledge, QMODEL requires the presence of vertex-extreme samples in the sample set, whereas EXTENDED QMODEL requires a set of edge-extreme samples. However, in either case, reliance on extreme data raises the possibility that a relatively few isolated data points might seriously affect the relative geometry of extreme samples with regard to a mixing polytope.

A more usual approach in conventional data analysis is not to rely on individual sample points, but to use aggregate properties of the data which tend to dissipate the effects of random variation (e.g., the greater stability of the mean compared to an individual data value). Another reason to deduce the nature of end members from the entire data structure rather than from a few outlying points is that outliers may, in many situations, not "belong" to the rest of the data. That is, they may represent either samples not belonging to the system of interest, or even incorrectly evaluated samples (i.e., errors in measurement or compilation). The alternative, "FUZZY QMODEL",[11] described herein, determines extremal end members based on the collective properties of all the data. Before the basic algorithm of FUZZY QMODEL is described, a brief overview of fuzzy clustering will be given.

In general, cluster analysis refers to a broad spectrum of methods which try to subdivide a data set X into c subsets (clusters), which are pairwise disjoint, all nonempty, and reproduce X via union. The clusters are then called a hard (i.e., nonfuzzy) c-partition of X. Many algorithms, each with its own mathematical clustering criterion for identifying "optimal" clusters, are discussed in the excellent monograph of Duda and Hart.[5] A significant fact about this kind of algorithm is the defect in the underlying axiomatic model — that each point in X is unequivocally grouped with other members of "its" cluster, and thus bears no apparent similarity to other members of X. One such way to characterize the similarity of an individual point to all the clusters was introduced in 1965 by Zadeh.[23] The key to Zadeh's idea is to represent the similarity a point shares with each cluster with a function (called the membership function) whose values (called memberships) are between zero and one. Each sample will have a high degree of similarity between the sample and a cluster; memberships close to unity signify a high degree of similarity between the sample and a cluster while memberships close to zero imply little similarity between the sample and that cluster. The history, philosophy, and derivation of such mathematical systems are well documented in Bezdek[2]. The net effect of such a function for clustering is to produce fuzzy c-partitions of a given data set. A fuzzy c-partition of X is one which characterizes the membership of each sample point in all the clusters by a membership function which ranges between zero and one. Additionally, the sum of the memberships for each sample point must be unity. If the function were to define values as either zero or one, then the memberships would define a hard c-partition (i.e., conventional clustering). In this light, the proportions derived from EXTENDED CABFAC-QMODEL,[17] EXTENDED QMODEL,[9] and FUZZY QMODEL define fuzzy c-partitions in themselves wherein the membership functions are linear.

Bezdek[2] discusses a number of ways to generate fuzzy c-partitions of data sets and contains many examples of data analysis drawn from a variety of applied fields, including taxonomy, medical diagnosis, and sprinkler design. The fuzzy c-means (FCM) algorithm[2] is one such fuzzy clustering technique that generates fuzzy c-partitions of a data set. The clustering criterion upon which FCM is based is the weighted sum of squared errors objective function $J_m$, defined as

$$J_m(U,V) = \sum_{k=1}^{N} \sum_{i=1}^{C} (u_{ik})^m \, ||X_k - v_i||^2 \qquad (1)$$

where

1. $U = (u_{ik})$ is a fuzzy c-partition of X (i.e., U is a matrix whose element $u_{ik}$ represents the membership of vector $X_k$ in the ith fuzzy cluster). The function which defines the $u_{ik}$ is a nonlinear inverse function.
2. $V = (u_1, \ldots, v_c)$ is the set of "cluster centers" (i.e., $v_i$ is the "center" or core of fuzzy cluster i). The location of the cluster centers is determined by a function which weights the contribution of each vector to a particular cluster by the memberships U.
3. $||\cdot||$ = any inner product norm. In the examples given, the Euclidean norm is used.
4. $m \epsilon [1,\infty)$ = a weighting exponent called the fuzziness exponent.

Outlying data have virtually no effect on the locations of the centers $v_i$ because $(u_{ik})^m$ (the weighted membership of $X_k$ in cluster i, raised to the nth power) approaches zero much faster than the $(u_{ij})^m$ of the other points $X_j$ (see Equation 1). In other words, the effects of noise can be essentially "filtered out" by adjusting m, the fuzziness weighting exponent. An example of the use of the FCM algorithm in geology can be found in Bezdek et al.,[3] while a synopsis and FORTRAN listing for FCM can be found in Bezdek et al.[4] Our present interest in FCM centers around three facts which are well established in both theory and practice:

1. The centers $v_i$ generated by FCM lie well "inside" the convex hull of data set X; this is important because the edge-adjustment DENEG procedure used in EXTENDED QMODEL always expands initial convex polytopes whose vertexes are putative end members.
2. The convex hull of the $v_i$s is, loosely speaking, a "skeleton" of the main structure of X itself; in particular, iterative expansion of the faces of this polytope via the DENEG procedure will proceed toward the type of unmixing solution described in Full et al.[9]
3. Noisy points (outliers in X) will exert little effect on the spatial location of any center $v_i$. This is one of the primary advantages of FCM — it is relatively insensitive to the presence of outliers.

The implications of (1) to (3) for the linear unmixing problem are simple: the centers $(v_i)$ generated by FCM always occupy spatial locations in end member space which are favorable for the initialization of the DENEG procedure. For this reason, the EXTENDED QMODEL procedure was further extended by incorporating the following algorithm:

1. Locate initial end members within normalized varimax space using the fuzzy c-means algorithm.
2. Determine the varimax loadings of these initial end members and submit them as end members defining the QMODEL oblique solution.
3. As these end members are contained within the data set (as opposed to the edges or vertexes of the hull of the data), the maximum edge translation (the DENEG value) must be enlarged for the first iteration only. After these steps are performed, the DENEG procedure is continued until either a solution is found, or a cutoff value is exceeded.

In the case where the data form a single cluster with internally uniformly distributed data, the FUZZY QMODEL algorithm is strongly influenced by the external configuration of the data. However, this influence is not derived from the outer edge of the data cluster (the convex hull) but from the geometry of an outer "zone" wherein the data density decreases. This is because of the nonlinear inverse weighting function used to define the fuzzy centers, upon which outlying points have much less influence than data closer to the centers. In the case of a data cloud that is arrayed as a clearly defined geometric figure, such as a triangle, the fuzzy centers will be imbedded in the data cloud in such a manner that the vertexes will define a similarly inscribed polygon. This appealing characteristic can become progressively destroyed as the data become more and more clustered. In such a case, the fuzzy cluster centers themselves would serve as vertexes and there is no guarantee that such vertexes would bear any relationship to the "true" mixing polytope. Strong clustering can be easily detected by simple inspection of the output from the algorithm. If the data has a definite cluster structure, then the mixing model for the data must be questioned or the sampling plan carefully analyzed.

The differences between QMODEL, EXTENDED QMODEL, and FUZZY QMODEL are illustrated in Figure 4. In the following example, ternary mixing in a multivariate data set can be visualized by examining the data structure as represented in normalized varimax space (Figure 4A). It is assumed for the following illustration that any point located on this triangular plane can serve as a potential end member for the mixing systems presented below. Figure 4B represents a ternary mixing space. Any of the QMODEL procedures (QMODEL, EXTENDED QMODEL, FUZZY QMODEL) will identify points $E_1$, $E_2$, and $E_3$ (the vertexes of the mixing triangle) as end members. Figure 4C represents a uniform data array that incompletely fills the mixing polytope depicted in Figure 4B. EXTENDED QMODEL could recover end members $E_1$, $E_2$, and $E_3$ as the proper end members of the mixing system whereas QMODEL could not. However, consider the same data configuration, but with outlier $E_4$ located external to the mixing polytope defined by $E_1$, $E_2$, and $E_3$ (Figure 4D). In this example, the QMODEL procedure will define point $E_4$ as a vertex; EXTENDED QMODEL, which uses the QMODEL solution as a starting position, would define a mixing polytope that will also contain point $E_4$. In both cases, the procedures cannot define triangle $E_1E_2E_3$ as the ternary mixing system when plotted in normalized varimax space. In this instance, the existence of point $E_4$ as an "aberrant" outlier can be readily observed. However, when the number of end members is four or more, or the edges of the mixing polytope are not clearly defined by the data, it is much more difficult to detect such a structure. The common occurrence of such configurations in geology motivated us to propose the alternative fuzzy means of data analysis. Whereas the initial iteration of previous methods relied heavily on the extremes of the convex hull of the data, initialization of the DENEG procedure with fuzzy clustering emphasizes the data configuration within the hull. The net effect of such a fuzzy analysis is that the effects of aberrant data are minimized.

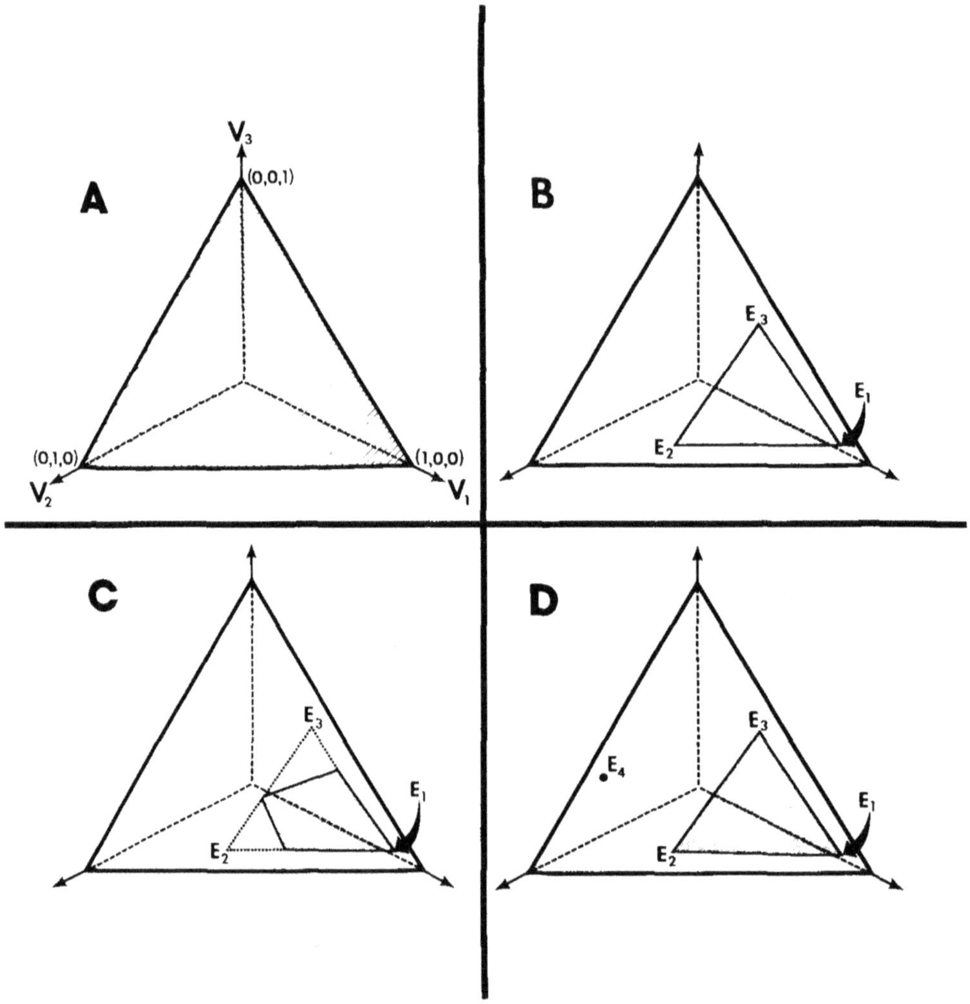


FIGURE 4. (A) Normalized varimax space where oblique end members can be conveniently visualized; (B) data structure (shaded) wherein end members $E_1$, $E_2$ and $E_3$, contained within the data set, can be properly located by both QMODEL and EXTENDED QMODEL; (C) portion of the data (shaded) section of the triangle represented in (B) wherein end members $E_2$ and $E_3$, now external to the data set, can be properly located by EXTENDED QMODEL; and (D) same data structure as (B) except an additional point $E_4$ is added, neither QMODEL nor EXTENDED QMODEL can recover $E_1$, $E_2$, $E_3$ in this case.

## V. SUMMARY

The primary steps in shape analysis are the following:

1. Particles are digitized and the two-dimensional edge points defined.
2. A finite Fourier series in polar form with the particle centroid used as the origin of the polar system is calculated.
3. The multiparticle data are summarized for each sample via shape frequency plots at each harmonic, using variable class interval widths based on entropy considerations.
4. The frequency diagrams in the form of sets of proportions are classified in terms of mixing proportions of end members by "unmixing" algorithms.

Our experience indicates that precise characterization of particle shape alone is not usually sufficient in itself, and further "downstream" analysis is required to solve most problems. Although most problems with which we have dealt require use of the CABFAC-QMODEL family of unmixing algorithms, others have required other multivariate procedures such as multiple discriminate analysis.

The algorithms described in (3) and (4) above have been useful with other types of multivariate data sets. The methods of creating frequency diagrams are useful for further analysis of any complex set of frequency distributions, as the defining of class intervals on the basis of maximum entropy ensures the least ambiguous results from the unmixing algorithms. The development of these unmixing algorithms (EXTENDED CABFAC, QMODEL, EXTENDED QMODEL, and FUZZY QMODEL) have allowed the interrelationships of large numbers of samples to be evaluated multivariately, using all variables simultaneously. The mixing proportions and end member compositions have proven to be of value with respect to a variety of sedimentological investigations using shape and size frequency data. Additionally, these algorithms have been used to successfully classify and analyze multichannel data from gas chromatography, gamma ray spectroscopy, major and trace element chemistry, and paleontologic assemblages.

## REFERENCES


1. **Bezdek, J. C.,** Numerical taxonomy with fuzzy sets, *J. Math. Biol.*, 1, 57, 1974.
2. **Bezdek, J. C.,** *Pattern Recognition with Fuzzy Objective Functions*, Plenum Press, New York, 1981.
3. **Bezdek, J. C., Ehrlich, R., Trivedi, M., and Full, W. E.,** Fuzzy clustering: a new tool for geostatistical analysis, Int. Syst. Means. Decis., 2, 13, 1981.
4. **Bezdek, J. C., Ehrlich, R., and Full, W. E.,** FCM: The fuzzy c-means clustering algorithms *Comput. Geosci.*, 10(2), in press.
5. **Duda, R. O. and Hart, P. E.,** *Pattern Classification and Scene Analysis*, J. Wiley & Sons, New York, 1973.
6. **Ehrlich, R. and Weinberg, B.,** An exact method for the characterization of grain shape, *J. Sediment. Petrol.*, 40, 205, 1970.
7. **Ehrlich, R., Brown, P. J., Yarus, J. M., and Przygocki, R. S.,** The origin of shape-frequency distributions and the relationship between size and shape, *J. Sediment. Petrol.*, 50, 475, 1980.
8. **Fico, C.,** Automated particle shape analysis: development of a microprocessor-controlled image analysis system, Masters Thesis, University of South Carolina, Columbia, 1980.
9. **Full, W. E., Ehrlich, R., and Klovan, J. E.,** EXTENDED QMODEL — objective definition of external end members in the analysis of compositional data, *Comput. Geosci.*, 7(4), 331, 1981.
10. **Full, W. E. and Ehrlich, R.,** Some approaches for location of centroids of quartz grain outlines to increase homology between Fourier amplitude spectra, *Math. Geol.*, 14, 43, 1982.
11. **Full, W. E., Ehrlich, R., and Bezdek, J. C.,** FUZZY QMODEL: a new approach for linear unmixing, *Math. Geol.*, 14, 257, 1982.
12. **Full, W. E., Ehrlich, R., and Kennedy, S. K.,** Optimal configuration and information content of sets of frequency distributions, *J. Sediment. Petrol.*, in press.
13. **Harmon, H. H., 1960,** *Modern Factor Analysis*, 2nd ed., University of Chicago Press, Chicago, 1960, 474.
14. **Imbrie, J. and Purdy, E. G.,** Classification of modern Bahamian carbonate sediments, in Mem. 1, Am. Assoc. Petrol. Geol., 253, 1962.
15. **Imbrie, J.,** Factor and Vector Analysis Programs for Analyzing Geologic Data, Tech. Report No. 6, ONR Task No. 389—135, Office of Naval Research, Geography Branch, Defense Technology Information Center, Alexandria, Va., 1963.
16. **Klovan, J. E. and Imbrie, J.,** An algorithm and FORTRAN IV program for large scale Q-mode factor analysis and calculation of factor scores, *J. Math. Geol.*, 3, 61, 1971.
17. **Klovan, J. E. and Miesch, A. T.,** EXTENDED CABFAC and QMODEL computer programs for Q-mode factor analysis of compositonal data, *Comput. Geosci.*, 1, 161, 1976.
18. **Kuenan, P. H.,** Experimental abrasion. III. Fluviatile action on sand, *Am. J. Sci.*, 257, 172, 1959.

19. **Miesch, A. T.,** 1976a, Q-mode factor analysis of geochemical and petrologic data matrices with constant row-sums, U.S. Geol. Surv. Prof. Paper 574-G, 47 p.
20. **Miesch, A. T.,** Interactive computer programs for petrologic modeling with extended Q-mode factor nalysis, *Comput. Geosci.*, 2, 439, 1976, 492.
21. **Shannon, C. E.,** *Bell System Tech.*, 379, 623, 1948; reprinted in Shannon, C. E. and Weaver, W., *The Mathematical Theory of Communication:* University of Illinois Press, Urbana, 1963.
22. **Smalley, I. J.,** Formation of quartz sand, *Nature (London)*, 211, 476, 1966.
23. **Zadeh, L.,** Fuzzy sets, *Inf. Control.*, 8, 338, 1965.

Chapter 8

# SMOOTHING A DIGITIZED PARTICLE PROFILE

**Russell V. Lenth**

## TABLE OF CONTENTS

## I. ABSTRACT

In particle morphology studies, the data are often obtained from some instrument which digitizes the particle profile. Then various parameters, such as Fourier coefficients, are estimated. In most cases, the resolution of the digitizing equipment is somewhat limited, and this introduces an error which must be taken into account. In particular, there is a danger of over-fitting the profile, that is, extracting "information" that is not actually present in the particle itself, but rather is a consequence of the digitization procedure. In this article, a method is given for deciding how many parameters can legitimately be estimated. An advantage of the procedure is that it is based on the values of certain shape parameters of interest. Thus the technique can be implemented conveniently, and the correct amount of smoothing is obtained more or less automatically.

## II. INTRODUCTION

The starting point in many particle morphology studies is the digitization of the profile of the particle. Various methods and equipment configurations can be used for this purpose. One example is the PIAS (Particle Image Analyzing System) apparatus at the University of Iowa,[1] which provides edge points on a rectangular coordinate system. Another configuration uses a grid consisting of radials and concentric circles, yielding data in polar coordinates.

Once the data are collected, we generally proceed to obtain some sort of parametric curve which fits the particle profile. This can be done, for example, by using a Fourier expansion. The parameters of these curves are then used in subsequent analysis as shape descriptors of the particle. In this article, we confine our attention to the Fourier approach, in particular the $(R, \theta)$ method.[2] This is especially well suited for the results developed here, although similar techniques can be found for other curve-fitting schemes.

No matter what curve-fitting method is being used, or whether the original data are rectangular or polar, there are certain inherent problems which must be considered. One is that the data are only as good as the photograph or microscopic image being digitized. There may be reflections of light, poor contrast, or portions of the profile out of focus. These can be alleviated to some degree by manual operations, such as a marking pen, but nonetheless they cause a certain amount of error in the data. We refer to this source of error as "fuzziness", not to be confused with fuzzy sets (see chapter following).

Another source of error is the lack of resolution of the equipment being used. This is due to mechanical limitations as well as the amount of computer memory available. The PIAS system, for example, currently uses a 128 × 128 grid, and the edge data so obtained contains a large number of small, but visible, "corners" not present in the original profile. This type of error is called "roundoff", because it is analogous to rounding numbers to the nearest integer.

It is clear that one thing that must *not* be done is to fit a curve passing through all the edge points. To see, this, refer to Figures 1 and 2. The smoothed profile shown in Figure 2 is the more believable of the two curves. The saturated fit in Figure 1 is the result of "over-fitting". Over-fitting does much more damage than merely producing a less than pleasing appearance. The parameters of this curve are used as descriptors of the particle itself, and thus if too much detail is incorporated, the particle will appear to possess properties that it does not actually have. Hence, it is important that the data be properly smoothed. In the use of Fourier analysis, this is easily accomplished by truncating the series at a suitably chosen number of terms. The manner in which this number is determined is the subject of this article.

Before developing the smoothing technique, it is necessary to introduce some notations and models. This is provided in Section III. In Section IV, we develop some statistical

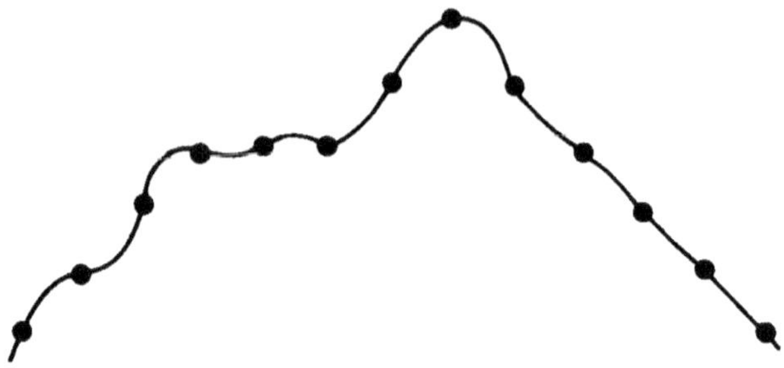

FIGURE 1. Result of over-fitting a portion of a particle profile.

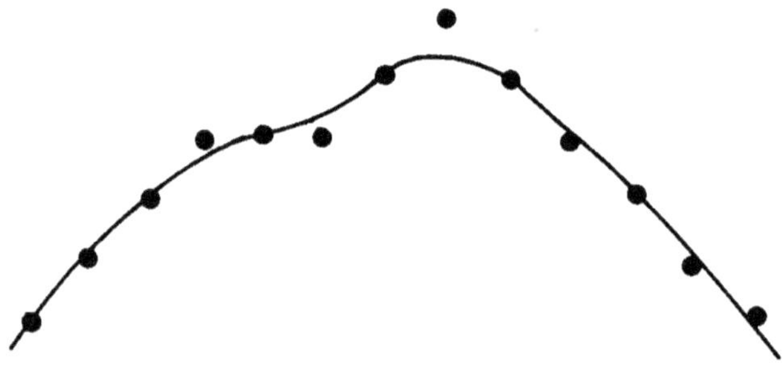

FIGURE 2. Smooth curve fitted to the data in Figure 1.

properties of roundoff and fuzziness in both the rectangular and polar cases. The smoothing technique is derived in the final section, and an example is given for illustration.

## III. MODEL AND BACKGROUND

To use the $(R,\theta)$ method of Fourier analysis, the data must be given in polar coordinates. Suppose that there are m distinct edge points in the data set. If the data are in polar form, we have the coordinates $(\theta_1, R(\theta_1)), (\theta_2, R(\theta_2)), \ldots, (\theta_m, R(\theta_m))$, and assuming that the $\theta_i$ are equally spaced, we can proceed to the next step. Otherwise, we have rectangular edgepoints $(x_1, y_1), (x_2, y_2), \ldots, (x_m, y_m)$, which must be transformed. In the course of doing so, we must be careful that we do not create any more than m points in polar coordinates, because otherwise we would be extracting more information from the data than is really there. With this in mind, let N be an integer no larger than $(m-1)/2$, and define angles $\theta_1, \ldots, \theta_{2N+1}$ by $\theta_i = 2\pi i/(2N+1)$. We now need to choose an origin. This can be done somewhat arbitrarily or by using the apparent center of gravity of the particle. Now define $R(\theta_i)$ as the distance between the origin and the edge along the radial whose angle is $\theta_i$. When necessary, linear interpolation between adjacent edge points can be used as an approximation to the "edge" of the profile along a given radial. The result is a suitable set of polar data.

For convenience, we assume that m is an odd number, i.e., $m = 2N + 1$. This makes things "come out right". However, even values of m can be handled by modifying the final term in the (finite, saturated) Fourier expansion. Define the parameters:

$$\hat{a}_0 = \sum_{i=1}^{2N+1} R(\theta_i) / (2N+1) = \text{average radius}$$

$$\hat{a}_n = \left[\sum_{i=1}^{2N+1} R(\theta_i) \cos n\, \theta_i\right] / N \qquad (n = 1, 2, \ldots, N)$$

$$\hat{b}_n = \left[\sum_{i=1}^{2N+1} R(\theta_i) \sin n\, \theta_i\right] / N \qquad (n = 1, 2, \ldots, N)$$

Then the finite, saturated Fourier expansion of the observed radii is given by

$$R_N(\theta) = \hat{a}_0 + \sum_{n=1}^{N} (\hat{a}_n \cos n\, \theta + \hat{b}_n \sin n\, \theta)$$

This is "saturated" since it fits the data perfectly, that is $R_N(\theta_i) = R(\theta_i)$ for each i.

Now consider the following shape parameters, developed by Luerkens[4]

$$L_2(n) = (a_n^2 + b_n^2) / (2R_0^2) \qquad (n = 1, 2, \ldots, N)$$

where $R_0$ is the radius of the circle having the same area as the particle silhouette. It can be shown algebraically that

$$\sum_{n=1}^{N} L_2(n) = \sum_{i=1}^{2N+1} (R(\theta_i) - \hat{a}_0)^2 / (2NR_0^2)$$

The importance of this is that the right-hand side of the above expression is proportional to the sample variance of the observed radii $R(\theta_i)$. Thus the quantity

$$L_2(n) / \sum_{j=1}^{N} L_2(j)$$

is a measure of the proportion of the radial variability of the particle that can be attributed to the n-th harmonic. In view of this, the $L_2(n)$ are useful descriptors of particle shape.

The problem we are concerned with here is deciding how many harmonics should be included in the Fourier expansion. In other words, the smoothed particle profile will be defined by

$$R_k(\theta) = \hat{a}_0 + \sum_{n=1}^{k} (\hat{a}_n \cos n\, \theta + \hat{b}_n \sin n\, \theta)$$

where k is chosen according to some criterion to be developed below. The $L_2(n)$ shape parameters are quite useful in this problem for two reasons. One is that the partial sum

$$S_k = \sum_{n=1}^{k} L_2(n)$$

can be used to describe how well the first k harmonics fit the observed radii. Secondly, the connection between the $L_2(n)$ and the radial variance allows us, after suitable justification, to employ classical statistical techniques to test the "goodness of fit" of the smoothed profile. The standard against which the $L_2(n)$ are tested is developed in the ensuing section.

## IV. FUZZINESS AND ROUNDOFF ERRORS

The shape parameters, Fourier coefficients, and so forth determined in a given analysis are based on measurements which are not perfectly precise. Clearly, the amount of smoothing that should be applied to a given particle image is dependent upon the size of the errors in measurement. In this section, we show how the error variability can be quantified. In order to do this, we must closely examine the method by which the data are obtained.

Suppose, for the moment, that we wish to measure the distance from the center to the edge of the particle along the radial at angle $\theta = 0$. Moving outward from the center, we have a series of measurements of gray levels from the image under study, and these occur at equally spaced points along the radial. Let c denote this distance between consecutive points on the grid. Assuming that "figure" cells (i.e., points inside the particle) correspond to low gray levels, we should observe a fairly rapid increase in gray levels somewhere along the radial. Thus we define a threshold value, $g_0$, such that a gray level less than $g_0$ corresponds to a grid point inside the profile. One way to estimate the distance to the edge is then to use the midpoint of the interval over which the gray level changes from below $g_0$ to above $g_0$.

Now, let G(r) denote the gray level at radius r. Within a small neighborhood of the actual edge of the particle, we may use the linear approximation $G(r) \doteq ar + b + E_G$. The line $ar + b$ describes the expected gray level at radius r, and $E_G$ is the error due to random fluctuations in gray level. Let R be the radius at which the level $g_0$ is achieved; then $R = [(g_0 - b) - E_G]/a$. Thus R is a random variable having mean $(g_0 - b)/a$ and variance $\sigma_G^2/a^2$, where $\sigma_G^2$ is the variance of the distribution of $E_G$. Note that the measured radius, $R^*$, is obtained by rounding R to the nearest odd multiple of c/2. Finally, we have

$$\begin{aligned}\text{Var}(R^*) &= E[R^* - E(R^*)]^2 \\ &= E[(R^* - R) + (ER - ER^*) + (R - ER)]^2\end{aligned}$$

Note that $R - R^*$ is roundoff error, which is uniformly distributed on the interval $(-c/2, c/2)$. Thus $ER - ER^* = 0$ and $E(R - R^*)^2 = c^2/12$. Furthermore, $R - R^*$ is independent of R, and hence the expected values of all cross products in the above expression are zero. Thus,

$$\text{Var}(R^*) = \text{Var}(R^* - R) + \text{Var}(R) = c^2/12 + \sigma_G^2/a^2$$

The first term is the result of roundoff and the second is due to fuzziness.

The development above is given only for a particular angle $\theta = 0$. In the case of a polar grid, the analysis would be the same for any other $\theta$. For a rectangular grid, these results apply only to horizontal and vertical radials, but it is reasonable to assume that one would achieve approximately the same level of accuracy along any other radial. A complication arises, however, in that the values of $\sigma_G^2$ and a may vary from angle to angle, regardless of which grid type is being used. In other words, the fuzziness does not remain constant. Also, the errors for two neighboring angles may be correlated. Taking such things into account makes fitting a curve much more complicated, and usually has little effect on the results. Consequently, we ignore these anomalies in our subsequent discussion.

What is needed now is a method for estimating the fuzziness parameters, a and $\sigma_G^2$. The latter can be estimated by computing the sample variances of observed gray levels inside and outside the profile. The average of these should give us a reasonable idea of the value of $\sigma_G^2$. Note that a is the slope of the line used for the linear approximation of gray levels across the boundary of the particle. Thus, it can be approximated by a difference quotient of the form $\Delta g/c$, where $\Delta g$ denotes the change in gray level between two adjacent grid points on opposite sides of the edge of the particle image. The average of all such quotients serves as an estimate of a.

In summary, the errors in measured radii have a distribution whose mean is approximately zero and variance that can be estimated by

$$\hat{\sigma}^2 = c^2\left[\frac{1}{12} + \frac{\text{avg(change in G not across edge)}^2}{(\text{avg(change in G across edge)})^2}\right]$$

This quantity can be used as a standard against which the smoothness of the fitted curve is compared, using the techniques of the next section.

## V. OPTIMAL SMOOTHING CRITERION

As is indicated in Section III, the quantity $S_k$, which is a partial sum of the $L_2(n)$, is a measure of how well the first k harmonics in the Fourier expansion fit the data. Similarly, we can define

$$\text{lof}(k) = \sum_{n=k+1}^{N} L_2(n) = \sum_{i=1}^{2N+1} (R(\theta_i) - \hat{a}_0)^{10} / (2NR_0^2) - S_k$$

as a measure of how badly $R_k(\theta)$ fits the data. Hence we call this quantity "lack of fit".

In the development which follows, we assume that $R(\theta_1), R(\theta_2), \ldots, R(\theta_{2N+1})$ are independent normal random variables having means equal to the actual radii and common variance $\sigma^2$ which can be estimated using the results of the preceding section.

The fact that $\theta_1, \theta_2, \ldots, \theta_{2N+1}$ are equally spaced implies that $\Sigma \cos n\theta_i = \Sigma \sin n\,\theta_i = 0$. Furthermore, the vectors whose components are $\cos n\theta_i$ and $\sin n\theta_i$ are mutually orthogonal. It follows that the random variables

$$\left[\sum_{i=1}^{2N+1} R(\theta_i) \cos n\,\theta_i\right]^2 / \left[\sigma^2 \sum \cos^2 n\,\theta_i\right] = N\hat{a}_n^2 / \sigma^2$$

and

$$\left[\sum_{i=1}^{2N+1} R(\theta_i) \sin n\,\theta_i\right]^2 / \left[\sigma^2 \sum \sin^2 n\,\theta_i\right] = N\hat{b}_n^2 / \sigma^2$$

are mututally independent. Their distributions are noncentral chi-square with one degree of freedom each and noncentrality parameters $Na_n^2$ and $Nb_n^2$, respectively. (Here, $a_n$ and $b_n$ denote the values of $\hat{a}_n$ and $\hat{b}_n$ that one would obtain if the exact radii were known.) Hence,

$$(2NR_0^2 / \sigma^2)\,\text{lof}(k) = (N/\sigma^2) \sum_{n=k+1}^{N} (\hat{a}_n^2 + \hat{b}_n^2)$$

is noncentral chi-square with $2(N-k)$ degrees of freedom and noncentrality parameter:

$$(N/\sigma^2) \sum_{k+1}^{N} (a_n^2 + b_n^2) = (2NR_0^2 / \sigma^2) \sum_{k+1}^{N} l_2(n)$$

where the $l_2(n)$ are the true shape parameters.

Now, the expected value of $R_0^2$ lof (k) is equal to

$$R_0^2 \sum_{k+1}^{N} l_2(n) + (N - k)\sigma^2/N.$$

Thus, if for some number k of harmonics, we observe a value of lof(k) that is not much greater than $(N-k)\ ^2/(NR_0^2)$, we would have reason to suspect that

$$\sum_{k+1}^{N} l_2(n)$$

is negligible. Since each term is positive, this would imply that $l_2(k+1)$, $l_2(k+2)$, . . ., $l_2(N)$ are all very small quantities. In other words, we should stop adding additional harmonics to the Fourier expansion as soon as the lack of fit is statistically insignificant.

The formal procedure for doing this is as follows. Suppose that k harmonics have been included in the fitted profile. We should include the (k + 1)st harmonic only if a test of the hypothesis that $l_2(k+1)$ and all subsequent $l_2$ (n) are equal to zero, is significant. The critical value for this test is obtained by noting that, if this hypothesis is true, then the noncentrality parameter for lof(k) is zero. Hence the hypothesis is rejected if

$$(2NR_0^2 / \sigma^2)\ \mathrm{lof}(k) > \chi^2_{1-\alpha}$$

where $\chi^2_{1-\alpha}$ denotes the $100(1-\alpha)$ percentile of the (central) chi-square distribution with $2(N-k)$ degrees of freedom, and $\alpha$ is the desired significance level for the test.

Note that the critical value changes with k. Since this procedure is most likely done by computer, it is helpful to find a closed-form approximation. This can be done using a normal distribution as long as k is not too close to N. In particular, for $\alpha = 0.05$ we have $\chi^2_{0.95} \doteq 2(N-k) + 2(1.645)\sqrt{N-k}$. While $\hat{\sigma}^2$, as given in Section IV, is only an estimate of $\sigma^2$, it is based on a large amount of data and hence we can use it in place of $\sigma^2$ without altering the validity of our procedure. In summary, then, the optimal amount of smoothing is obtained by fitting the function

$$R_k(\theta) = \hat{a}_0 + \sum_{n=1}^{k} (\hat{a}_n \cos n\,\theta + \hat{b}_n \sin n\,\theta)$$

where k is the smallest number such that

$$\mathrm{lof}(k) < \hat{\sigma}^2\ [N - k + 1.645\ \sqrt{N - k}]\ /\ (NR_0^2)$$

As an illustration of how this procedure performs, refer to Figures 3 and 4. Figure 3 shows the result of fitting $R_N(\theta)$ to a set of data from the PIAS. In this case, the saturated model contains N = 170 harmonics. Note the numerous corners and flat regions resulting from the discreteness of the edge data. Using the lack-of-fit technique described above, only k = 13 harmonics were fitted to the profile, and the result is shown in Figure 4. The rough edges have been smoothed out while preserving the basic shape characteristics of the particle. This smoothed profile compares favorably with the original photograph from which the data were obtained.

## VI. CONCLUSIONS

We have developed a criterion for determining how to smooth a particle profile in an optimal manner. Such smoothing is necessary in order to avoid over-fitting a set of data which contains errors of various types. Otherwise, the validity of subsequent analysis becomes questionable, because many of the parameters estimated from the over-fit are a result of noise in the data rather than characteristics of the particle in question.

The method given is practical in that the lack of fit, critical values, and error variance

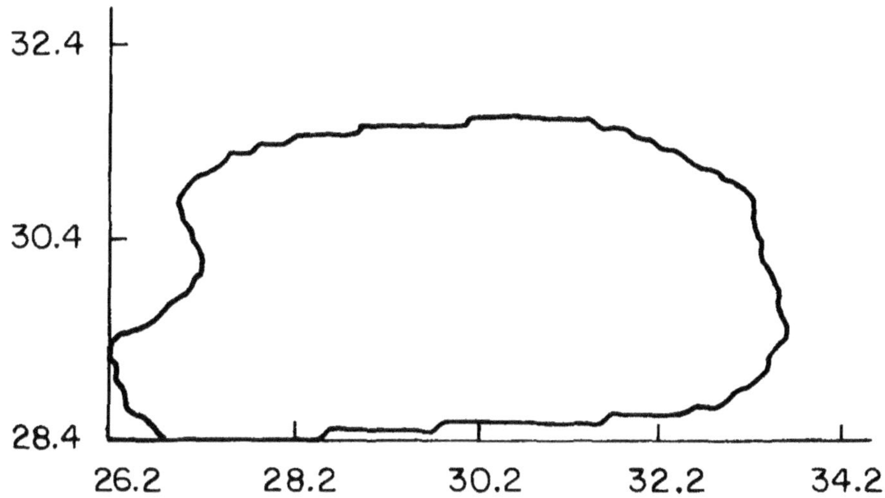


FIGURE 3. Saturated curve fitted to a particle profile, N = 170.

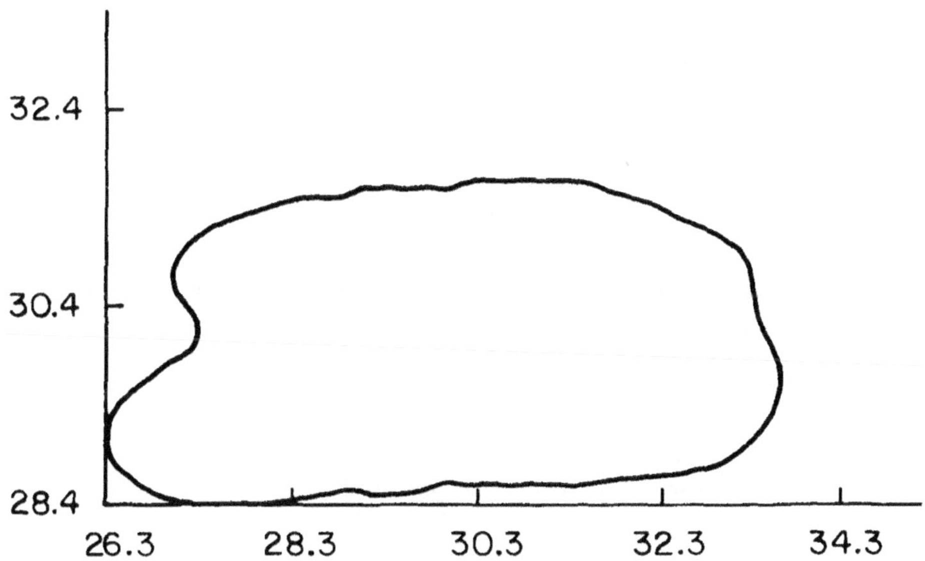


FIGURE 4. Smoothed profile using k = 13 harmonics.

are easily computed. An additional advantage is that most of the computation involves the $L_2(n)$ terms, which are of interest in their own right for describing the particle shape. Thus the procedure can be implemented efficiently, and it lends itself to a straightforward interpretation.

## ALPHABETICAL SYMBOLS

| | |
|---|---|
| a | Rate of change in gray levels across edge |
| $\hat{a}_0$ | Estimated average radius |
| $a_n$ | Coefficient of cos n θ if no errors are present |
| $\hat{a}_n$ | Estimate of $a_n$ |
| $b_n$ | Coefficient of sin n θ if no errors are present |
| $\hat{b}_n$ | Estimate of $b_n$ |
| c | Distance between two adjacent grid points on a rectangular grid, or distance between two adjacent rings on a polar grid |
| E | Mathematical expectation |
| $E_G$ | Random error in gray level |
| $g_0$ | Threshold value for gray levels for determining edge |
| G(r) | Gray level at radius r |
| $l_2(n)$ | n-th shape descriptor |
| $L_2(n)$ | Estimate of $l_2(n)$ |
| lof(k) | Lack of fit when $R_k(\theta)$ is used, i.e., the sum of $L_2(n)$ for all $n > k$ |
| m | Number of edge points in data |
| N | Number of harmonics in saturated fit |
| R | Radius at threshold gray level, $G(R) = g_0$ |
| $R_0$ | Radius of circle having same area as particle |
| R* | Rounded value of R |
| $R(\theta_i)$ | The i-th measured radius |
| $R_k(\theta)$ | Fourier expansion containing the first k harmonics |
| $R_N(\theta)$ | Saturated Fourier expansion |
| $S_k$ | Sum of first k values of $L_2(n)$ |
| Var | Variance operator |
| $x_i, y_i$ | Coordinates of i-th observed edge point on a rectangular grid |

## GREEK SYMBOLS

| | |
|---|---|
| $\Delta g$ | Change in gray level across threshold between two adjacent grid points |
| θ | An angle |
| $\theta_i$ | The i-th angle in the polar data |
| $\sigma^2$ | Error variance of the $R(\theta_i)$ |
| $\sigma_G^2$ | Variance of $E_G$ |
| $\hat{\sigma}^2$ | Estimate of $\sigma^2$ |
| $\chi_{1-\alpha}^2$ | The $1-\alpha$ fractile of a chi-square distribution |

## REFERENCES

1. **Lenth, R. V., Chang, C.-R., Beddow, J. K., and Vetter, A. F.,** A research mode of a particle image analyzing system, *Particle Characterization in Technology,* Vol. 2, CRC Press, Boca Raton, Fla., 1984.
2. **Luerkens, D. W., Beddow, J. K., and Vetter, A. F.,** The (R,Θ) method, *Powder Technol.*, in press.
3. **Bezdek, J. D.,** *Particle Characterization in Technology,* Vol. 2, CRC Press, Boca Raton, Fla., 1984.
4. **Luerkens, D. W.,** Morphological Analysis of Fine Particles Using the Fourier Method, Ph.D. thesis, University of Iowa, Iowa City, 1980.

Chapter 9

# SHAPE PREDICTION USING PROTOTYPES AND MEMBERSHIP INTERPOLATION*

**James C. Bezdek and Kenneth H. Solomon**

## TABLE OF CONTENTS



* This research was supported by NSF Grant No. MCS80-02856.

## I. INTRODUCTION

The "Fourier" method for analyzing particulate mixtures via two-dimensional contours of grain shapes is a well-established technique in particulate morphology.[1,2] The usual procedure entails several transformations of the raw data before proceeding to the objective of a particular investigation. While individual studies sometimes demand deviations from the norm, it seems fair to assert that the overall objectives usually include detection and characterization of a few prototypical grain shapes, with a subsequent view toward prediction and/or classification of points not in the data. While the Fourier method is undeniably useful for these applications, the multi-stage transformation of raw data prior to final analysis inevitably evokes the thought that worthwhile information for these purposes may be lost before it is used. Furthermore, the Fourier method is largely confined to classification of measured samples: it does not readily lend itself to the problem of *predicting* the shape of an unmeasured sample on the basis of variables implicit in the shaping process. This note proposes a new technique for shape analysis and prediction which operates directly on the raw data. The proposal method is based on the "Fuzzy c-Means" (FCM) clustering algorithms. The basis for the technique described below was presaged by one author at a 1977 National Science Foundation workshop on Advanced Particulate Morphology.[3] Consequently, this note brings to fruition at least one small ambition of that workshop, viz., to infuse some novel methodologies into the mainstream of particulate morphology.

Section II contains a short description of the Fourier method. The proposed method is briefly described in Section III. Section IV illustrates the new technique with a numerical example. The data discussed are shape data generated by irrigation sprinklers of various types. While this type of data is hardly typical of particulate morphology, its characteristics are in fact exactly the same as those encountered in grain shape analysis. Hence, the example given is relevant in the general context of particulate morphology.

## II. THE FOURIER METHOD

Let $g_k$ denote the k-th grain in a sample of n grains (the "grains" may be sand grains, lumps of coal, Rorschach stains, or representatives of any process which results in closed two-dimensional shapes). Thus, at its most elementary level, the data in physical form may be symbolized as

$$G = \{g_1, g_2, \ldots, g_n\} = \text{physical "grains"}. \tag{1}$$

Each $g_k$ has a shape, say $s_k$, with boundary $\delta s_k$. Geometrically, $s_k$ and $\delta s_k$ are sets in 2-space ($\mathbb{R}^2$). From $(s_k, \delta s_k)$ we seek a mathematical representation of $g_k$. The usual approach is to locate the first moment or centroid of area of $s_k$, say $(\overline{x}_k, \overline{y}_k)$, and place an origin of (polar) coordinates at it. Next, the geometric representation of G, i.e.,

$$S = \{s_1, s_2, \ldots, s_n\} = \text{geometric "grains"} \tag{2}$$

is converted to a first mathematical form, by dividing 360° into d equal angles $\phi_d = 360/d$, and measuring the radial distance to boundary $\delta s_k$ along the rays emanating from $(\overline{x}_k, \overline{y}_k)$ at the angles $0, \phi_d, 2\phi_d, \ldots (d\text{-}1)\phi_d$. This process generates d 2-vectors for each $g_k$, i.e., the polar coordinate pairs $((0, r_{k0}), (\phi_d, r_{k1}), (2\phi_d, r_{k2}), \ldots ((d\text{-}1)\phi_d, r_{k,d-1}) = \underline{z}_k$, a vectorial digitization in $\mathbb{R}^{2d}$ of boundary $\delta s_k$. The collection of $\underline{z}_k$'s is the first mathematical representation of $g_k$:

$$Z = \{\underline{z}_1, \underline{z}_2, \ldots, \underline{z}_n\} = \text{digitized "grains"}. \tag{3}$$

Letting $\theta_j = j\phi_d$, j = 0, 1, 2, . . . , (d-1), the measured values $r_{kj}$ are points on the graph of a continuous function $\rho_k$: $[0, 2\pi] \rightarrow (0, \infty)$, with $\rho_k(\theta_j) = r_{kj}$, j = 0, 1, 2, . . . ,(d-1). This points up the fact that each $\underline{z}_k$ really contains only d features, namely, the dependent variable values $r_{kj}$. In other words, the "unrolled" shape of $\delta s_k$ embodied in measurements $\{\underline{z}_k\}$ is actually contained entirely in the $r_{kj}$'s, provided the $\theta_j$'s remain fixed (as is usually the case in practice) from grain to grain. Let us denote this reduced set as

$$R = \{\underline{r}_1, \underline{r}_2, \ldots, \underline{r}_n\} = \text{digitized "grains"}, \tag{4}$$

where $\underline{r}_k = (r_{k0}, r_{k1}, \ldots, r_{k,d-1}) \in \mathbb{R}^d$, $1 \leq k \leq n$. This distinction is quite important for the method proposed below.

In the Fourier method, data set Z is now processed again by computing the discrete Fourier transform (DFT) of each $\underline{z}_k \in Z$. This results in the periodic functional approximation $h_k$:[0, $2\pi] \rightarrow \mathbb{R}$ given by

$$h_k(\theta) = a_{k0} + \sum_{j=1}^{d} (a_{kj} \cos j\,\theta + b_{kj} \sin j\,\theta); \qquad 1 \leq k \leq n. \tag{5}$$

As d increases, $h_k(\theta_j) \approx \rho_k(\theta_j) = r_{kj}$ for all k and j. The functions $\{h_k\}$ transform data set Z into one represented by the Fourier coefficients in Equation 5. Letting $\underline{w}_k = (a_{k0}, a_{k1}, b_{k1}, \ldots a_{kd}, b_{kd}) \in \mathbb{R}^{2d+1}$, the new data is

$$W = \{\underline{w}_1, \underline{w}_2, \ldots, \underline{w}_n\} = \text{Fourier "grains"}. \tag{6}$$

Generally speaking, a further transformation of Z is realized by conversion of Equation 5 to its amplitude-phase angle form, i.e.,

$$h_k(\theta) = \alpha_{k0} + \sum_{j=1}^{d} \alpha_{kj} \cos(\theta - \delta_{kj}). \tag{7}$$

From Equation 7 there follows an alternate form for W:

$$T = \{\underline{t}_1, \underline{t}_2, \ldots, \underline{t}_n\} = \text{Fourier "grains"}, \tag{8}$$

where $\underline{t}_k = (\alpha_{k0}, \alpha_{k1}, \delta_{k1}, \ldots, \alpha_{kd}, \delta_{kd}) \in \mathbb{R}^{2d+1}$. Data set T is usually parsed by ignoring the phase angles $\delta_{kj}$. Letting $\underline{\alpha}_k = (\alpha_{k0}, \alpha_{k1}, \ldots, \alpha_{kd})$, this set becomes, after elimination of the constant term,

$$A = \{\underline{\alpha}_1, \underline{\alpha}_2, \ldots, \underline{\alpha}_n\} = \text{Fourier "grains"}. \tag{9}$$

Finally, the Fourier amplitudes $\{\alpha_{kj}\}$ are converted to frequency percentages over p specified amplitude intervals, where p may or may not equal d. In this way, the nxd array A is compressed into a single vector $\underline{f} = (f_1, f_2, \ldots, f_p)$. If the entire procedure (Equations 1 to 9) is then repeated N times, the resultant Nxp array

$$F = \{\underline{f}_1, \underline{f}_2, \ldots, \underline{f}_n\} = \text{N shape samples}, \tag{10}$$

is the point of departure for shape analysis via the QMODEL-CABFAC technique discussed in Reference 1. The entire process is illustrated schematically in Figure 1. The view espoused here is that the overall effect of such extensive data transformations may delete the quality of shape information contained in the original data. In Section III we outline an alternative

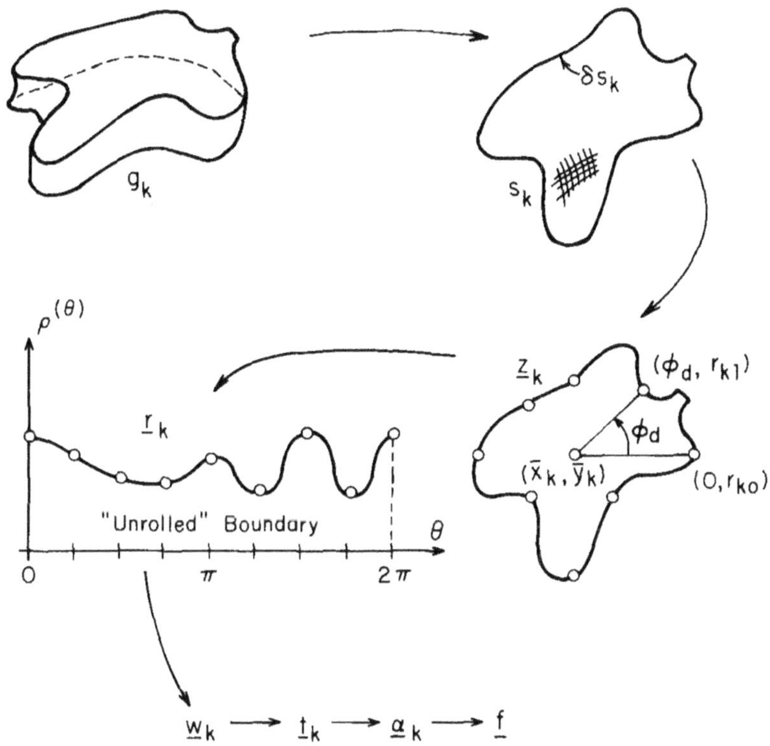


FIGURE 1. Evolution of Fourier grain shape data.

to the Fourier method that operates directly on data set R, the digital form of each grain boundary.

The ultimate goal of most Fourier shape analysis is to describe the given data in terms of c "end-members" ($2 \leq c < n$) or shape prototypes, from which the given data can be reconstructed via convex recombination. The convex weights used are completely determined geometrically once the end members are known. The general problem is called linear unmixing (see Reference 1). In this note, however, we investigate a slight variant of the usual analysis: viz., prediction of the shape of an unmeasured grain (or suite of grains) based on variables implicit but not used as the original data (geographical coordinates of sampling sites are an example of such variables), and on the information condensed from the measured grains (or suite thereof). The convex hull of c end members, say $\{\underline{e}_1, \underline{e}_2, \ldots, \underline{e}_c\} \subset \mathbb{R}^d$, is just

$$\text{conv}(\{\underline{e}_i\}) = \left\{ \underline{r} \in \mathbb{R}^d \mid \underline{r} = \sum_{i=1}^{c} \beta_i \underline{e}_i; \sum_{i=1}^{c} \beta_i = 1, 0 < \beta_i < 1 \right\}. \tag{11}$$

If each $\underline{e}_i$ is an implicit function of a vector $(\underline{x})$, then so is every $\underline{r} \in \text{conv}(\{\underline{e}_i(\underline{x})\})$. We seek a means for estimating $\underline{r}(\underline{x}) = \Sigma_i \beta_i \underline{e}_i(\underline{x})$. Although the Fourier method will yield each $r \in \text{conv}(\{\underline{e}_i\})$, $\underline{r}(\underline{x})$ cannot be predicted directly using the information available with Fourier shape analysis. It is this problem to which we now turn.

## III. SHAPE PROTOTYPES AND CLASS MEMBERSHIPS

The algorithm we propose to use is discussed extensively in Reference 4 in its general context, and its potential utility for particulate morphology was described in Reference 3. Relying on References 3 and 4 for the details then, we give a brief outline of the basic method. Regarding $R = \{\underline{r}_1, \underline{r}_2, \ldots, \underline{r}_n\} \subset \mathbb{R}^d$ as the raw data, we suppose the existence of c, $2 \leq c < n$, prototypical shape vectors, say $\{\underline{v}_1, \underline{v}_2, \ldots, \underline{v}_c\} \subset \mathbb{R}^d$, which suffice to compress the n $\underline{r}_k$'s into c paradigmatic representatives. Further, we seek an assignment of membership for each $\underline{r}_k$ to each of the c classes epitomized by the $\underline{v}_i$'s, which measures the affinity or degree of belonging $\underline{r}_k$ seems to demand for identification with each of the i prototypes. Let $u_i(\underline{r}_k) = u_{ik}$ denote this membership, and let $U = [u_{ik}]$ denote the array of $u_{ik}$ values as a cxn matrix. A little reflection on the meaning of this model suggests the following constraints for U.

$$\forall\, i, k : 0 \leq u_{ik} \leq 1 \text{ (Memberships are ``percentages'')} \tag{12a}$$

$$\forall\, k : \sum_{i=1}^{c} u_{ik} = 1 \quad \text{(Each } \underline{r}_k \text{ has a total affinity of 1)} \tag{12b}$$

$$\forall\, i : \sum_{k=1}^{n} u_{ik} > 0 \quad \text{(Each } \underline{v}_i \text{ is really needed by some } \underline{r}_k\text{)} \tag{12c}$$

Letting $M_{fc} = \{U|\ u_{ik} \text{ satisfies (12)}\}$, and $v = (\underline{v}_1, \underline{v}_2, \ldots, \underline{v}_c)$, the model we envision for shape analysis of R is a vector $(U, \underline{v}) \in M_{fc} \times \mathbb{R}^{cd}$. There are several methods currently under development for generating "optimal" U's and $\underline{v}$'s. The usefulness of the results, of course, depend entirely on how well $(U, \underline{v})$ explains the substructure of R. Different criteria of optimality yield different "best" $(U, \underline{v})$'s, for the same data, so one should be forewarned that the technique described below is only as good as empirical results suggest.

The criterion of optimality discussed at length in References 3 and 4 is the weighted sum of squared errors criterion $J_m$: $M_{fc} \times \mathbb{R}^{cd} \rightarrow (0, \infty)$ defined by

$$J_m(U, \underline{v}) = \sum_{k=1}^{n} \sum_{i=1}^{c} (u_{ik})^m \|\underline{r}_k - \underline{v}_i\|^2 ; \qquad 1 \leq m < \infty, \tag{13}$$

where m is a weighting exponent, and $\|\cdot\|$ is any norm on $\mathbb{R}^d$. Assuming that local minima of $J_m$ are appealing candidates for $(U, \underline{v})$ as discussed above, it has been shown that (U, v) must satisfy the necessary conditions (with $m > 1$, $d_{ik} = \|\underline{r}_k - \underline{v}_i\| > 0\ \forall\, i, k$):

$$u_{ik} = \left( \sum_{j=1}^{c} \left( \frac{d_{ik}}{d_{jk}} \right)^{2/(m-1)} \right)^{-1} \forall\, i, k ; \tag{14a}$$

$$\underline{v}_i = \left( \sum_{k=1}^{n} (u_{ik})^m \underline{r}_k \right) \left( \sum_{k=1}^{n} (u_{ik})^m \right)^{-1} \forall\, i. \tag{14b}$$

The fuzzy c-means (FCM) algorithms (one for each m) simply loop through conditions (Equation 14): starting with an initial guess for $U \in M_{fc}$, compute $\underline{v}_i$'s with Equation 14b; update U with Equation 14a; and stop if iterates of U are "close" (e.g., $\max_{i,k} |u_{ik}^{(new)} - u_{ik}^{(old)}| \leq \epsilon$); otherwise, loop. Extensive discussion of this procedure is available in References 3 and 4. For our purposes here, it suffices to know that the vectors $\{\underline{r}_k\}$ at Equation 4 are thereby "supplanted" by terminal protypical shape vectors $\{\underline{v}_i\}$ from Equation 14b; and that with these prototypes there come, for each $r_k$, c memberships $\{u_{ik}\}$ from Equation 14a which, with the $\underline{v}$'s, are an approximate local minimum of $J_m$, c and m fixed. Choices for c and m, both crucial to the model being used, are discussed in Section IV. In the interests of brevity, the distances $\{d_{ik}\}$ in Equation 14a will be Euclidean (Formula 14a holds for any inner product induced norm). The cutoff criterion mentioned above is used with $\epsilon = 0.01$ in the numerical example to follow.

At this point, we have used the n grains $\{\underline{r}_k\}$ to find $(U, \underline{v})$. One must now ask how R might be regenerated using $(U, \underline{v})$. In the Fourier model, function $h_k$ at Equation 7 regenerates the shape $\underline{r}_k$ exactly as $d \rightarrow \infty$: for $d < \infty$, $h_k$ provides an approximation to $\underline{r}_k$, $h_k(\theta_j) \approx r_{kj}$.

In the proposed model, we have $\underline{v}_i$, a shape prototypical of class i, and $u_{ik}$, the membership of $\underline{r}_k$ in this class. If $u_{ik}$ represents the "percentage" or extent to which $\underline{r}_k$ has class i features, it seems plausible to suppose that $\Sigma_i\, u_{ik}\, \underline{v}_i$ might be a reasonable reconstruction of $\underline{r}_k$. Based on this supposition, we define, for $1 \leqslant k \leqslant n$

$$H_k(U, \underline{v}) = \sum_{i=1}^{c} u_{ik}\, \underline{v}_i. \tag{15}$$

To extend $H_k$, we note that U and $\underline{v}$ are both continuous functions of the data (exactly our assumption concerning $\rho_k$). If there are variables implicit in shaping each $\underline{r}_k$ which are not realized numerically in R, then each row of U specifies n points on a continuous membership surface in the implicit variables. For example, each grain $\underline{r}_k$ (although more probably each suite of N grains) may have associated with it $\underline{x}_k = (x_k, y_k, z_k)$, the coordinates of the sample site. In such cases, it seems reasonable to assume that U is a continuous function of $\underline{x}$ over the hull or mesh of the sample sites used (provided, of course, this is physically plausible). If the prototypes $\underline{v}_i$ are fixed end members, then Equation 15 can be used to estimate any vector in the convex hull of the $\{\underline{v}_i\}$ (i.e., any shape which is a linear mixture of the c end members) once its implicit variables are known, provided we have an interpolation surface $u_i(\underline{x})$ fitting each row i of U. Thus, the predicted shape of grain $\underline{r}$ at $\underline{x}$ might be defined as

$$\underline{r}\,(\underline{x}) = \sum_{i=1}^{c} u_i(\underline{x})\, \underline{v}_i. \tag{16}$$

As with all mathematical models, the "proof is in the pudding". Having cooked up our delicacy, we now turn to its consumption.

## IV. A NUMERICAL EXAMPLE

The data set R discussed below was first published as Table 1 of Reference 5. It consists of 19 "grains" $\underline{r}_k \in \mathbb{R}^{20}$. Each $\underline{r}_k$ is actually the normalized application rate of water applied by a given sprinkler (the Rain Bird 103 Rain Gun) at equi-interval radial distances from the center of the sprinkler. Figure 2 depicts grain $\underline{r}_8$; it corresponds precisely to the digitized grains $\underline{r}_k = \rho_k(\theta_j)|\, j = 1, 2, \ldots, d)$ in the "unrolled" form of Figure 1, with the exception that application rate is zero at $\theta = 0$, $\theta = 2\rho$. This cannot occur with real grains, of course, because it would require the centroid of $s_k$ to lie on the boundary. This small difference aside, our sprinkler data is certainly a shape data set. The actual values for R appear in Reference 5: $\underline{r}_8$ is listed explicitly in Table 1. There are two implicit variables attached to each $\underline{r}_k$, namely, nozzle size $N_k$ and operating pressure $P_k$. In the (N, P) plane, these variables form a mesh as shown in Figure 4, above which we hope to construct the $u_i(N, P)$'s.

Using the protocols established above, R was processed by the FCM algorithm with c = 2 at various values of m. The first fact established by our calculations was that $\underline{v}_1$ and $\underline{v}_2$, the terminal shape prototypes arrived at via Equation 14b, were quite indifferent to small changes in m. This surprising fact lends credence to the hypothesis that the application shape of the 19 sprinklers can be characterized as a continuous function of two fixed prototypical shapes on the (N,P) grid.

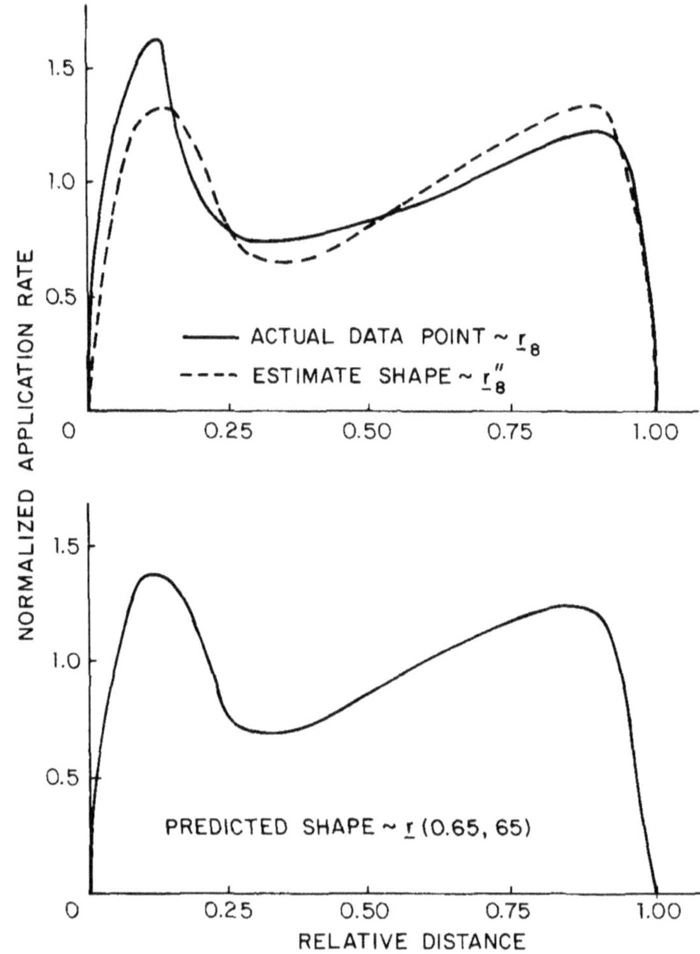


FIGURE 2. Shape prediction of real and hypothetical data.

The viability of Equation 16 for shape prediction rests entirely on empirical results. Anticipating a need to test the proposed scheme with known shapes, the variation of $\underline{v}_1$ and $\underline{v}_2$ due to deletions from data set R was studied. Three data sets were processed:

$$R = \{\underline{r}_1, \underline{r}_2, \ldots, \underline{r}_{19}\}; \quad |R| = 19 = n \tag{17a}$$

$$R' = \{\underline{r}_1, \underline{r}_3, \ldots, \underline{r}_{19}\}; \quad |R'| = 10 = n' \tag{17b}$$

$$R'' = \{\underline{r}_1, \underline{r}_4, \ldots, \underline{r}_{19}\}; \quad |R''| = 7 = n'' \tag{17c}$$

We denote by (U, $\underline{v}$), (U′, $\underline{v}'$), and (U″, $\underline{v}''$), respectively, the terminal pairs found by FCM using R, R′, and R″. Tables 2 and 3 list the shape prototypes and membership functions derived by FCM for each of these sets at $m = 2.2$. Note that the prototypes seem relatively stable to (rather drastic) deletions in the data upon which they are based. We take this as further evidence of the physical existence of (hypothesized) prototypes. Figure 3 shows graphs of $\underline{v}_1''$ and $\underline{v}_2''$, corresponding to columns 4 and 7, respectively, of Table 2. Although the shape prototypes shown in Figure 3 are not related to "grains" in the usual sense, these graphs are directly comparable to the "unrolled" digital grain $\underline{r}_k$ in Figure 1.

**Table 1**
**ESTIMATION OF SHAPE DATA BY CONVEX COMBINATIONS OF WEIGHTS $\{u_{ik}\}$ AND PROTOTYPES $\{\underline{v}_i\}$**

| Relative distance | #8: N = 0.6, P = 70 Actual | n′ = 10 | n″ = 7 | Hypothetical: N = 0.65 P = 65 |
|---|---|---|---|---|
| 0.025 | 0.59 | 0.38 | 0.35 | 0.39 |
| 0.075 | 1.20 | 1.03 | 0.90 | 1.07 |
| 0.125 | 1.63 | 1.40 | 1.27 | 1.39 |
| 0.175 | 1.30 | 1.28 | 1.18 | 1.23 |
| 0.225 | 0.90 | 0.94 | 0.97 | 0.93 |
| 0.275 | 0.72 | 0.77 | 0.77 | 0.78 |
| 0.325 | 0.78 | 0.72 | 0.65 | 0.71 |
| 0.375 | 0.80 | 0.73 | 0.67 | 0.73 |
| 0.425 | 0.80 | 0.79 | 0.72 | 0.78 |
| 0.475 | 0.87 | 0.83 | 0.76 | 0.82 |
| 0.525 | 0.91 | 0.89 | 0.83 | 0.88 |
| 0.575 | 0.91 | 0.96 | 0.91 | 0.95 |
| 0.625 | 0.92 | 1.05 | 0.98 | 1.03 |
| 0.675 | 0.99 | 1.11 | 1.04 | 1.09 |
| 0.725 | 1.01 | 1.14 | 1.12 | 1.14 |
| 0.775 | 1.21 | 1.20 | 1.16 | 1.20 |
| 0.825 | 1.19 | 1.21 | 1.21 | 1.21 |
| 0.875 | 1.19 | 1.27 | 1.34 | 1.26 |
| 0.925 | 1.34 | 1.18 | 1.14 | 1.19 |
| 0.975 | 0.53 | 0.42 | 0.52 | 0.45 |

**Table 2**
**SHAPE PROTOTYPES FOR THE RB #1013 RAIN GUN**

| Relative distance | Cluster $\underline{v}_1$ | Center $\underline{v}_1'$ | For u $\underline{v}_1''$ | Cluster $\underline{v}_2$ | Center $\underline{v}_2'$ | For $u_2$ $\underline{v}_2''$ |
|---|---|---|---|---|---|---|
| 0.025 | 0.45 | 0.44 | 0.46 | 0.31 | 0.28 | 0.27 |
| 0.075 | 1.28 | 1.20 | 1.27 | 0.73 | 0.68 | 0.64 |
| 0.125 | 1.57 | 1.56 | 1.52 | 1.10 | 1.08 | 1.07 |
| 0.175 | 1.30 | 1.35 | 1.27 | 1.12 | 1.13 | 1.11 |
| 0.225 | 0.91 | 0.92 | 0.92 | 0.98 | 1.00 | 1.02 |
| 0.275 | 0.80 | 0.78 | 0.80 | 0.75 | 0.77 | 0.75 |
| 0.325 | 0.78 | 0.77 | 0.79 | 0.60 | 0.60 | 0.54 |
| 0.375 | 0.81 | 0.80 | 0.82 | 0.60 | 0.60 | 0.55 |
| 0.425 | 0.87 | 0.87 | 0.87 | 0.64 | 0.63 | 0.60 |
| 0.475 | 0.91 | 0.91 | 0.93 | 0.68 | 0.67 | 0.63 |
| 0.525 | 0.96 | 0.96 | 0.98 | 0.75 | 0.75 | 0.71 |
| 0.575 | 1.03 | 1.02 | 1.03 | 0.83 | 0.84 | 0.81 |
| 0.625 | 1.09 | 1.12 | 1.08 | 0.93 | 0.92 | 0.90 |
| 0.675 | 1.15 | 1.17 | 1.13 | 1.00 | 0.99 | 0.97 |
| 0.725 | 1.18 | 1.18 | 1.19 | 1.09 | 1.06 | 1.07 |
| 0.775 | 1.22 | 1.22 | 1.19 | 1.17 | 1.16 | 1.14 |
| 0.825 | 1.19 | 1.19 | 1.19 | 1.26 | 1.26 | 1.24 |
| 0.875 | 1.16 | 1.17 | 1.18 | 1.45 | 1.48 | 1.47 |
| 0.925 | 1.01 | 1.01 | 1.01 | 1.52 | 1.54 | 1.69 |
| 0.975 | 0.37 | 0.34 | 0.38 | 0.59 | 0.59 | 0.63 |
| | | Shape 1 = $\underline{v}_1$ | | | Shape 2 = $\underline{v}_2$ | |

*Note:* m = 2.2; c = 2.

**Table 3**
**SHAPE CLASS MEMBERSHIPS FROM FCM**

| Test #k | Nozzle N, in. | Press P, psi | Membership $u_{1k}$ (N,P) in Shape 1 $u_{2k}(N,P) = 1 - u_{1k}(N,P)$ n = 19 | n′ = 10 | n″ = 7 |
|---|---|---|---|---|---|
| 1 | 0.5 | 50 | 0.17 | 0.16 | 0.19 |
| 2 | 0.5 | 60 | 0.54 | — | — |
| 3 | 0.5 | 70 | 0.83 | 0.87 | — |
| 4 | 0.5 | 80 | 0.83 | — | 0.87 |
| 5 | 0.6 | 40 | 0.18 | 0.16 | — |
| 6 | 0.6 | 50 | 0.07 | — | — |
| 7 | 0.6 | 60 | 0.40 | 0.44 | 0.55 |
| 8 | 0.6 | 70 | 0.72 | — | — |
| 9 | 0.6 | 80 | 0.85 | 0.86 | — |
| 10 | 0.7 | 40 | 0.20 | — | 0.15 |
| 11 | 0.7 | 50 | 0.08 | 0.07 | — |
| 12 | 0.7 | 60 | 0.43 | — | — |
| 13 | 0.7 | 70 | 0.48 | 0.53 | 0.58 |
| 14 | 0.7 | 80 | 0.80 | — | — |
| 15 | 0.8 | 40 | 0.19 | 0.18 | — |
| 16 | 0.8 | 50 | 0.18 | — | 0.13 |
| 17 | 0.8 | 60 | 0.15 | 0.16 | — |
| 18 | 0.8 | 70 | 0.53 | — | — |
| 19 | 0.8 | 80 | 0.85 | 0.87 | 0.86 |

Turning now to the questions of empirical verification, we consider the construction of vectors which were not in R″ (thus, not part of the information used to find (U″, $\underline{v}''$)). Sprinkler $\underline{r}_8$ was not in R′ or R″, so we constructed its shape vector using Equation 16 with $\underline{v}''$ and the membership surface $u_1''(N, P)$ shown in Figure 4. Surface $u_1''$ was fitted to the eight membership values ($u_{11}''$, $u_{12}''$, . . . , $u_{18}''$) taken from the first row of U″(the last column of Table 3). The fit was done by eye, and the eight (N, P) grid points used are clearly not optimal, since their convex hull does not cover the (N,P) net for R. Both of these facts are circumstantial rather than intentional: both suggest areas for further study of the proposed technique. Although not shown here, membership surfaces for all 19 (N, P) points were plotted, and had very similar general shapes — with greater detail of course — to the one depicted in Figure 4. Comparison of columns 4 to 6 in Table 3 verifies this.

Note in Figure 4 the regions denoted as "1" and "2" in the (N, P) plane. Physical reasoning suggests that region 1 corresponds to $\underline{v}_1''$ — high pressure, low nozzle aperture; conversely, low pressure and high nozzle size are associated to $\underline{v}_2''$ and region 2. We cannot assign real (N, P) values to $\underline{v}_1''$ and $\underline{v}_2''$. It is useful, however, to envision the characteristics of shapes such as $\underline{r}_8$, at $(n_8, P_8) = (0.6, 70)$, as continuous convex combinations of the shape prototypes. This also emphasizes an obvious shortcoming of using c = 2 prototypes on an essentially 4-sided grid: Equation 16 can be exact only for vectors in the convex hull of the prototypes — in this case, on a line in data space $\mathbb{R}^{20}$ connecting $\underline{v}_1''$ and $\underline{v}_2''$. The projection of this hull onto the (N, P) grid is presumably (roughly) the diagonal from region 1 to region 2 in Figure 4. Since $\underline{r}_8$ is on this diagonal, our expectation is that Equation 16 should be fairly accurate, provided (U″, $\underline{v}''$) is indeed the entity.

In order to implement the method in the present instance, we write for Equation 16.

$$\underline{r}_8(N_8, P_8) = r_8(0.6, 70) \approx u_1''(0.6, 70)\,\underline{v}_1'' + u_2''(0.6, 70)\,\underline{v}_2''.$$

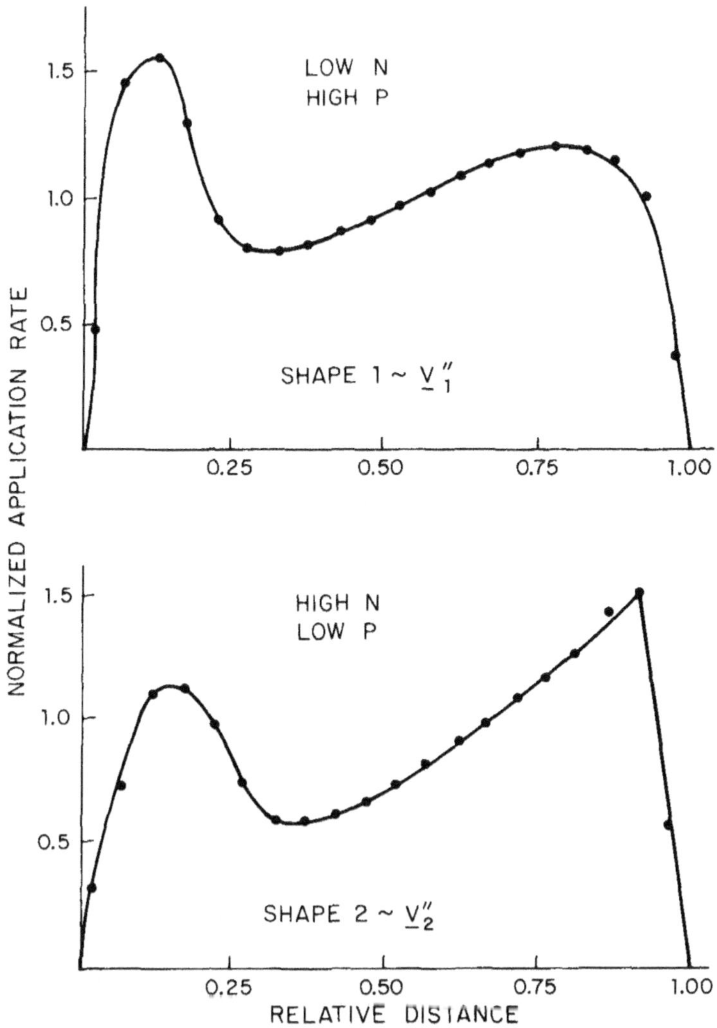


FIGURE 3. Shape prototypes derived from data R″.

Locating (0.6, 70) in Figure 4 and interpolating by eye, we find

$$u_1''(0.6, 70) \approx 0.44 \quad ; \quad u_2''(0.6, 70) \approx 0.56.$$

These values yield the estimate

$$\underline{r}_8 \approx 0.44\, \underline{v}_1'' + 0.56\, \underline{v}_2'' = \underline{r}_8'' \tag{18}$$

where $\underline{v}_1''$ and $\underline{v}_2''$ come from Table 2. Column 3 of Table 1 lists the estimate of $\underline{r}_8$ produced by Equation 18. The predicted values are, in general, quite close to the actual ones: this is evident in Figure 2, where $\underline{r}_8''$ is graphed along with $\underline{r}_8$. Test results with $r_{18}$ (N = 0.8, P = 70) not reported here were equally encouraging, and $\underline{r}_{18}$ is not on the "convex diagonal" discussed above.

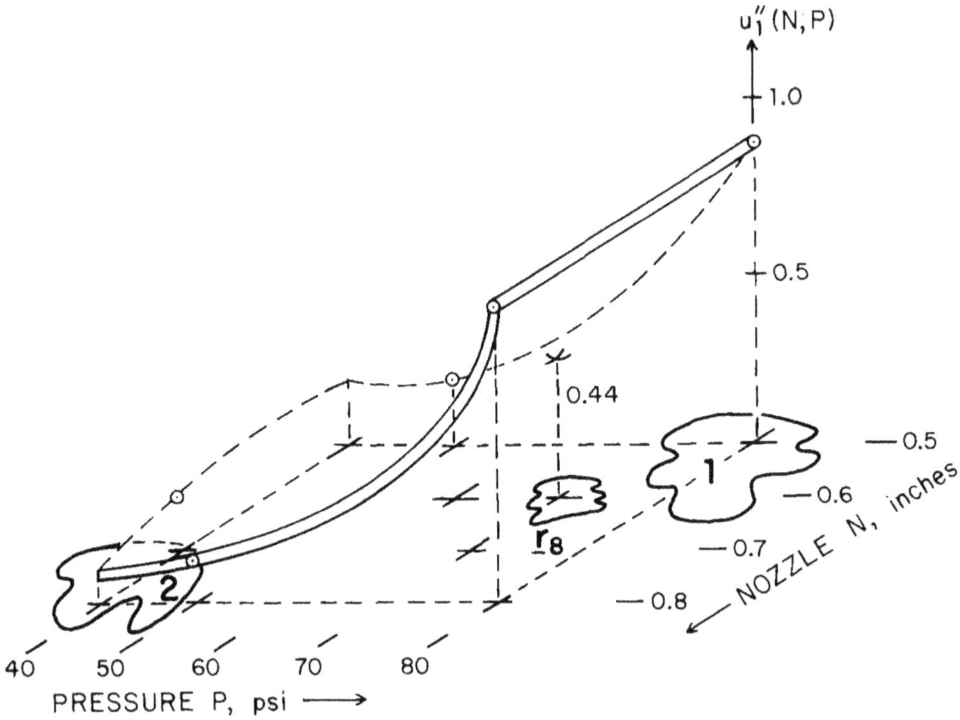


FIGURE 4. A membership surface derived from data R″.

The derivation of Equation 18 suggests one general procedure for shape prediction:

1. Use of data set R with FCM to find (U, $\underline{v}$) which "decomposes" R into c convex factors or prototypes.
2. Fit each row of U with an interpolation surface $u_i(\underline{x})$, $1 \leq i \leq c$, where $\underline{x}$ are the variables implicit in R, so that $u_{ik} \approx u_i(\underline{x}_k)$.
3. Given $\underline{x}$, estimate shape $\underline{r}(\underline{x})$ as $\underline{r}(\underline{x}) = \Sigma_i\, u_i(\underline{x})\underline{v}_i$.

As an example of the general method, we take $\underline{x}$ = (0.65, 65), (N, P) values not included in the 19 point grid defined by R. Using the membership values in U associated with $\underline{v}_1$ and $\underline{v}_2$ in Table 2 led to interpolated values for $u_1$ and $u_2$ at (0.65, 65) which, when used in Equation 16, produced the predicted shape $\underline{r}(\underline{x})$ shown in Figure 2, and listed as column 4 of Table 1. There is, of course, no way to verify the accuracy of this prediction: whatever confidence one has in its reliability must be drawn from tests on sprinklers such as $\underline{r}_8$ above.

The results of this study are not exhaustive enough to justify general usage of the proposed method without further investigation. Many interesting questions have arisen, however, and the example given seems successful enough to encourage further research. Among the larger issues: how shall the parameters m and c for FCM be chosen? What method of surface fitting is best for samples of small size? What implicit variables ($\underline{x}$) should be associated to a shape data set R to maximize its utility for shape prediction? Finally, what method of empirical verification would serve generally to "prove" an hypothetical model?

## ALPHABETICAL SYMBOLS

| | |
|---|---|
| $g_k$ | k-th (Real) grain in a sample |
| n | Number of observations in a sample |
| G | Grain sample |
| $s_k$ | Shape of $g_k$ |
| $(\bar{x}_k,\bar{y}_k)$ | Centroid of $s_k$ |
| S | Grain shape sample |
| d | Number of samples taken from $\delta s_k$ |
| $r_{kj}$ | Radial distance to j-th digital coordinate on $\delta S_k$ |
| $\underline{z}_k$ | Vector of digitized coordinates along $\delta s_k$ |
| Z | Digitized shape sample |
| R | Unrolled digitized grains |
| $\underline{r}_k$ | Vector of radial coordinates |
| $h_k$ | Trigonometric polynomial |
| $a_{kj}$ | Fourier coefficients |
| $b_{kj}$ | Fourier coefficients |
| $\underline{w}_k$ | Vector of Fourier coefficients |
| W | Fourier grain sample |
| T | Fourier grain sample |
| $\underline{t}_k$ | Vector of Fourier coefficients |
| p | Number of class intervals |
| A | Fourier grain sample |
| $f_i$ | Number of grains in interval |
| $\underline{f}$ | Vector of frequencies |
| N | Number of samples |
| F | Shape samples |
| c | Number of prototypes |
| $\underline{e}_k$ | End-member vector in $\mathbb{R}^d$ |
| $\underline{r}$ | Vector in $\mathbb{R}^d$ |
| $\underline{x}$ | Implicit vector of data parameters |
| $\underline{v}_k$ | Prototype vector in $\mathbb{R}^d$ |
| $u_i$ | Membership function of cluster i |
| $u_{ik}$ | Membership of $r_k$ in cluster i |
| U | Membership matrix (cxn) |
| $M_{fc}$ | Set of membership matrices |
| $\underline{v}$ | Vector of prototype vectors |
| m | Weighting exponent |
| $J_m$ | Criterion function |
| $d_{ik}$ | Distance from $r_k$ to $v_i$ |
| $H_k$ | Reconstruction function |
| $x_k$ | Implicit coordinate |
| $y_k$ | Implicit coordinate |
| $z_k$ | Implicit coordinate |
| N | Nozzle diameter |
| P | Operating pressure |

## GREEK SYMBOLS

| | |
|---|---|
| $\delta s_k$ | Boundary of shape $s_k$ |
| $\phi_d$ | Sampling angle for digitizing $s_k$ |
| $\theta_j$ | Aggregated sampling angle |
| $\pi$ | "pi" |
| $\rho_k$ | Shape function of $\delta s_k$ |
| $\alpha_{kj}$ | Fourier coefficients |
| $\delta_{kj}$ | Fourier coefficients |
| $\alpha_k$ | Vector of Fourier amplitudes |
| $\Sigma$ | Summation |
| $\beta_i$ | Convex coefficients |

## MATHEMATICAL SYMBOLS

| | |
|---|---|
| $\{\ \}$ | Set brackets |
| $\mathbb{R}^2$ | 2-Dimensional Euclidean space |
| $^\circ$ | Degrees |
| $[0, a]$ | Square brackets: numbers between and including 0 and a |
| $\infty$ | Infinity |
| $\mathbb{R}^d$ | d-Dimensional Euclidean space |
| $\epsilon$ | "Belongs to" |
| $\approx$ | Approximately |
| $\forall$ | "For all" |
| $\leqslant$ | "Less than or equal to" |
| $=$ | "Equals" |
| $>$ | "Greater than" |
| X | Cartesian product |
| $\rightarrow$ | "Maps" |
| $\|\ \|$ | Norm |
| $\mid\ \mid$ | Absolute value |
| $+$ | Plus |

## REFERENCES

1. **Ehrlich, R. and Weinberg, B.,** An exact method for characterization of grain shape, *J. Sediment. Petrol.*, 40, 205, 1970.
2. **Beddow, J. K. and Vetter, A. F.,** On the use of classifiers in morphological analysis of particulates, *J. Powder Bulk Technol.*, 1(1), 42, 1977.
3. **Bezdek, J. C.,** Particle and grain shape analysis with fuzzy sets, in *Advanced Particulate Morphology*, Beddow, J. K. and Meloy, T. P., Eds., CRC Press, Boca Raton, Fla., 121, 1980.
4. **Bezdek, J. C.,** *Pattern Recognition with Fuzzy Objective Function Algorithms*, Plenum Press, New York, 1981.
5. **Solomon, K. and Bezdek, J. C.,** Characterizing sprinkler distribution patterns with a clustering algorithm, *Trans. ASAE*, 23, 899, 1980.

Chapter 10

# MORPHOLOGICAL ANALYSIS OF UNKNOWN SAMPLES

**Louise Hua, John Keith Beddow, and Arthur F. Vetter***

## TABLE OF CONTENTS



* The authors are grateful to the National Science Foundation Particulate Multiphase Processing Program for partial support under grant CPE-80-23868. The authors are also grateful to the International Fine Particle Research Institute.

## I. INTRODUCTION

A test was made to establish the reliability of methods for comparing and contrasting morphologies of different samples of unknown origin. The two samples, marked ALN and ALO, were subject to morphological analysis in an optical microscope and statistical analysis using invariant Fourier descriptors[1] as features. The result of the analyses indicated no significant difference in morphology between the two samples.

To test the reliability of the statistical technique being used to differentiate the two unknown samples, a third sample (Ottawa sand, mesh size no. 50/60) with a shape different from the unknowns, was used as a contrast. The result showed a good separation between sand and either ALN or ALO, thus confirming the adequacy of this analysis technique.

## II. SAMPLE PREPARATION

Representative photomicrographs of the three samples are shown in Figure 1. They are

| | | |
|---|---|---|
| ALN | unknown sample | mean size $< 33.8$ μm |
| ALO | unknown sample | mean size 48.7 μm |
| OTW1 | Ottawa sand (mesh no. 50/ 60) | mean size 169.8 μm |

One hundred particles were picked randomly from each sample, and digitized through PIAS (Particle Image Analyzing System)[2] to get sets of (x,y) coordinates, one set for each profile. The (x,y) pairs were then converted to polar coordinates $(R,\theta)$ using the center of gravity as the origin. By applying Fourier transformation, a set of Fourier coefficients ($A_n$s and $\alpha_n$s) was obtained from $(R,\theta)$ pairs for each particle. The invariant Fourier descriptors were finally calculated from Fourier coefficients as described in the previous chapter.

## III. STATISTICAL ANALYSIS

The technique used was the multivariate linear discriminant analysis based on Bayesian decision theory.[3,4] The probability density function for each group is assumed to be multivariate normal and is written as

$$f_k(\chi) = \frac{1}{(2\pi)^{d/2}|\Sigma_k|^{1/2}} \exp\left[ -\frac{1}{2}(\chi - m_k)^t \Sigma_k^{-1} (\chi - m_k)\right] \quad (1)$$

where $m_k$ is the d-component mean vector for k-th group, $\Sigma_k$ is the dxd covariance matrix for k-th group, $(\chi - m_k)^t$ is the transpose of $(\chi - m_k)$, $\Sigma_k^{-1}$ is the inverse of $\Sigma_k$, and $|\Sigma_k|$ is the determinant of $\Sigma_k$; $m_k$ and $\sigma_k$ are founded by taking expectations of $\chi_k$ and $(\chi_k - m_k)^t(\chi_k - m_k)$, respectively, where $\chi_k$ is a d-component column vector storing the features of a particle from k-th group. To be more specific, if $\chi_{i,k}$ is the i-th component of $\chi_k$, $m_{i,k}$ is the i-th component of $m_k$ and $\sigma_{ij,k}$ is the component ij-th of $\Sigma_k$ then

$$m_{i,k} = E[\chi_{i,k}]$$

$$\sigma_{ij,k} = E[(\chi_{i,k} - m_{i,k})^t(\chi_{j,k} - m_{j,k})]$$

in which E[ ] is the expectation of whatever is inside the brackets. The covariance matrix is always symmetric and positive semidefinite. The diagonal element $\sigma_{ii,k}$ is the variance of $\chi_{i,k}$, and the off-diagonal element of $\sigma_{ij,k}$ is the covariance of $\chi_{i,k}$ and $\chi_{j,k}$.

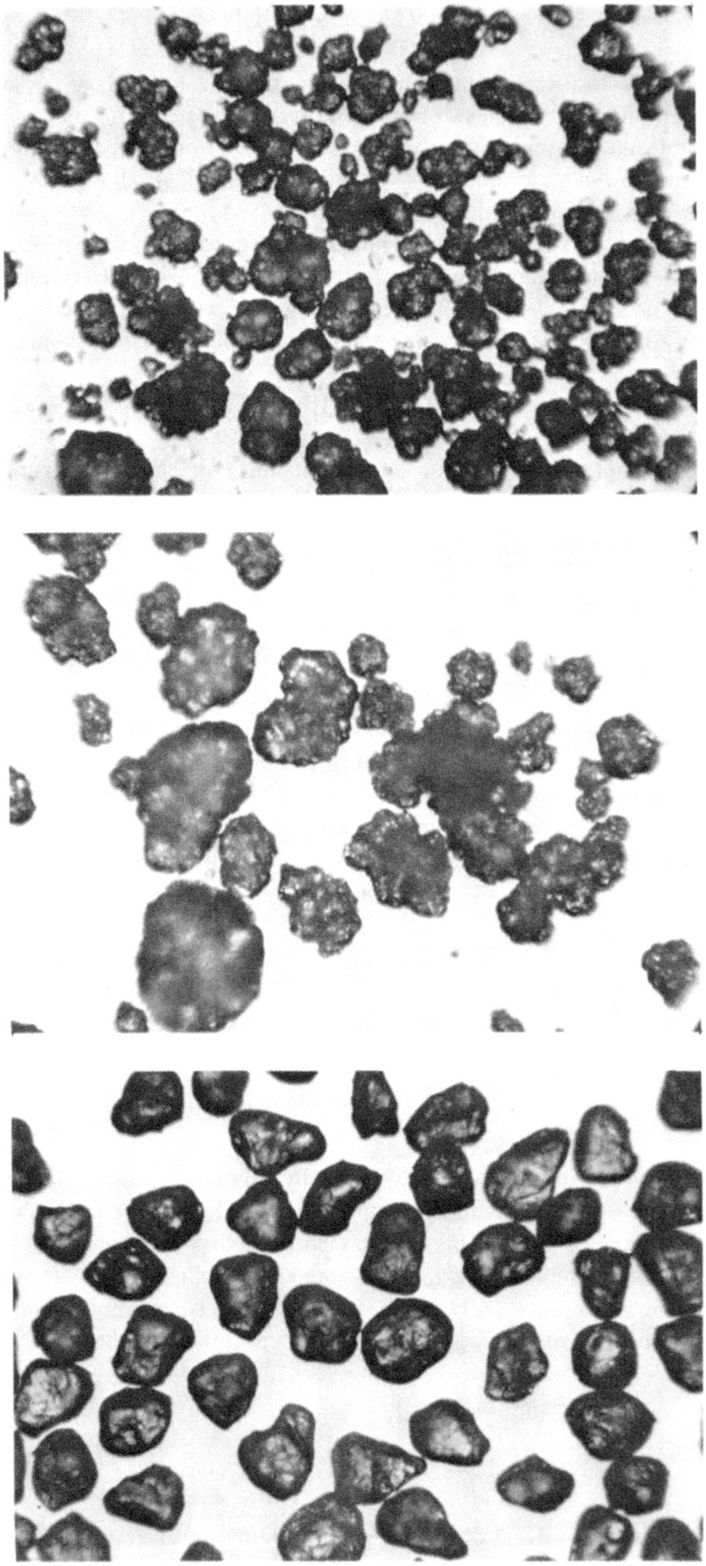

FIGURE 1. Representative photomicrographs of the three samples. (Top) ALN, unknown sample; mean size < 33.8 μm. (Center) ALO, unknown sample; mean size 48.7 μm. (Bottom) OTW1, Ottawa sand (mesh no. 50/60); mean size 169.8 μm.

Given the a priori probability for each group ($P_k$), the posterior probabilities of an observation ($\chi$) belonging to each group can be estimated. The linear discriminant function ($g_k(\chi)$) is the natural logarithm of the posterior probability:

$$g_k(\chi) = \ell n(f_k(\chi)P_k)$$

The above equation can be readily evaluated if the densities $f_k(\chi)$ are multivariate normal. Then

$$g_k(\chi) = -\frac{d}{2}\ell n2\pi - \frac{1}{2}\ell n|\Sigma_k| - \frac{1}{2}\chi'\Sigma_k^{-1}\chi$$

$$+ m_k'\Sigma_k^{-1}\chi - \frac{1}{2}m_k'\Sigma_k^{-1}m_k + \ell nP_k \qquad (2)$$

Since the $-d/2\ell n2\pi$ term in Equation 2 is independent of k, it is an unimportant additive constant that can be ignored. If we define

$$W_k = -\frac{1}{2}\Sigma_k^{-1}$$

$$\omega_k = \Sigma_k^{-1}m_k^t$$

$$\omega_{ko} = -\frac{1}{2}m_k'\Sigma_k^{-1}m_k - \frac{1}{2}\ell n|\Sigma_k| + \ell nP_k$$

then Equation 2 becomes

$$g_k(\chi) = \chi'W_k\chi + \omega_k^t\chi + \omega_{ko} \qquad (3)$$

The Bayesian classification rule assigns $\chi$ to the group associated with the largest $g_k(\chi)$.

A routine used in this study, which performs a multivariate linear discriminant analysis among several groups, is available from the IMSL routine ODNORM. However, ODNORM assumes the density for each group to be multivariate normal with common covariance matrix and uses pooled covariance matrix ($\Sigma$ pooled) instead of individual covariance matrix ($\Sigma_k$) as in Equation 1. Therefore, the linear discriminant function (Equation 3) reduces to Equation 4 by ignoring the constant terms.

$$g_k(\chi) = \omega_k^t\chi + \omega_{ko} \qquad (4)$$

where

$$\omega_k = \Sigma_{pooled}^{-1}m_k$$

$$\omega_{ko} = -\frac{1}{2}m_k'\Sigma_{pooled}^{-1}m_k + \ell nP_k$$

## IV. RESULTS

Because the discreteness of the digital data ((x,y) pairs) produces an edge composed of a number of square corners not actually present in the profile, a technique for smoothing the profile[5] was developed by omitting high-order harmonics (i.e., choosing the best value of k in Equation 1) using the statistical properties of the rounding error (discretization). On investigating the data, it was found that 10% of OTW1 has the k value less than 10. For this reason, only 10 L2s with n from 1 to 10 were chosen as variables. In addition, 36 L3s ($m + n = 2,3,----,12$), together with 5 other morphic terms ($L_0$, PER, IRR, VAR, SKE), were also used as variables. The 51 variables are listed in Table 1.

In some systems, it is not at all unusual to encounter a dimensionality problem, not only because too many features are involved, but also because of the nature of the feature. If a dimensionality problem is encountered or an economy factor is concerned, the Newman-Keul multiple range test[6] can be applied to reduce the number of features, i.e., to select those features showing significant discriminating power. Table 2 summarizes the variables selected for $\alpha = 0.01$. They are organized in decreasing order of discriminating power. This order is consistent with the result of human-eye judgment according to the histograms of those variables (not given here).

The resulting Bayesian classifier was tested on the original data with different variables sets and the performances were summarized in matrix D (see Table 3) in which the ij-th component of D is the fraction of samples in the i-th group that were classified in the j-th group. Table 3 shows that the result obtained by using reduced variables is acceptable when compared with that using 51 variables. Furthermore, if computing time is of concern, the latter cost about four times more than the former, thus confirming the validity and worthiness of the Newman-Keul multiple range test.

## V. DISCUSSION

It was found that no matter how many variables were used, samples ALN and ALO were inseparable. This is because the profile optical images of the two samples are so similar to each other morphologically that they may be confounded (See Figure 1), and the method used is actually based on the morphological factor. Introducing the Ottawa sand confirms that the Fourier descriptors do represent the particle morphology and can be used as features in discriminant analysis.

The basic assumption in this analysis is that the populations of descriptors have multivariate normal distribution with common covariance structures. From the histograms of these descriptors (not given here), it can be found that they are not far away from normal distribution. However, one is unable to tell whether they have equal covariance matrices. If the populations have a common covariance matrix $\Sigma$, then $\Sigma_{pooled}$ is an unbiased estimate of $\Sigma$. A method of multivariate analog of Bartlett's test for homogeneity of variance[7] can be applied to test whether the populations have equal covariance matrix or not, but this test is not used in this study. From the good separation between OTW1 and either ALN or ALO, it might be concluded that the assumption of common covariance matrix is reasonable.

As mentioned in the previous section, only those descriptors with $n \leq 10$ or $m + n \leq 12$ were chosen as variables. It is not implied that the other descriptors are not important, but comparatively, their use is not necessary since using only those 51 variables already makes the fraction classified for OTW1 to be 1.

Since the particle morphology is considered as the only factor of interest for investigation, the size term $R_0$ was isolated and ignored although it does show significant discriminating power between ALN and ALO.

**Table 1**
**THE 51 VARIABLES (VARIABLE SET I)**

**Individual shape terms**

| L3s | | | | | |
|---|---|---|---|---|---|
| **Abbreviation** | **Convention** | **Abbreviation** | **Convention** | **L2s and $L_0$** | **Morphic terms** |
| L3(1) | L3(1,1) | L3(19) | L3(2,9) | L2(1) | PER |
| L3(2) | L3(1,2) | L3(20) | L3(2,10) | L2(2) | IRR |
| L3(3) | L3(1,3) | L3(21) | L3(3,3) | L2(3) | VAR |
| L3(4) | L3(1,4) | L3(22) | L3(3,4) | L2(4) | SKE |
| L3(5) | L3(1,5) | L3(23) | L3(3,5) | L2(5) | |
| L3(6) | L3(1,6) | L3(24) | L3(3,6) | L2(6) | |
| L3(7) | L3(1,7) | L3(25) | L3(3,7) | L2(7) | |
| L3(8) | L3(1,8) | L3(26) | L3(3,8) | L2(8) | |
| L3(9) | L3(1,9) | L3(27) | L3(3,9) | L2(9) | |
| L3(10) | L3(1,10) | L3(28) | L3(4,4) | L2(10) | |
| L3(11) | L3(1,11) | L3(29) | L3(4,5) | $L_0$ | |
| L3(12) | L3(2,2) | L3(30) | L3(4,6) | | |
| L3(13) | L3(2,3) | L3(31) | L3(4,7) | | |
| L3(14) | L3(2,4) | L3(32) | L3(4,8) | | |
| L3(15) | L3(2,5) | L3(33) | L3(5,5) | | |
| L3(16) | L3(2,6) | L3(34) | L3(5,6) | | |
| L3(17) | L3(2,7) | L3(35) | L3(5,7) | | |
| L3(18) | L3(2,8) | L3(36) | L3(6,6) | | |

**Table 2**
**VARIABLES SHOWING SIGNIFICANT DISCRIMINATING POWER[a]**

| Variable set II (Samples ALN and OTW1 $\alpha$ = 0.01) | Variable set III (Samples ALO and OTW1 $\alpha$ = 0.01) |
|---|---|
| PER | PER |
| L2(9) | L2(9) |
| L2(10) | L2(10) |
| L2(7) | L2(8) |
| L2(8) | L2(7) |
| L2(6) | L2(6) |
| VAR | L2(5) |
| $L_0$ | VAR |
| IRR | $L_0$ |
| L2(5) | IRR |
| L2(4) | L2(2) |
| L2(2) | L2(4) |
| | L3(11) |

[a] None of the 51 variables show significant discriminating power between ALN and ALO at $\alpha$ = 0.01.

**Table 3**
**MATRIX D (SUMMARY OF RESULTING BAYESIAN CLASSIFIER TESTED ON ORIGINAL DATA)[a]**

| | Set I (51 Variables) | Set II (α = 0.01) (12 Variables) | Set III (α = 0.01) (13 Variables) |
|---|---|---|---|
| ALN & ALO (I = 1) (I = 2) | $\begin{bmatrix} 0.70 & 0.30 \\ 0.30 & 0.70 \end{bmatrix}$ | | |
| OTW1 & ALN (I = 1) (I = 2) | $\begin{bmatrix} 0.99 & 0.01 \\ 0.05 & 0.95 \end{bmatrix}$ | $\begin{bmatrix} 0.96 & 0.04 \\ 0.17 & 0.83 \end{bmatrix}$ | |
| OTW1 & ALO (I = 1) (I = 2) | $\begin{bmatrix} 1.00 & 0.00 \\ 0.09 & 0.91 \end{bmatrix}$ | $\begin{bmatrix} 0.97 & 0.03 \\ 0.15 & 0.85 \end{bmatrix}$ | $\begin{bmatrix} 0.97 & 0.03 \\ 0.15 & 0.85 \end{bmatrix}$ |
| OTW1, ALN, & ALO (I = 1) (I = 2) (I = 3) | $\begin{bmatrix} 1.00 & 0.00 & 0.00 \\ 0.10 & 0.66 & 0.24 \\ 0.08 & 0.30 & 0.62 \end{bmatrix}$ | $\begin{bmatrix} 0.96 & 0.04 & 0.00 \\ 0.13 & 0.47 & 0.40 \\ 0.11 & 0.36 & 0.53 \end{bmatrix}$ | |

[a] D(I,J) is the fraction of the samples in I-th group that were classified in the J-th group.

## VI. CONCLUSION

Three conclusions may be drawn from this investigation:

1. This is the first time the newly developed Fourier descriptors have been used in discriminant analysis.
2. As long as the samples of interest show distinguishable morphologies, this analysis is effective and reliable.
3. Newman-Keul's multiple range test is recommended for reducing the number of variables so that one can save computing time and avoid dimensionality problems.

## SYMBOLS

| | |
|---|---|
| $d$ = | Number of variables (features) |
| $\chi_k$ = | d-Component feature vector of a particle from k-th group |
| $\chi_{i,k}$ = | i-th Component of $\chi_k$ |
| $m_k$ = | d-Component mean vector of k-th group |
| $m_{i,k}$ = | i-th Component of $m_k$ |
| $f_k$ = | Probability density function for k-th group |
| $\Sigma_k$ = | dxd Covariance matrix for k-th group |
| $\sigma_{ij,k}$ = | i-th Row, j-th column element of $\Sigma_k$ |
| $\chi$ = | d-Component feature vector for an observation |
| $P_k$ = | A priori probability for k-th group |
| $g_k$ = | Linear discriminant function for k-th group |
| $c$ = | Number of groups |
| $D$ = | cxc Matrix |
| $\alpha$ = | Significant level |

## REFERENCES

1. **Luerkens, D. W.,** Morphological Analysis of Fine Particles Using the Fourier Method, Ph.D. thesis, University of Iowa, Iowa City, 1980; *Powder Technol.*, 13(2), 209, 1982.
2. **Lenth, R., Chang, C. R., Beddow, J. K., and Vetter, A. F.,** Particle image analyzing system, *Particulate Systems: Application and Fundamentals*, Hemisphere Corporation, New York, 1983.
3. **Duda, R. O. and Hart, P. E.,** *Pattern Classification and Scene Analysis*, John Wiley & Sons, New York, 1973.
4. **Kendall, M. and Stuart, A.,** *The Advanced Theory of Statistics*, Vol. 3, Charles Griffin, London, 1976.
5. **Lenth, R.,** Smoothing a digitized particle profile, in *Particle Characterization in Technology, Vol. 2: Morphological Analysis*, CRC Press, Boca Raton, Fla., 1984.
6. **Burr, I. W.,** *Applied Statistical Methods*, Academic Press, New York, 1974.
7. **Winer, B. J.,** *Statistical Principles in Experimental Design*, 2nd ed., McGraw-Hill, New York, 1971.

Chapter 11

# OPTIMAL DEFINITION OF CLASS INTERVALS OF HISTOGRAMS OR FREQUENCY PLOTS

**William E. Full, Robert Ehrlich, and Stephen Kennedy**

## TABLE OF CONTENTS

## I. ABSTRACT

Data sets are often analyzed in the form of collections of histogram frequencies or percentiles derived from equivalent cumulative frequency distributions. Decisions concerning the number of intervals and interval width obviously affect the quality of the data in subsequent analysis. Relying on the basic concepts of information theory, a procedure is presented which evaluates the relative information content of a set of frequency data when subdivided in various manners. Maximum information is always preserved when "maximum entropy" histograms (with unequal class intervals) are used. Evaluation of several schemes of histogram subdivision (phi-based arithmetic, log arithmetic, Z-score, log Z-score, maximum entropy) indicate that, surprisingly, collections of equal-interval phi-based histograms contain the least information. Additionally, the concept of the relative entropy of a given collection of histograms is defined. The relative entropy is useful as a feature extractor wherein several collections of data with potentially similar information can be compared. An example of using the relative entropy as a feature extractor is given in shape analysis where the choice of which harmonic(s) that represents the greatest shape differences needs to be defined.

## II. INTRODUCTION

Particle scientists have traditionally cast data in the form of frequency histograms or their equivalent, cumulative frequency distributions. After data are put in this form, the frequencies in each class interval are used as data for further analysis such as moment measures,[7,8,14] discriminant functions,[24,26,32,33,34] and factor analysis.[2,12,17,31] Obviously, this is done because it is assumed that the data manifest in this manner contain information. However, the quality of the information expressed by histogram frequencies must be affected by the choice of the number of class intervals. Extreme situations range from the case where all the observations fall in one class interval (i.e., intervals too wide), to the case where the data are vanishingly sparse in all intervals (i.e., intervals too narrow and numerous). From this example, it can be seen that the width of a class interval affects the quality of information we can glean from histograms. Examples of the effect of the width of the histogram intervals can be seen in Folk,[6] Jaquet and Vernet,[14] Swan et al.,[30] and Kennedy et al.[16]

Even with an "appropriate" number of class intervals, some parts of a frequency distribution will have large frequencies ("modes") and other portions will be relatively bare. Where class frequencies are low, the effect of random perturbations are strongly felt, whereas in high-frequency portions of a histogram the effect of these random perturbations are reduced. Conversely, high-frequency portions of the histograms may contain more observations than needed in terms of precision, and might be profitably subdivided in order to test for the presence of polymodality in these intervals.

If, therefore, one compares a set of samples, each in the form of a frequency histogram with equal-width class intervals, certain intervals may not contribute significantly to the comparison because of the greater probability of random perturbations in intervals of low frequency in all samples, whereas in others, information is lost because high-frequency intervals are too broad. One could remedy this by creating class intervals of unequal width. For example, percentile-based size measures[7,13,22] calculated from cumulative frequency curves assume that variable width intervals are desirable in size analysis. The graphic mean of Folk and Ward,[13] for instance, is calculated by summing the cumulative percent represented by the 16th percentile, 50th percentile, and the 84th percentile and dividing this sum by 3. In effect, the graphic mean can be considered to be derived from a three-class interval histogram whose interval widths are unequal.

The objective of the paper is to demonstrate that such unequal-width class intervals can maximize the amount of information that can be extracted from frequency distributions. These conclusions draw heavily on the fundamental relations of information theory which were originally derived for analysis of electronic signal propagation and detection. Based on these fundamental relationships of information theory, a different method of defining interval boundaries has been designed. This new "maximum entropy method" defines interval boundaries based on the aggregate properties of the entire sample set. Examples from size and shape analysis will be used to demonstrate the usefulness of such a method to increase the potential of any analysis (discrimination, clustering, unmixing) performed on data cast in such a manner. In the process of the above demonstration, questions will be raised concerning presently used techniques in size analysis. Before this is done, a brief overview of the fundamental measure of information will be given.

## III. THE CONCEPT OF ENTROPY

The concept of entropy is one that has been generally shrouded in mystery. This is due, in part, to the fact that the word entropy is used in many disciplines with seemingly different connotations. There is "chemical entropy", "thermal entropy", and a "probabilistic entropy", to name a few. Shannon,[27] however, described a measure called "information entropy", which has been shown to be the primitive concept which could explain the other entropy measures (see Jaynes[15]). Henceforth, the word entropy will refer to the information entropy as described by Shannon.

Shannon defined entropy (E) by the following formula:

$$E = -\sum_{i=1}^{n} p_i \operatorname{Log}_a p_i$$

where $p_i$ is the probability of an event occurring in interval i, $\underline{a}$ is any base of logarithm (base $\underline{e}$ in this paper), and n is the number of intervals. The probability of an event occurring in interval i is calculated by dividing the frequency within that interval by the total frequency count. This unitless form of entropy has been applied in geology to histograms and cumulative curves by Sharp and Fan[29] in order to define a sorting index for size analysis. Later, Sharp[28] applied information entropy to define a measure of parity between the mean and standard deviation of a distribution and how much information is expressed by these values. Sharp's earlier studies on entropy represent some of the first applications of entropy in sedimentary geology.

Entropy, as applied to frequency plots, is a measure of the contrast between intervals. Low entropy values represent frequency plots with large differences between intervals (Figure 1A) and high entropies characterize frequency plots with relatively slight contrast between intervals (Figures 1B and 1C). Frequency plots exhibiting no contrast between intervals (Figure 1C) are referred to as maximum entropy plots. However, inasmuch as isolated frequency plots are rarely used in sedimentological analysis such as unmixing, discrimination, or clustering, other properties of entropy can be used to measure the information contained within a system of frequency plots.

One such property is that it is additive; the total entropy contained within a system of frequency plots is equal to the sum of the entropies of the individual frequency plots. The total entropy of a set of frequency plots is a measure of the total contrast between intervals and hence is a measure of the total amount of useful information contained within that system. The entropy of the total system is a direct measure of the potential of the frequency plots to provide reasonable solutions when further analysis (such as unmixing, discrimination, or clustering) is performed on that set of data.

Entropy, as a measure of the information contained within a particular set of data, can be used to measure the efficiency of several schemes for determining class intervals of

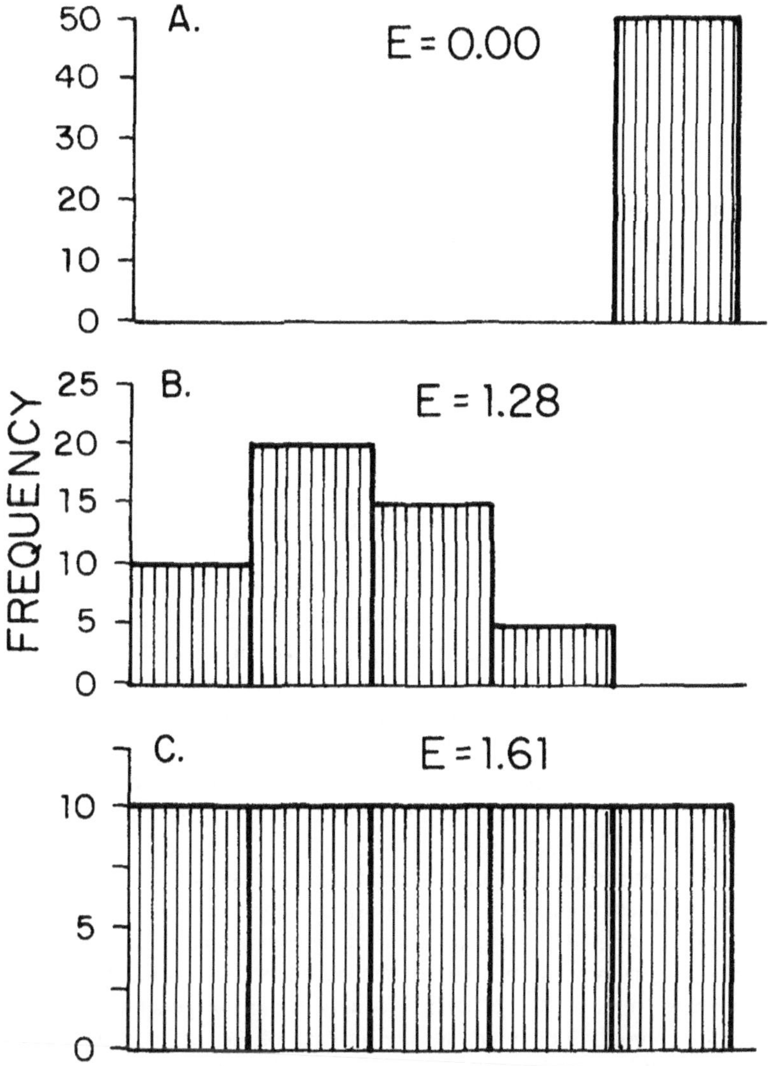


FIGURE 1. Examples of frequency plots and resulting entropies(E).

histograms. The importance of such an evaluation is that some commonly used methods of defining the interval boundaries may be better than others in displaying differences or similarities between samples and, hence, can drastically affect the efficacy of further analysis performed on that data set. Included in this comparison is a new approach that maximizes the entropy within a particular set of frequency plots. The approach that produces such plots is presented in the next section.

## IV. MAXIMUM ENTROPY METHOD

Maximum entropy is attained when any single event has equal probability of occurring in any interval.[27] Relative to a particular set of samples (histograms), maximum entropy can be obtained by defining the interval boundaries, using the following method:

1. Pool all the observations in the sample set.
2. Order the observations from smallest to largest.

3. Let N be the total number of pooled observations and I, the number of intervals. Then, starting at the smallest (or largest) observation, the (N/I)-th value defines the end of the first interval.
4. Repeat step (3) for the $2 \times$ (N/I)-th, $3 \times$ (N/I)-th, ..., $(I - 1) \times$ (N/I)-th values and use these to define the remaining interval boundary values.

The above method is easily programmable. However, because step (2) can be very time- and space-consuming for large data sets, modifications to the approach can be made wherein the procedure becomes practical for large data sets. A computer program designed to handle large data sets is currently being prepared for publication. Obviously, such an approach assumes that size distributions, for instance, are measured continuously (either grain by grain, or via cumulative frequency curves derived from settling tubes or interpolated from sieving. It is the availability of such data that permits definition of class intervals of unequal width.

The above approach (involving the pooling of the entire data set) allows differences between samples to be expressed without forcing the samples to be different. This can be seen by examining the two extreme examples shown in Figure 2. The first data set (Figure 2a) represents four samples that are exactly alike, while the second data set (Figure 2b) represents four samples that are completely different from each other. Four intervals were chosen, although in this case any number of intervals greater than four would produce similar results. The maximum entropy method was performed on each of the data sets, producing the frequency plots given (Figure 2). In Figure 2a, the similarity between samples can be seen while samples in Figure 2b exhibit maximum differences, hence demonstrating that differences between samples are not forced. In this light, the maximum entropy approach represents the "maximally least biased" method of interval definition.[21] Additionally, maximum entropy frequency plots increase the potentially useful information that, when available, should be used in further analysis.[21]

Conceivably, conventional schemes of determining class intervals in sedimentology may already approach a maximum entropy subdivision of the frequency distributions. However, this can only be determined by comparing the entropy generated by the conventional methods with the entropy generated by histograms whose intervals are formally determined by the maximum entropy method. The efficiency of any system of generating class intervals can be evaluated by comparing the total entropy for each of the methods. The larger this value, the greater is the efficiency in capturing the information contained within the system. The following section compares the efficiency of various styles of generating class intervals.

## V. EFFICIENCIES OF VARIOUS HISTOGRAMMING APPROACHES

In addition to the previously defined maximum entropy scheme, five additional methods will be compared using grain size data (Figure 3). The first three are similar in that some measure of size is divided into equal intervals. The first method considered is the traditional (in geology) phi method ($-\log_2$ mm) (TP) where the intervals are each quarter-phi units wide (Figure 3). The second is an arithmetic (A) method wherein millimeter-based intervals are each a fixed distance wide (Figure 3). The third is the log arithmetic method (LA) where the log (base 10) of the diameters is used to define equally spaced log millimeter intervals (Figure 3). The latter method, functionally related to the TP method, has been used in spectral analysis and has been recently applied to the analysis of quartz particle shapes.[1]

The fourth method, the Z-scores method (ZS), was used in many shape studies (Porter et al.,[25] Brown et al.,[2] and Mazzullo and Ehrlich,[23] to name a few). Its underlying assumption is that the pooled data have a higher density closer to the mean. The intervals are defined in such a way as to create equal frequency counts in each interval by assuming the pooled

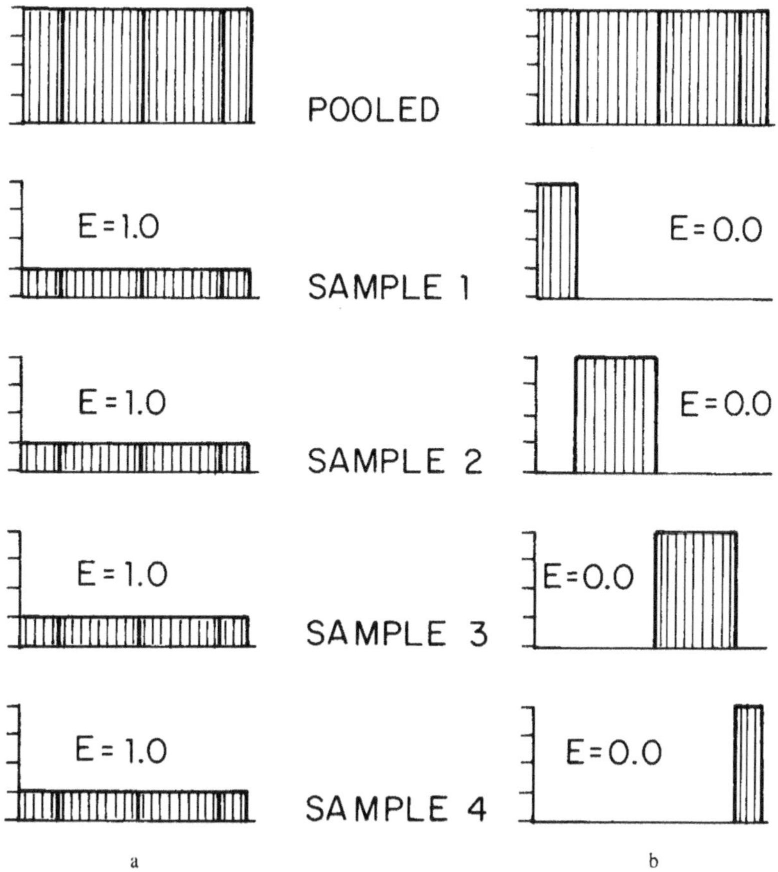


FIGURE 2. Example of two sample sets: alike when pooled, but expressing (a) similarity between samples and (b) differences between samples when the same interval definition is applied to the individual samples.

data are normally distributed. The net effect of such a scheme is to create narrower intervals where the bulk of the data are situated (in the middle of the pooled data set), and progressively wider intervals toward the tails (Figure 3).

The fifth method that is to be compared is the log Z-score method (LZS) wherein the ZS algorithm is applied to the pooled data set created by taking the log (base 10) of each individual observation. This method is included for the sake of comparing a particular method to its log-transformed counterpart (i.e., A vs. LA, ZS vs. LZS). The log-maximum entropy method does not have to be compared in such a way because the log transformation will not result in a change of information relative to the maximum entropy method.[20] That is, for most common transformations (log, square root, square) the maximum entropy method is transformation independent. Therefore, there is no a priori need to transform the data if the data are cast in maximum entropy format.

A data set of 12 grain size distributions was analyzed in order to compare the various methods mentioned. The distributions were obtained on a grain-by-grain basis using a video-digitizing microprocessor system (ARTHUR II) described elsewhere.[5] The average diameter was calculated based on the maximum projection of each particle and converted to volume by assuming sphericity.[16] Approximately 200 grains comprised each sample. The results of the entropy comparisons are shown in Table 1. The calculated entropy for each sample is listed by method; the total entropy is given for each method at the bottom of the table.

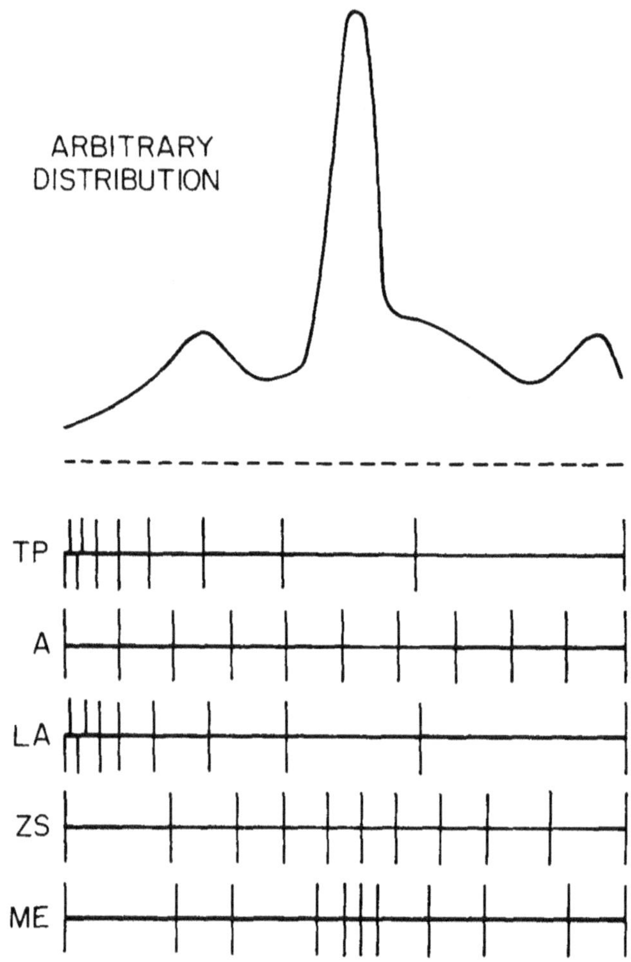


FIGURE 3. Example of the partition of an arbitrary pooled collection of distributions using the traditional phi (TP), arithmetic (A), log (base 10) arithmetic (LA), Z-scores (ZS), and maximum entropy methods.

Surprisingly, the equal-interval phi-based frequency plot can be seen, in this instance, to be the poorest carrier of information. Interestingly, even the arithmetic method produced a higher total entropy than either case of equal-width interval histograms of log-transformed data—demonstrating that, at least in the present instance, the log transformation can reduce intersample information. This suggests that data transformations must be done with the greatest of caution unless the maximum entropy technique is used.

Less surprising is the fact that the log Z-score method produced relatively high entropy values. This is due to the fact that the data are close to lognormality with relatively equal means.[16] As a matter of fact, if all the samples are lognormal with the same means and standard deviations, the log Z-score method would produce maximum entropy plots. Many size distributions are neither as normal or lognormal as the samples used here. Therefore, the Z-score or log Z-score method is not universally better than the other methods discussed above, especially when the data consist of nonsymmetrical or polymodal distributions.

The least surprising result is that the maximum entropy method produced the largest entropy, showing that this method conveys more information than the previously defined methods. Generally, the increased information content of the maximum entropy plots has

**Table 1**
**LISTINGS OF THE INDIVIDUAL SAMPLE COMPARISONS FOR DIFFERENT METHODS OF INTERVAL DEFINITION**

| | Method | | | | | |
|---|---|---|---|---|---|---|
| Sample | TP | A | LA | ZS | LZS | ME |
| 1 | 1.5 | 1.8 | 1.7 | 2.2 | 2.3 | 2.3 |
| 2 | 1.6 | 1.9 | 1.6 | 2.2 | 2.2 | 2.3 |
| 3 | 1.5 | 1.8 | 1.5 | 2.2 | 2.2 | 2.3 |
| 4 | 1.5 | 1.8 | 1.7 | 2.2 | 2.2 | 2.2 |
| 5 | 1.5 | 1.8 | 1.6 | 2.1 | 2.3 | 2.3 |
| 6 | 1.5 | 1.7 | 1.5 | 2.2 | 2.2 | 2.3 |
| 7 | 1.5 | 1.8 | 1.5 | 2.2 | 2.2 | 2.2 |
| 8 | 1.5 | 1.8 | 1.5 | 2.2 | 2.2 | 2.3 |
| 9 | 1.5 | 1.8 | 1.5 | 2.1 | 2.3 | 2.3 |
| 10 | 1.6 | 1.9 | 1.5 | 2.0 | 2.2 | 2.3 |
| 11 | 1.6 | 1.9 | 1.7 | 2.1 | 2.3 | 2.2 |
| 12 | 1.6 | 1.9 | 1.7 | 2.1 | 2.3 | 2.2 |
| Total system entropy | 18.4 | 21.9 | 18.9 | 25.8 | 26.9 | 27.3 |

produced higher-quality solutions, relative to quartz grain shape analysis,[4] when used as input data into the Q-mode EXTENDED CABFAC-QMODEL family of algorithms.[10,11,18,19]

## VI. ENTROPY AS A FEATURE EXTRACTOR

The previous sections introduced the idea of using the total system entropy as a means to compare the information captured by various interval definition schemes. The maximum entropy method was shown to capture the largest amount of information. Now that the method of interval definition has been chosen, entropy can serve as a measure of the total information contained within a particular data set. This measure of information becomes important when two or more data sets consisting of different observations may contain similar geologic information. For example, an investigator may have access to separate data sets such as size, shape, chemical, and hydrologic analyses from the sample array or a time series. There may be no practical way (or justification) to analyze all these data. A question to be answered is which variable(s) offers the best opportunity for a clear, unambiguous analysis. Entropy becomes a numerical index by which the geologist can decide which variable(s) holds the most potential for further analysis. Such a criterion is called a ''feature extractor''.

In order to use entropy as a feature extractor, the concept of relative entropy must be introduced. Relative entropy is the ratio of the calculated sample entropy and the maximum possible entropy. The maximum possible entropy for an individual sample is equal to the natural logarithm of the number of intervals. Therefore, the relative entropy of an entire data set is the sum of the calculated entropies divided by the product of the maximum possible entropy and the number of samples. Given a particular method of interval definition (hopefully the maximum entropy method), low relative entropies (closer to zero) indicate data sets with greater contrast between samples. This can be seen in Figure 2 wherein samples in Figure 2b represent a system whose relative entropy is zero. In contrast, high relative entropies (closer to one) reflect data sets with little intersample variation (data set A of Figure 2). The data set that contains the lowest relative entropy among several data sets contains the highest potential for a clear, unambiguous solution in any subsequent analysis. This is in contrast to the previous section wherein low entropies indicated a reduced potential to carry sample-by-sample contrast when we were discussing the various interval definition schemes. After the data set with the greatest potential is identified, attempts at maximizing the information contained within that data set can be attempted (i.e., raise the total entropy by either combining intervals or increasing or decreasing the number of intervals). The following example will demonstrate the use of the relative entropy as a feature extractor.

In shape analysis, the maximum projection shapes of particles are quantified via a Fourier series.[3,9] Each shape is quantified by calculating 20 shape components (called harmonics); the greater the magnitude (amplitude) of the harmonic, the greater its contribution to the total shape. Each sample of 200 grains is represented by the distributions of 20 sets of harmonic amplitudes. A study may involve 100 or more samples and analysis of each variable (harmonic) separately would be cumbersome. Because each harmonic measures a different scale of shape variability, different amounts and types of information can occur at different harmonics. For instance, if one is concerned with the effects of diagenesis or abrasion, the higher harmonics measuring small-scale perturbations on the grain outline are more efficient carriers of information than the lower harmonics, which measure the grosser aspects of shape (elongation, triangularity, etc.). In addition, among the higher harmonics, some may be more inherently sensitive to the system of interest than others. The problem, then, is to find one or a few harmonics which are most likely to carry the most information. The class frequencies of the amplitudes of these harmonics will be used in further analysis.

An example is shown in Figure 4. The data set represents stream sediment samples from drainage basins on the west flank of the Big Horn Mountains in Wyoming.[35] The streams traversed Precambrian through Mesozoic crystalline and sedimentary rocks. The relative entropy is plotted for harmonics 2 through 20; (the first harmonic is an error measurement and is not used for further analysis). The plot in Figure 4 indicates that harmonic 15 shows the lowest relative entropy, followed by harmonics 14, 16, and 17. All of these harmonics, when individually analyzed using the EXTENDED CABFAC-FUZZY QMODEL algorithm,[10] produced similar four end member (four source) solutions which clearly reflect the four important source terranes in the region.

## VII. POTENTIAL PROBLEMS

There may be occasions where the maximum entropy formalization may pose some difficulties. One of the potential difficulties is that the data themselves determine interval definition, thus making direct comparisons between data from two independent studies more difficult. Another problem, one which is equal to the problem of interval width definition, is the choice of the number of intervals. Using the maximum entropy approach, all samples are subdivided into the same number of intervals. Too few intervals will obscure the information present by submerging inherent variability, whereas too many intervals will obscure information by increasing the "signal to noise ratio". The subdivisions are based on the average properties of the data set. However, this fixed number of intervals can conceivably not be optimal for any of the samples. That is, relatively simple frequency distributions might be adequately served by a few intervals whereas complex ones might be better served with a larger number of intervals. However, a sample set may contain samples containing both types of distributions. In such a case, less or more than the optimal number of intervals for any given distribution may be accepted in order to retain the potential definition of all the distributions. Therefore, what is needed, complimentary to the maximum entropy approach for optimal interval width definition, is a method for determining the optimal number of intervals. Unfortunately, to our knowledge such a procedure has not yet been developed.

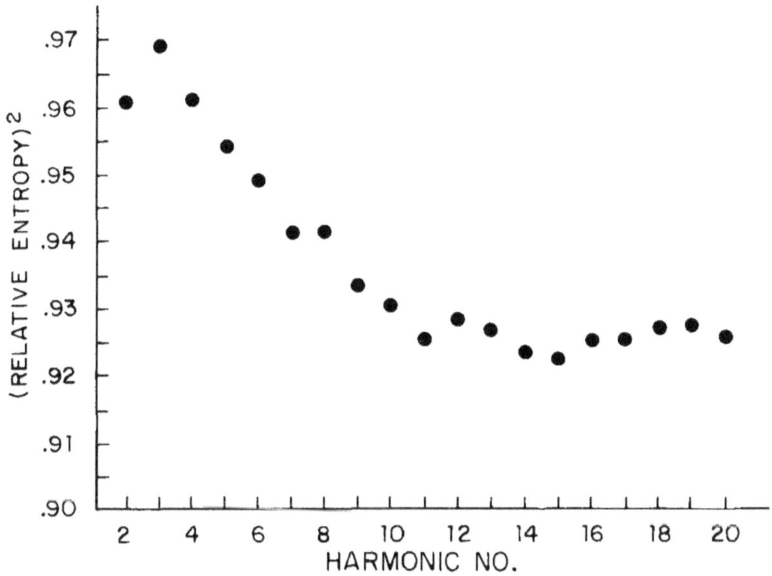


FIGURE 4. Plot of the relative entropy (squared to enhance separation) vs. harmonic number.

## VIII. CONCLUSIONS

The concept of entropy, as defined by Shannon,[27] can be a powerful tool for geologists. With entropy measurements, the geologist has an index of the information represented by a particular data set. In addition to measuring information, the principle of entropy can be used to define interval definition for histograms and frequency distribution plots via the maximum entropy method, wherein the maximum amount of information contained within a specific data system can be assumed to have been captured. The maximum entropy method maximizes the possibility of extracting additional information in terms of unmixing, clustering, scaling and discrimination, and minimizes any bias introduced by the a priori definition of interval values or by assuming a priori knowledge of the distributions present.

The original reason for performing the above entropy analysis was to optimize the unmixing algorithms EXTENDED QMODEL[11] and FUZZY QMODEL[10] for shape analysis. However, applications of the principle of entropy in size and other types of analyses promise to increase the potential use of such data to solve more complex geologic problems. In terms of realized potential, the maximum entropy method has already resulted in higher-quality solutions in Fourier shape analysis.[4]

In addition to optimizing the definition of intervals in frequency plots, the entropy can be used as a feature extractor for a particular system of data sets. This has been applied to shape analysis to point out which harmonics carry the largest amount of information, and helps to define which harmonics should be analyzed.

## ACKNOWLEDGMENTS

This project was sponsored by the Office of Naval Research under Contract No. N00014-78-C-0698. Carleen Sexton typed and helped edit the final draft.

## ALPHABETICAL SYMBOLS

E = Entropy
pi = Probability of event occurring in interval i
i = Interval
q = Base of log
n = Number of intervals
N = Number of pooled observations

## REFERENCES


1. **Boon, J. D., III, Evans, D. A., and Hennigar, H. F.,** Interpretation of grain shape information from Fourier analysis of digitized two-dimensional images, *Int. J. Math. Geol.*, 14(6), 589, 1982.
2. **Brown, P. J., Ehrlich, R., and Colquhoun, D.,** Origin of patterns of quartz sand types on the southeastern United States continental shelf and its implication on contemporary shelf sedimentation-Fourier grain shape analysis, *J. Sediment Petrol.*, 50, 1095, 1980.
3. **Ehrlich, R., Brown, P. J., Yarus, J. M., and Przygocki, R. S.,** The origin of shape frequency distributions and the relationship between size and shape, *J. Sediment. Petrol.*, 50, 475, 1980.
4. **Ehrlich, R., Full, W. E., Judson, M., and Umbehagen, G.,** Analysis of quartz detritus of complex provenance via analysis of shape data, in press.
5. **Fico, C.,** Automated Particle Shape Analysis — Development of a Microprocessor Controlled Image Analysis System, Masters thesis, University of South Carolina, Columbia, 1980.
6. **Folk, R. L.,** A review of grain-size parameters, *Sedimentology*, 6, 73, 1966.
7. **Folk, R. L. and Ward, W. C.,** Brazos River bar: a study in the significance of grain size parameters, *J. Sediment. Petrol.*, 27, 3, 1957.
8. **Friedman, G. M.,** Dynamic processes and statistical parameters compared for size frequency distributions of beach and river sands, *J. Sediment. Petrol.*, 37, 327, 1967.
9. **Full, W. E. and Ehrlich, R.,** Some approaches for location of centroids of quartz grain outlines to increase homology between Fourier amplitude spectra, *J. Math. Geol.*, 14, 43, 1982.
10. **Full, W. E., Ehrlich, R., and Bezdek, J. C.,** FUZZY QMODEL: a new approach for linear unmixing, *J. Math. Geol.*, 14(3), 259, 1982.
11. **Full, W. E., Ehrlich, R., and Klovan, J. E.,** EXTENDED QMODEL — objective definition of external end members in the analysis of mixtures, *J. Math. Geol.*, 13, 331, 1981.
12. **Hudson, C. B. and Ehrlich, R.,** Determination of relative provenance contributions in samples of quartz sand using Q-mode factor analysis of Fourier grain shape data, *J. Sediment. Petrol.*, 50, 1101, 1980.
13. **Inman, D. L.,** Measures for describing the size distribution of sediments, *J. Sediment. Petrol.*, 22, 25, 1952.
14. **Jaquet, J.-M. and Vernet, J.-P.,** Moment and graphic size parameters in sediments of Lake Geneva (Switzerland), *J. Sediment. Petrol.*, 46, 305, 1976.
15. **Jaynes, E. T.,** Electrodynamics today, *Proc. 4th Rochester Conf. Coherence and Quantum Optics*, Mandel, L. and Wolk, E., Eds., Plenum Press, New York, 1978.
16. **Kennedy, S. K., Ehrlich, R., and Kana, T. W.,** The nonnormal distribution of intermittent suspension sediments below breaking waves, *J. Sediment. Petrol.*, 51, 1103, 1981.
17. **Klovan, J. E.,** The use of factor analysis in determining depositional environments from grain-size distributions, *J. Sediment. Petrol.*, 36, 115, 1966.
18. **Klovan, J. E. and Imbrie, J.,** An algorithm and FORTRAN-IV program for large scale Q-mode factor analysis and calculation of factor scores, *J. Math. Geol.*, 3, 61, 1971.
19. **Klovan, J. E. and Miesch, A. T.,** EXTENDED CABFAC and QMODEL computer programs for Q-mode factor analysis of compositional data, *Comput. Geosci.*, 1, 161, 1976.
20. **Kullback, S.,** *Informaton Theory and Statistics*, John Wiley & Sons, New York, 1959.
21. **Lee, R.,** *Entropy Models in Spatial Analysis*, Univ. Toronto Dep. Geogr. Discuss. Pap. Ser., Toronto, 1974.
22. **Mason, G. C. and Folk, R. L.,** Differentiation of beach, dune, and aeolian flat environments by size analysis, Mustang Island, Texas, *J. Sediment. Petrol.*, 28, 211, 1958.
23. **Mazzullo, J. and Ehrlich, R.,** A vertical pattern of variation in the St. Peter sandstone — Fourier grain shape analysis, *J. Sediment. Petrol.*, 50, 63, 1980.

24. **Moiola, R. J., Spencer, A. B., and Weiser, D.,** Differentiation of modern sand bodies by linear discriminant analysis, *Trans. Gulf Coast Assoc. Geol. Soc.*, 24, 324, 1974.
25. **Porter, G. A., Ehrlich, R., Combellick, R. A., and Osborn, R. H.,** Sources and nonsources of beach sand along southern Monterey Bay, California — Fourier Shape analysis, *J. Sediment. Petrol.*, 49, 727, 1978.
26. **Reed, W. E., Le Fever, R., and Moir, G. J.,** Depositional environment interpretation from settling velocity (PSI) distributions, *Geol. Soc. Am. Bull.*, 86, 1305, 1975.
27. **Shannon, E.,** *Bell System Tech. J.*, 379, 623, 1948; reprinted in **Shannon, C. E. and Weaver, W.,** *The Mathematical Theory of Communication*, University of Illinois Press, Urbana, 1963.
28. **Sharp, W. E.,** Entropy as a parity check, *Earth Res.*, 1, 27, 1973.
29. **Sharp, W. E. and Fan, P.,** A sorting index, *J. Geol.*, 71, 76, 1963.
30. **Swan, D., Clague, J., and Leternauer, J. L.,** Grain-size statistics. II. Evaluation of grouped moment measures, *J. Sediment. Petrol.*, 49, 487, 1979.
31. **Soluhob, J. T. and Klovan, J. E.,** Evaluation of grain-size parameters in Lacustrine environments, *J. Sediment. Petrol.*, 40, 81, 1970.
32. **Taira, A. and Scholle, P. A.,** Discrimination of depositional environments using settling tube data, *J. Sediment. Petrol.*, 49, 787, 1979.
33. **Tillman, R. W.,** Multiple group discriminant analysis of grain size data as an aid in recognizing environments of deposition, 8th Int. Sediment. Congr. Abstr., Heidelberg, Germany, 1971, 102.
34. **Tucker, R. W. and Vacher, H. L.,** Effectiveness of discriminating beach, dune, and river sands by moments and the cumulative weight percentages, *J. Sediment. Petrol.*, 50, 165, 1980.
35. **Kennedy, S. K.,** Provenance and Dispersal of Sand and Silt in a High Gradient Stream System on the West Flank of the Bighorn Mountains, Wyoming — Fourier Shape Analysis, Ph.D. dissertation, University of South Carolina, Columbia, 1983, unpublished.

# *Part III*
# *Applications*

Chapter 12

# RECENT APPLICATIONS OF MORPHOLOGICAL ANALYSIS*

**John Keith Beddow**

## TABLE OF CONTENTS



* The author is grateful to the National Science Foundation Particulate Multiphase Processing Program for partial support under grant CPE-80-23868. The author is also grateful to the International Fine Particle Research Institute.

# I. ABSTRACT

The need for and the principles of morphological analysis as applied to particulate matter are briefly discussed. Applications of morphological analysis to a number of problems are cited. Topics include relating morphology to the mechanisms of powder production, wear debris analysis by morphological analysis, and kinetic modeling of crystallization processes. Relationships between the behavior of particulate materials and morphology of the particles are discussed in connection with abrasive wear of metals, sedimentation in low flow regimes, measuring of bulk properties including internal angle of friction and flow times, dry separation technology, and scrap recovery techniques. The adaptation of the Particle Image Analyzing System (PIAS) to analyze color as a morphic feature is presented. A number of examples of mixture analysis — classification studies are detailed and finally future work in packed and fluidized beds and slurry transportation is outlined.

# II. INTRODUCTION

During the last 10 years the morphological analysis research program has been developed along the following lines: the objectives have been clearly defined, the theory has been developed, needed equipment has been built, and the utility of the analysis technique has been demonstrated. The program is now in what might best be termed the exploitation state, in which applications of morphological analysis are being studied. Particle morphology may be defined as that part of fine particle science and technology concerned with the form of finely divided matter and the structures, homologies, and processes of change which govern that form. Morphological analysis is a technique for converting the image of the profile of a particle into a set of mathematical descriptors. These descriptors include a size term and shape terms. The technique of morphological analysis is being extended to include color as a morphic feature.[1]

There are many properties of particles which can be considered under the general heading of particle characterization. Some of the more important ones have been given in Table 1.

The particle characteristics which are the focus of this report include shape, size, and color.

# III. MORPHIC DESCRIPTORS

Particle shape has been defined as "the pattern of all the points on the boundary or surface of the particle profile."[2] There has been a rather steady interest in the measurement of shape of matter in finely divided form during the last 30 to 40 years. Some of the earlier methods of measurement are included in Table 2.

For example, Heywood defined the length, the breadth, and the thickness as shown in Figure 1, and then developed the flatness and elongation ratios given in Table 2.[3] Bird, who was a graduate student of Heywood, studied the flatness and elongation ratios of crushed glass particles and some of Bird's results are shown in Figure 2.[4] The ratio of Martin's diameter over Feret's diameter has been used as a shape factor, and both diameters are defined in Figure 3.

Both sphericity and circularity have been used for many years and they tend to be more popular in geological applications. Sphericity is defined as the surface of the equivalent volume sphere divided by the actual surface of the particle, and circularity is defined as circumference of a circle of area equal to the particle projected area divided by the actual perimeter of the particle.[5] The surface area shape factor and the volume shape factor should also be noted in Table 2. Heywood's analysis of lunar particles is shown in Figure 4.[6] It is a very good example of the use of these classical shape parameters in the characterization

**Table 1**
**PARTICLE CHARACTERIZATION**

1. Size
2. Shape
3. Chemistry: surface, subsurface, interior
4. Physics: electrical, magnetic
5. Structure: porosity, defects, boundaries
   cracks, fissures

**Table 2**
**SHAPE**

Earlier Methods of Measurement Including

1. Heywood, L, B, T:
   M = flatness ratio = B/T
   N = elongation ratio = L/B
2. $D_m/D_f$
3. $(PER)^2/2(area)$
4. Sphericity
5. Circularity
6. Surface area shape factor $R = C_2D^2$
7. Volume shape factor $V = C_3D^3$

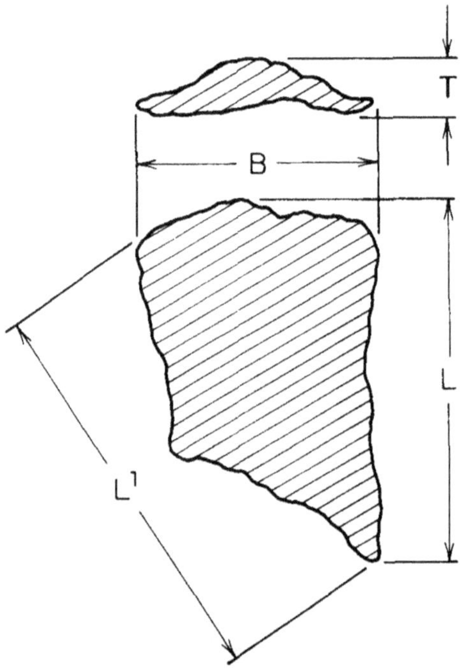


FIGURE 1. Heywood's method.

of particles. Note that r, the rugosity coefficient, is defined as the perimeter of the particle profile divided by the perimeter of the smooth circumscribing curve.

Keeping in mind the definition of particle shape given earlier, it will be apparent that the information content of the various shape parameters is not the same. For example, if we define the information content of a parameter on the basis of the number of x,y points of the profile that were used to define that parameter, then we may develop the information

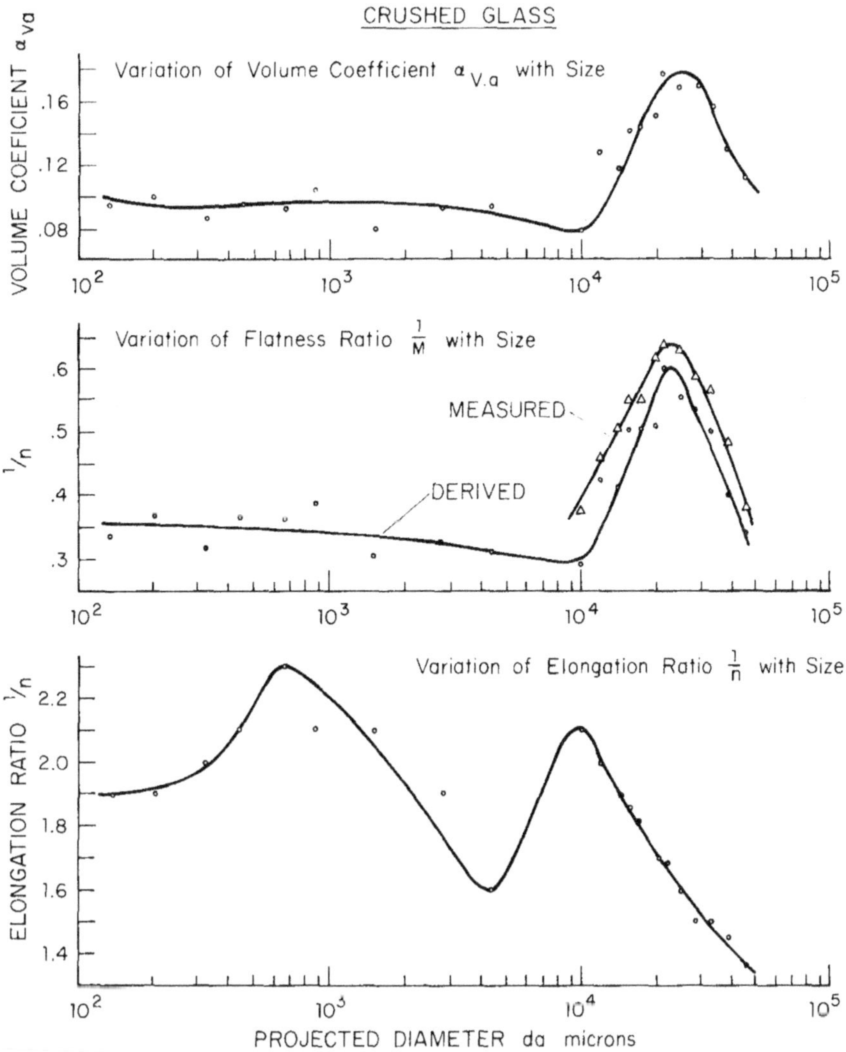


FIGURE 2. Bird's results.

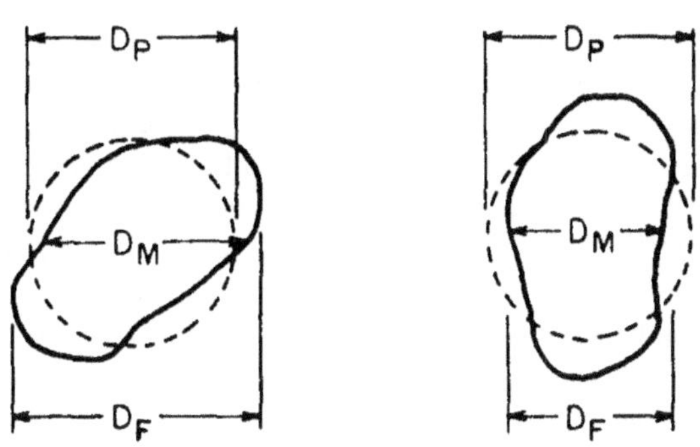


FIGURE 3. $D_m$, $D_f$ Defined.

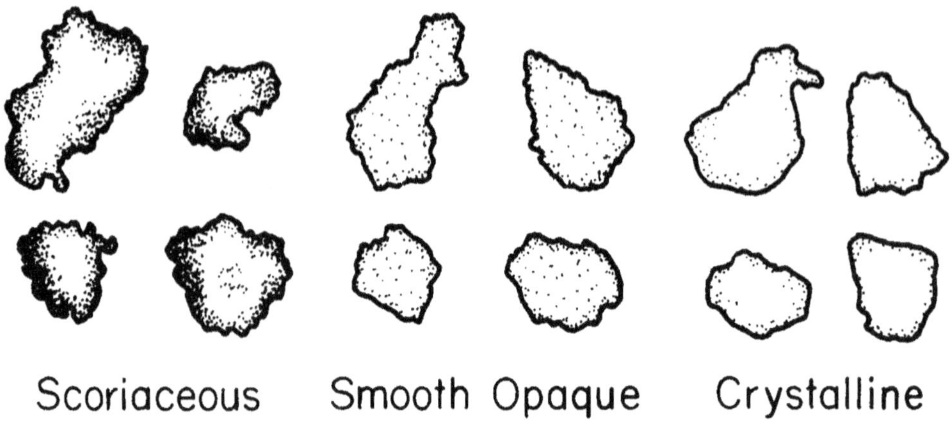


**SHAPE COEFFICIENTS MEASURED ON 30 PARTICLES 700 μm PROJECTED AREA DIAMETER[40]**

| Shape coefficient | Particle type | | |
|---|---|---|---|
| | Scoriaceous | Smooth opaque | Crystalline |
| n | 1.33 | 1.38 | 1.32 |
| m | 1.37 | 1.17 | 1.20 |
| $\alpha_a$ | 0.70 | 0.72 | 0.74 |
| $\alpha_{v,a}$ | 0.32 | 0.35 | 0.34 |
| r | 1.10 | 1.04 | 1.04 |
| $\phi^*$ | 0.93 | 0.93 | 0.93 |
| $\phi_r^*$ | 0.84 | 0.89 | 0.90 |

FIGURE 4. Lunar particle data.

of the sort given in Table 3, which contains some examples of these shape descriptors in relation to their information content on the basis of the number of x,y points used to define them. Note that the content of Table 3 is selective. Rugosity uses all of the x,y points whereas the ratio of Martin's diameter over Feret diameter uses only four. The flatness and elongation ratios likewise use only four. Isometry and bulkiness use few x,y points, and therefore they have a low information content. On the other hand, sphericity, roundness, and perimeter squared over twice the area contain all of the x,y points and therefore have a higher information content. Note, also, that the morphic descriptors also contain all of the x,y points and therefore have a high information content. If one can generalize by saying that information content should be equivalent to utility, then it is interesting to note that of all of the shape descriptors shown in Table 3, only the set of morphic descriptors can be used to regenerate the original two-dimensional shape. Therefore, the set of morphic descriptors which is obtained in morphological analysis is a unique set.

The set of morphological descriptors referred to is shown in Table 4. The following size and shape terms are defined.[7]

**Table 3**
**INFORMATION CONTENT OF SHAPE DESCRIPTORS**

| | Info. content No. of (X,Y) points |
|---|---|
| Rugosity | All |
| $D_m/D_f$ | 4 |
| Flatness ratio M = (B/T) | 4 |
| Elongation ratio N = (L/B) | 4 |
| Sphericity | All |
| Roundness | All |
| Anisometry | Few |
| Bulkiness | Few |
| Perimeter²/2(area) | All |
| Morphic descriptors | All |

## A. The Set of Morphic Descriptors

The following size and shape terms are defined. These are rotationally invariant and related to the amplitudes, $a_n$ and $b_n$, as follows.

The size term is defined as

$$R_0 = \sqrt{a_0^2 + \frac{1}{2}\sum_{n=1}^{\infty} (a_n^2 + b_n^2)} \tag{1}$$

in which $R_0$ is termed "equivalent radius" and is the radius of a circle having the same area as that of the particle profile, and n is the order of the coefficient.

The shape terms are defined as

$$L_0 = a_0/R_0 \tag{2}$$

$$L_{1,n} = 0 \qquad \text{for all n}$$

$$L_{2,n} = \frac{1}{2R_0^2}(a_n^2 + b_n^2) \tag{3}$$

$$L_{3,n} = \frac{3}{4R_0^3}(a_n^2 a_{2n} - b_n^2 a_{2n} + 2a_n b_n b_{2n}) \tag{4}$$

## B. Statistical Properties

$$\mu_0 = L_0 R_0 \qquad \text{(mean radius)}$$

$$\mu_1 = R_0 \sum_{n=1}^{\infty} L_{1,n} = 0 \qquad L_{1,n} = 0 \qquad \text{for all n} \tag{5}$$

$$\mu_2 = R_0^2 \sum_{n=1}^{\infty} L_{2,n} \tag{6}$$

$$\mu_3 = R_0^3 \sum_{n=1}^{\infty} \sum_{m=1}^{\infty} L_{3,m,n} \tag{7}$$

where $\mu_2$ is the radial standard deviation (the radance) and $\mu_3$ is the radial skewness.

**Table 4**
**SET OF MORPHOLOGICAL DESCRIPTORS[a]**

| Morphic descriptors | |
|---|---|
| $R_0$ | the size term |
| $L_0$ | the shape terms |
| $L_2(n)$ | the shape terms |
| $L3(m,n)$ | the shape terms |
| $\mu_2$, $\mu_3$ | the shape terms |

[a] These are called *morphic features*.

The morphic descriptors and the statistical properties are collectively called the morphic features. In effect, they represent features of the pattern of the profile of the particle.

## IV. RESEARCH PROGRAMS

The underlying strategy in the morphological research program outlined in this chapter is shown in Figure 5. The morphic descriptors derived from the morphological analysis procedure can be related to the mode of origin of the particle. This type of research can lead to a deeper understanding of the mechanisms of particle formation and can lend itself to process control, for example. Relating the morphic properties of particles to their physical or chemical behavior is of importance in understanding the processing characteristics of bulk solids. The third main research category in morphological analysis deals with the analysis of particle mixtures which lead directly to the development of classifiers.

In general, we believe that particulate science and technology is a multidisciplinary field and, accordingly, we work with others wherever and whenever possible.

## V. TECHNIQUES OF MORPHOLOGICAL ANALYSIS

In order to convert a particle image into useful mathematical information, PIAS is required. An example of such a system is given in Figure 6. In its present form, this consists of a high-quality TV camera connected to a computer via an AD converter with an accompanying TV monitor. The output may be manipulated in the Ramtek device. Thus, for example, the Ramtek permits the operator to examine the morphological information in the computer. In its present form, the output data are transmitted to a larger computer for further processing. The original image may be viewed as a photomicrograph, for example, from a SEM or TEM source. If convenient, the original particles themselves may be viewed, or if the particles are of intermediate size, an optical microscope of high quality is used to present the particle image to the camera.

The outline of the morphological analysis steps is given in Table 5. The set of x,y pairs is converted to polar coordinates which are then further processed to the Fourier coefficients and these, in turn, are calculated to produce the invariant morphological descriptors as shown. These morphological descriptors have properties that correspond to the desirable properties of descriptors as listed in Table 6.

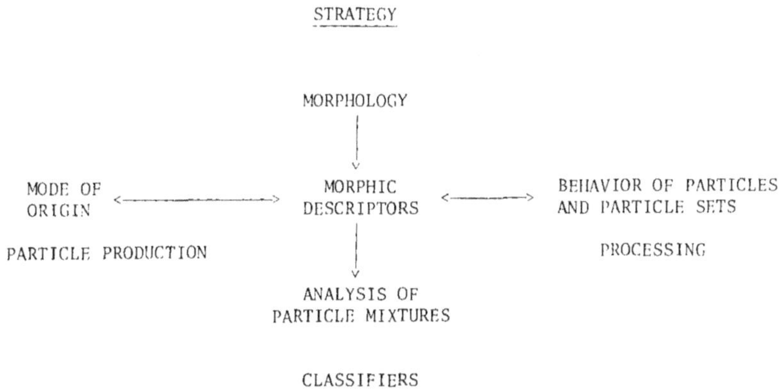


FIGURE 5. Morphological research strategy.

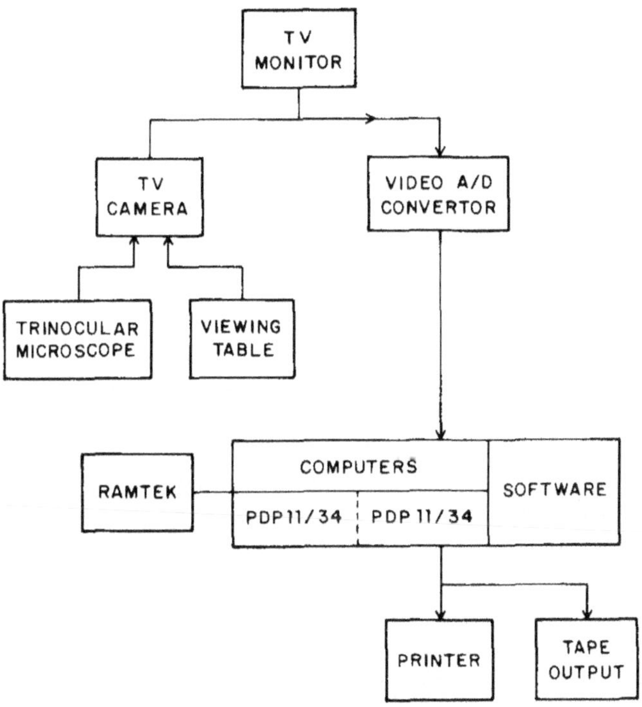


FIGURE 6. Particle image analyzing system.

## VI. MEANING OF MORPHOLOGICAL DESCRIPTORS

$R_0$ is defined as the equivalent radius; it is the radius of a circle having the same area as that of the particle profile. That is to say, $R_0$ measures the size of the particle. This is clearly demonstrated in Figure 7.

The meaning of the L2(n) terms is not quite so clear-cut. For example, we constructed an equilateral triangle and obtained the L2(n) terms as shown in Table 7. It is clear that L2(3), L2(6), L2(9), L2(12), etc. are very large compared with all the other L2(n) terms. Next we increased the roughness of the equilateral triangle, and as shown in the second

**Table 5**
**STEPS IN MORPHOLOGICAL ANALYSIS**

Analysis

$$(x,y) \rightarrow (R,\theta) \rightarrow (A_n,B_n) \rightarrow R_0,\ L_0,\ L2(n),\ L3(m,n)$$
$$\mu_2,\ \mu_3$$

**Table 6**
**DESIRABLE PROPERTIES OF DESCRIPTORS**

1. The descriptors should be derivable from first principles and they should relate to the inherent properties of the Fourier series; they should not be empirical formulas
2. The descriptors should be easily calculated from the input data
3. They should be demonstrated to be invariant
4. They should be capable of treatment with the usual statistical methods, preferably standard, widely accepted methods so that a broad spectrum of workers can follow and use the analysis technique
5. Any physical interpretations of the meanings of the descriptors should be unequivocal and preferably capable of rigorous definition, if at all possible
6. The descriptors should be capable of relating to the fundamental physical-chemical properties of particles and their surfaces, and ultimately to the mode of formation of the particle (i.e., relating to the mechanisms occurring during particle formation), and also ultimately to the chemical and mechanical behavior of the particle in specific environments
7. One should be able to describe sets of particles in terms of the invariant descriptors without losing precision inherent in the data, using, if possible, standard statistical techniques
8. It should be possible, in principle at least, to recalculate the particle profile from the set of invariant descriptors
9. The descriptors should be invariant to a change of image size; this is especially important with regard to the use of modern image analyzing instruments

column of Table 7, the L2(3), L2(6), etc. terms remained about the same order of magnitude as for the straight-sided equilateral triangle, but the other L2(n) terms increased in magnitude substantially. Yet again, for an isosceles triangle, the L2(3),..., terms are of a similar order of magnitude to those for the roughened equilateral triangle but the L2(2) term, for example, is very large indeed. What this means is that although the human observer notes that the equilateral triangle is a triangle, the morphological analysis notes that it is primarily a figure with a large aspect ratio. Finally, if one looks at the extreme right column in Table 7, it will be obvious that it is not possible from the numbers to determine that the scalene triangle is indeed a scalene triangle, or that it is a triangle at all. Thus, although the L2(n) terms indicate the relative contribution of the second, the third, the fourth harmonic, etc. it is somewhat misleading to relate L2(3), for example, to the angularity (specifically triangularity) in each and every case.

The radance is the sum of the L2(n)s and the skewness is the sum of the L3(m,n)s. Thus $\mu_2$ and $\mu_3$ are statistical properties of the L2(n)s and L3(m,n)s, respectively. Is there any additional meaning in these statistical terms? In the case of a plot of $\mu_2$ vs. $\mu_3$ for regular equal dimensional shapes, it appears that there may be some relationship between the shape of the particles and their position in the $\mu_2$, $\mu_3$ space. However, this is an incorrect conclusion because, as is shown in Figure 8, a Maltese cross can be placed in various locations in the $\mu_2$, $\mu_3$ space and it remains forever a Maltese cross, which is to say its shape does not change as it moves around the $\mu_2$, $\mu_3$ space.[8] The reason for this is very clear if one considers the meaning of the term standard deviation and skewness. If one measures the radii of a particular shaped profile, then the L2(n) and L3(m,n) terms will indicate the order in which these radii are arranged. However, the $\mu_2$, $\mu_3$ terms contain no information about the sequence

$R_0 = 21.6$

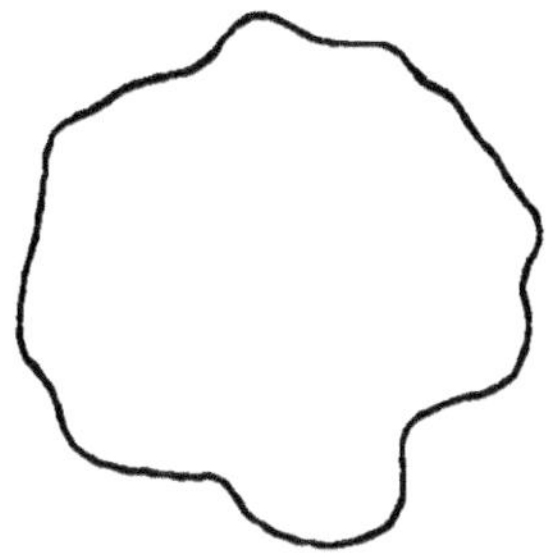

$R_0 = 48.6$

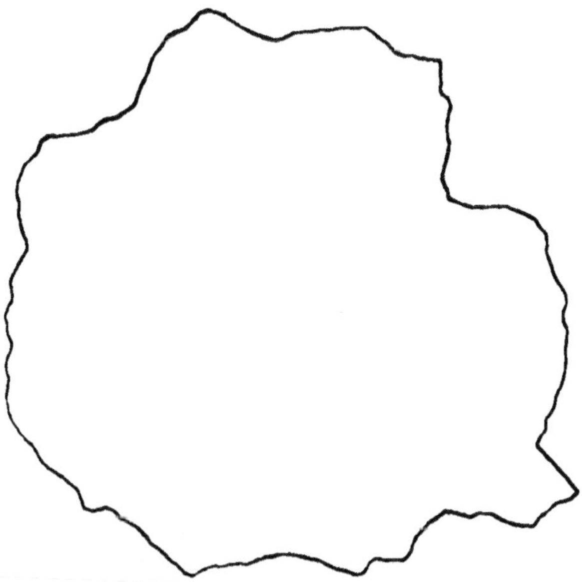

FIGURE 7. $R_0$ measures particle size.

or order in which the radii are arranged. Such information is lost when the L2(n) terms, for example, are summed up to produce $\mu_2$. What this means is that the $\mu_2$, $\mu_3$ space contains less shape information than the L2(n) and L3(m,n) terms contain.

## VII. OTHER METHODS

The R-Theta method which has been described above is useful for particle profiles which contain no reentrants. For particle profiles which contain reentrants, other methods are necessary. The $(\phi,\ell)$ method, which has been reported elsewhere,[9] reparameterizes the particle profile in terms of the change of slope of the tangent with distance moved around the profile. The R,S method is a very general method which may be used for any particulate shape.[10] Both of these methods have been reported elsewhere.

**Table 7**
**L2(n) TERMS FOR TRIANGULAR FIGURES**

| | Equilateral triangle |  DEMO0001 |  DEMO0003 |  DEMO0004 |
|---|---|---|---|---|
| L2(1) | 2.945E−9 | 7.501E−6 | 3.195E−3 | 6.349E−3 |
| L2(2) | 3.166E−9 | 2.688E−5 | 5.430E−2 | 9.867E−2 |
| L2(3) | 3.619E−2 | 3.725E−2 | 1.863E−2 | 8.959E−3 |
| L2(4) | 3.351E−9 | 4.953E−5 | 2.596E−4 | 1.155E−2 |
| L2(5) | 3.740E−9 | 5.623E−5 | 1.240E−2 | 1.236E−2 |
| L2(6) | 5.202E−3 | 5.298E−3 | 6.917E−4 | 2.737E−3 |
| L2(7) | 3.650E−9 | 1.362E−5 | 1.389E−3 | 5.380E−3 |
| L2(8) | 4.068E−9 | 2.854E−5 | 2.378E−3 | 5.648E−4 |
| L2(9) | 1.377E−3 | 1.260E−3 | 4.552E−5 | 2.613E−3 |
| L2(10) | 3.634E−9 | 2.170E−5 | 1.170E−3 | 7.040E−4 |
| L2(11) | 3.939E−9 | 6.491E−5 | 3.324E−4 | 1.218E−3 |
| L2(12) | 5.004E−4 | 6.123E−4 | 1.052E−4 | 6.922E−4 |
| L2(13) | 3.316E−9 | 1.302E−4 | 5.628E−4 | 2.094E−4 |
| L2(14) | 3.579E−9 | 1.042E−5 | 4.371E−5 | 8.073E−4 |
| L2(15) | 2.212E−4 | 3.316E−4 | 2.274E−4 | 3.122E−5 |
| L2(16) | 2.753E−9 | 2.628E−4 | 1.659E−4 | 4.611E−4 |
| L2(17) | 3.577E−9 | 1.620E−6 | 2.635E−5 | 1.224E−4 |
| L2(18) | 1.118E−4 | 8.887E−6 | 2.140E−4 | 1.402E−4 |
| L2(19) | 2.611E−9 | 2.341E−5 | 3.880E−5 | 2.168E−4 |
| L2(20) | 4.244E−9 | 9.089E−5 | 4.573E−5 | 5.174E−5 |
| LO | 9.779E−1 | 9.762E−1 | 9.504E−1 | 9.194E−1 |
| VAR | 4.379E−2 | 4.694E−2 | 9.666E−2 | 1.546E−1 |
| SKE | 9.675E−3 | 9.934E−3 | 2.940E−2 | 7.113E−2 |

## VIII. APPLICATIONS

### A. General

The production of material in finely divided form is an important industrial activity. Examples of such processes include atomizing, spray drying, comminution, crystallization. Furthermore, many industrial processes involve stages in which particulate materials are dissolved into liquids. Some examples of growth and diminution processes involving chemical change are given in Table 8. Traditional modeling processes idealize particle profiles as circles and the use of morphological analysis allows us to break away from this practice so that we can incorporate both size and shape into the models used.

Some commonly measured bulk particulate properties include apparent density, tap density, and mechanical strength properties. Again for the most part, modeling of these properties is done using spherical particles. The use of morphological analysis should enable us to incorporate both size and shape in studies of static and dynamic situations, some examples of which are given in Table 9. In the following sections we discuss some of the experiments that are currently being conducted in morphological analysis.

### B. Morphology in Relation to the Mode of Origin of the Particle

As is shown in Table 10, a number of experiments are underway including relating morphological characteristics to the production process, wear debris analysis, kinetic modeling of crystallization, and comminution experiments.

We produced adipic acid powders by three different methods: atomizing from the melt, comminution, and crystallization from aqueous solution. The form of typical particles from the three processes are shown in Figure 9 and a plot of $\mu_2$ vs. $\mu_3$ for the three types of material is shown in Figure 10. It can be observed from Figure 10 that one can distinguish

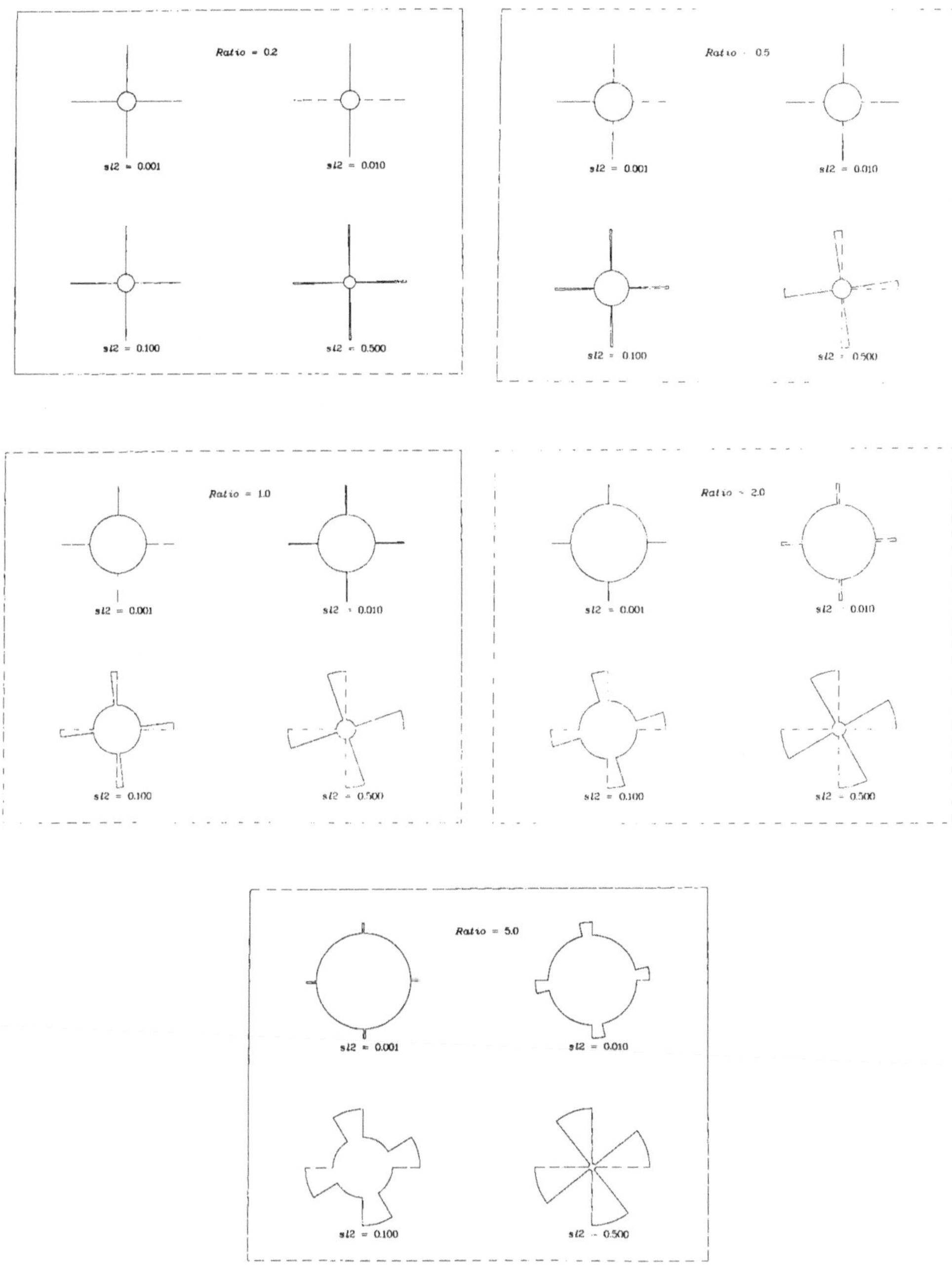


FIGURE 8. Maltese crosses $\mu_2$, $\mu_3$ space; ratio = $\mu_2/\mu_3$ $s\ell 2$ read $\mu_2$.

between adipic acid powders from the three modes of origins by using the $\mu_2$, $\mu_3$ morphology plot.

Iron powders were made by reduction of the oxide produced from steel scrap (this was made at Indian Institute of Technology, Bombay, by Professor P. Ramakrishnan), by atomization, and by electrodeposition. In general, manufacturing processes to produce metal powders are much more complicated than the processes that we used to produce the adipic acid powders. For example, the production of atomized iron powders involves atomizing followed by an annealing process, which tends to sinter particles together, and this is then

**Table 8**
**CHEMICAL PROCESSES INVOLVING PARTICULATE MATERIAL**

| | Process | |
|---|---|---|
| **Matrix** | **Growth** | **Diminution** |
| Gas | Vapor deposition | Sublimation |
| Liquid | Crystallization | Dissolution |
| Solid | Precipitation | Reprecipitation |

**Table 9**
**STATIC AND DYNAMIC BEHAVIOR OF PARTICULATE MATERIAL**

| Matrix | Static | Dynamic |
|---|---|---|
| Gas | Apparent density<br>tap density<br>Shear strength | Cyclone<br>Pneumatic transport<br>Mixing |
| Liquid | Sedimentation | Hydraulic transport<br>Mixing |
| Solid | Compaction<br>Sintering | Balling |

**Table 10**
**MORPHOLOGY VS. MODE OF ORIGIN EXPERIMENTS**

| Morphology | Mode of origin |
|---|---|
| Production of powders | Atomizing<br>Comminution<br>Reduction of oxide<br>Crystallization<br>Electrodeposition |
| Wear debris analysis | Metallic debris from sliding wear<br>Fatigue and others |
| Kinetic modeling | $(R,\theta)$ approach unrealistic<br>Facial growth rates independent |
| Comminution | Breakage function<br>Selection function |

followed by a hammer milling to break up the sinter cake. In these circumstances it is not surprising that the complexity of the manufacturing processes is reflected in the complexity of the particle morphologies. Reference to Figure 10 will show that the reduced iron powder and the atomized iron powder are in the comminuted zone of the adipic acid plot and the electrolytic iron powder is on the borderline between the crystallized and comminuted zone.[11]

The analysis of wear debris is potentially a very fruitful application of morphological analysis. It should be possible to differentiate between wear debris produced by different mechanisms. Indeed, as shown in Figure 11, some work along these lines has already been

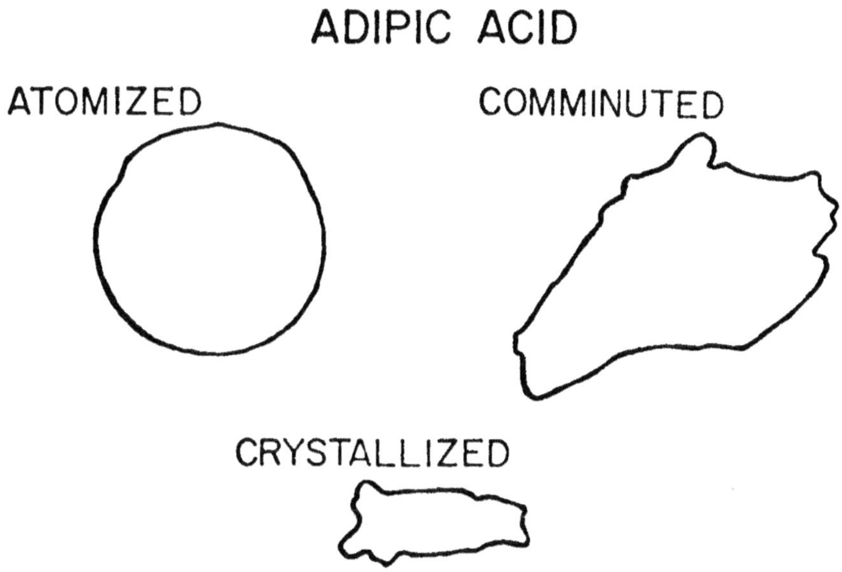


FIGURE 9. Adipic acid powder types.

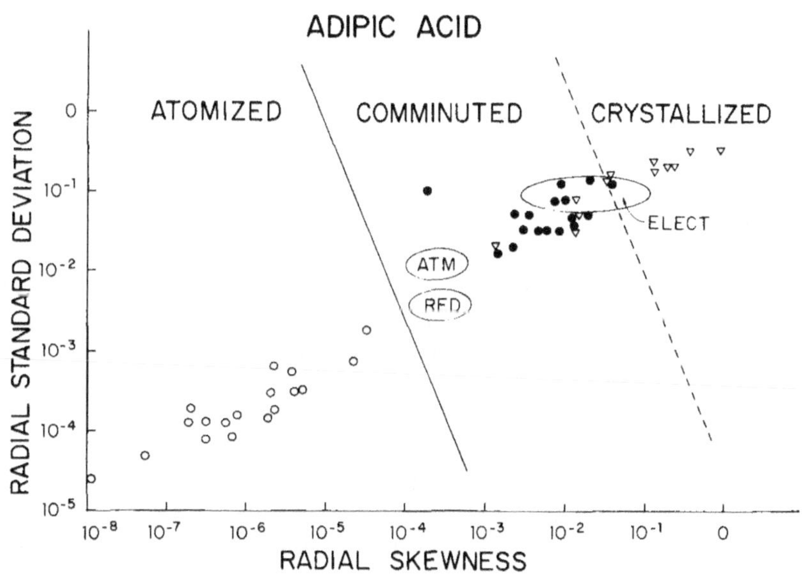


FIGURE 10. Morphology plot adipic acid; ATM — atomized iron, RED — reduced iron, ELECT — electrolytic iron.

carried out to demonstrate this possibility.[12] Furthermore, it is likely that the stage that a particular wear process has reached in the life of a component should be capable of being determined from morphological analysis of the appropriate wear debris.

Kinetic modeling of crystallization and dissolution processes using the R-Theta method has been carried out[13] but it suffers from the disadvantages that the process is assumed to be isotropic, whereas, we know in the case of crystallization, for example, that the facial growth rates are different for different crystallographic faces. With this in mind, we have developed a model of crystal growth that permits different facial growth rates.[14] An example

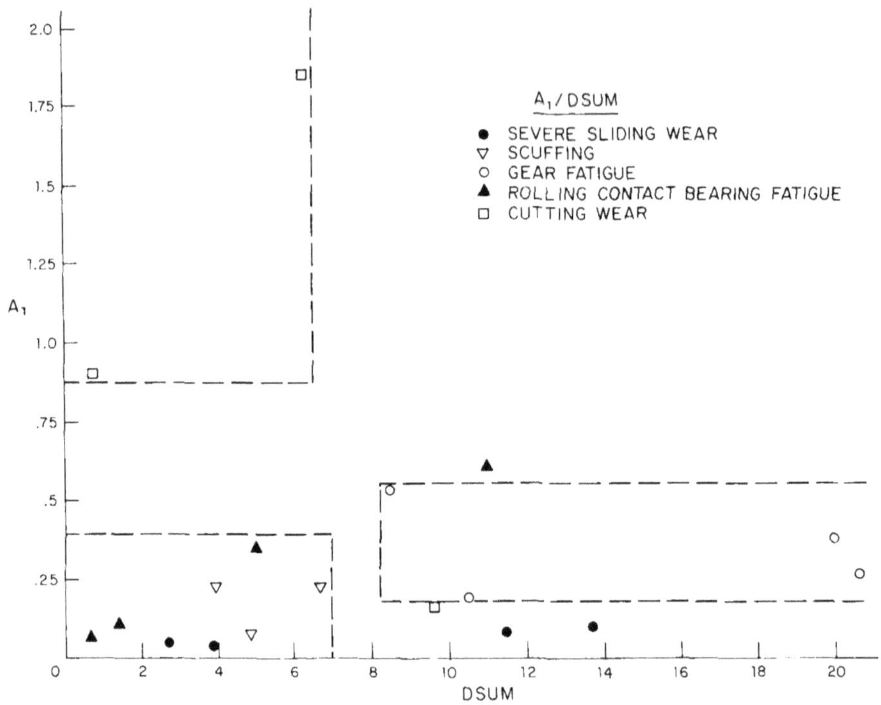


FIGURE 11. Wear debris classifications.

of some of this modeling work is given in Figure 12. In addition, the criterion for face elimination has been determined and this has been confirmed by using experimental data previously reported in the literature.

A comminution experiment is being conducted at the moment in which the materials being comminuted are adipic acid and propyl paraben samples made by atomization, crystallization, and comminution, respectively, and the objective of the experiment is to see if and in what way the breakage functions and selection functions differ depending upon morphology. These two materials were chosen because they both have a low melting point and they are nontoxic. They can be formed into crystals and they fracture without much difficulty.

## C. Relating Behavior to Particle Morphology

The behavior of particulate materials covers a very wide range of activities, and experiments have been performed or are being conducted currently in a wide variety of areas, as is shown in Table 11.

It has been known for some time that the size and also the shape of abrasive particles is important, but it is only recently, due to the application of morphological analysis, that it has been possible to differentiate between the abrasive wear due to the size of the particle and the abrasive wear due to the shape of the particle. These very interesting results are shown in Figure 13.[15]

In morphological analysis we measure the shape as that of a two-dimensional profile but, of course, shape is actually a three-dimensional characteristic. In order to measure shape in three dimensions, a variety of procedures may be adopted. One such procedure has been used in the study of the effect of shape on sedimentation in low-flow regimes. Because the particles used in these sedimentation experiments were regular geometrical shapes, it was possible to take three orthogonal profiles and develop a correlation that not only includes

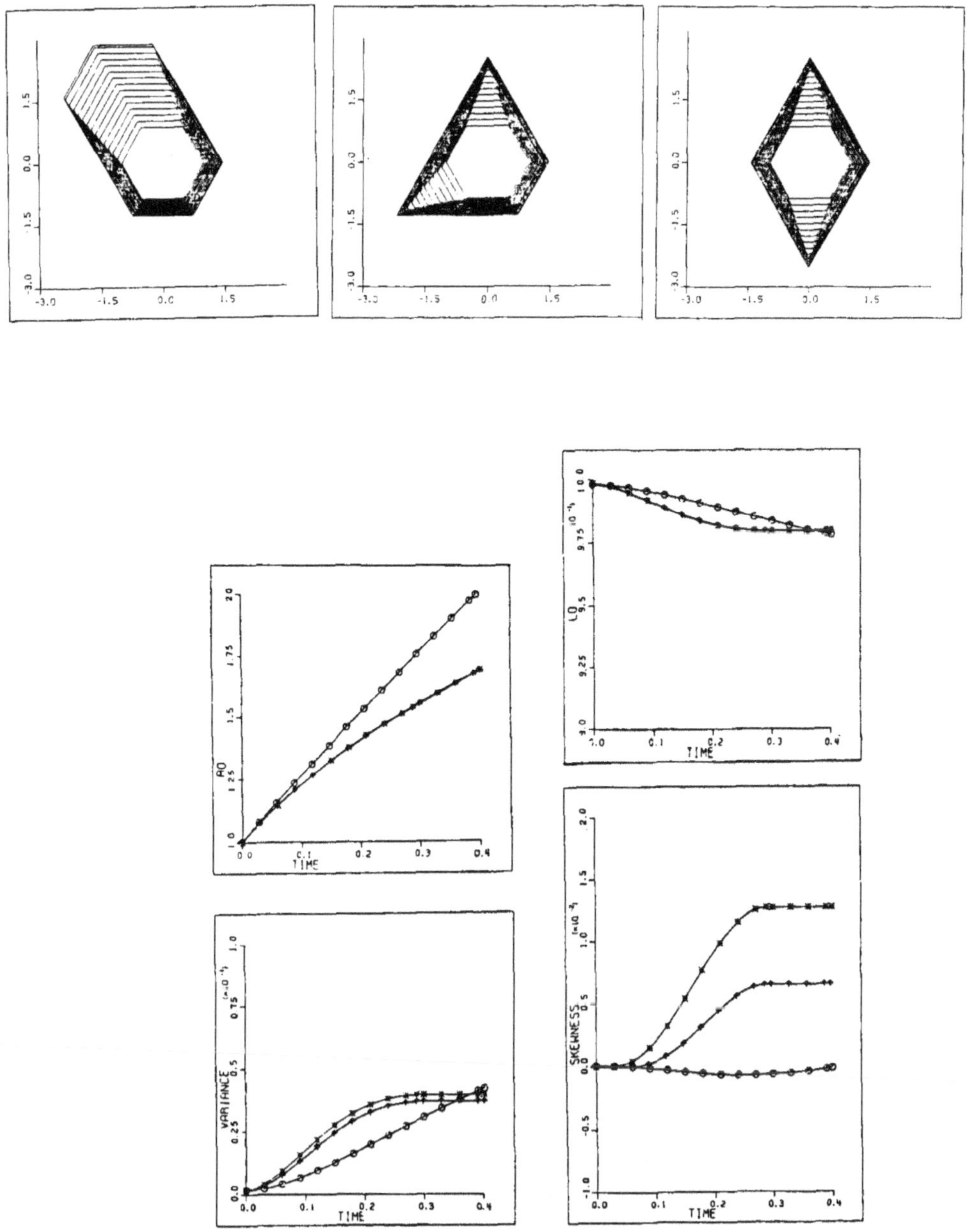


FIGURE 12. Growth of hexagonal crystal by different facial growth rates.

the shape of the particle but also their orientation during sedimentation. These data are illustrated in Figure 14.[16]

Plastic parts may be fabricated from a variety of input materials which the manufacturer may wish to characterize using morphological analysis. Figure 15 shows an empirical correlation which has been obtained between the angle of internal friction and some morphological features in the case of a number of samples of polyethylene powders made by different processes. Yet another relationship has been developed in the case of a manufacturing process parameter for PVC powders, as shown in Figure 16.[17]

**Table 11**
**BEHAVIOR VS. MORPHOLOGY EXPERIMENTS**

| Behavior | Morphology |
|---|---|
| Abrasive wear | Effects of size and shape of abrasive particles on wear rate |
| Sedimentation | 3-D analysis of particle morphology in slow flow regime |
| Internal angle of friction | Plastic powders |
| Manufacturing process parameters | Molding of plastic parts from powders |
| Dry separation technology | Sieving, magnetic separation |
| Storage and flow of grains | Bulk properties |
| Slurry flow study | $T_1O_2$ slurry |
| Comminution study | Phosphate rock |
| Scrap recovery | Dry separation of different material |

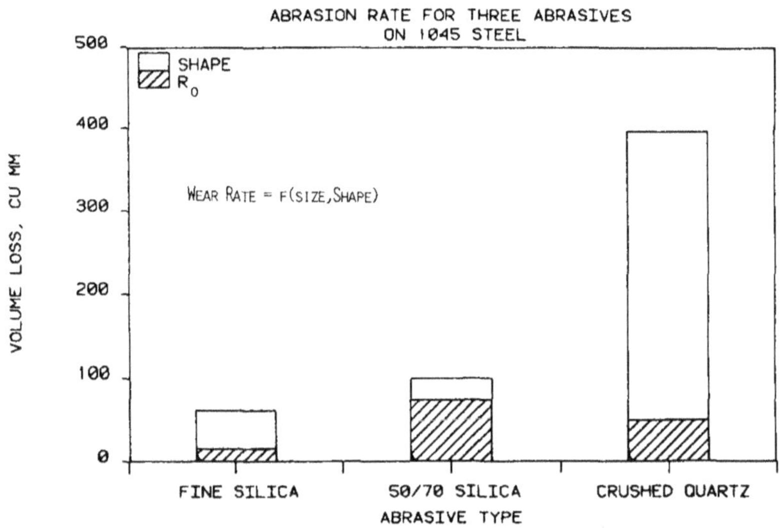


FIGURE 13. Effect of abrasive particle (sand) size and shape on metal wear.

## IX. COLOR AS A MORPHIC FEATURE

For many years microscopists have used size, shape, and also color of fine particles in order to differentiate them. To be used successfully, these techniques require a high level of education and skill on the part of the observer and also involve subjective judgments. Although morphological analysis has been developed initially to include size and shape parameters as particle characteristics, there is no fundamental reason why color should not be included as a morphic feature, if indeed one can carry out the analysis with color. The PIAS is intrinsically a black and white system, but with the incorporation of the Ramtek device into the system as shown in Figure 6, the PIAS has a capability of analyzing for color because the Ramtek can electronically ''reproduce'' colors.

In the first instance, there are three experiments being conducted.[18] One, to try to use the PIAS to reproduce a colored image presented to the TV camera. The second, to use the system to identify and differentiate between idealized particles and some self-colored real

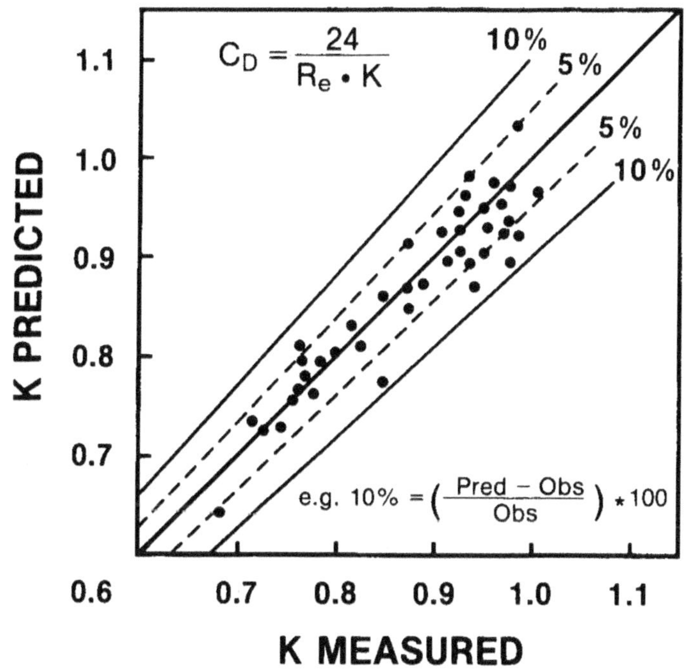


FIGURE 14. Sedimentation in low flow regimes.

particles. The third, to increase the number of color filters being used to, perhaps, six so that a large number of different substances can be readily identified and differentiated on the basis of their color.

Figure 7 in Leong's chapter shows the large range of colors that the PIAS can handle and Figure 8 in Leong's chapter gives an illustration of the reproduced image of a colored logo. It should be noted that the reproduction is quite inferior to the original logo, but it has the merit of being consistent, which is important to the present study.

## X. ANALYSIS OF MIXTURES — CLASSIFICATION

There are many important technical problems which can benefit from the use of morphological analysis and, as a group, the problems of classification and mixture analysis are very important. Some of the work that has been done in this area is summarized in Table 12.

Two unknown samples, O and N, were found to be agglomerations of smaller particles, as is shown in the scanning electromicrographs in Figure 17. The very close similarity between these two materials was confirmed when the histograms of the morphic features were examined. To all intent and purposes, the materials seemed indistinguishable. As shown in Table 13, using the size term $R_0$ as the discriminating feature proved unacceptable, for the samples N and O were wrongly allocated in 34 and 27% of the time, respectively. Using the shape features only (the 51 variables) and excluding the size term failed to improve the discriminant situation appreciably, for in this case the misclassification of N and O was 30% in each case. However, there was some improvement in the classification process when both the size and the shape terms were used together. In this case, the misclassifications were reduced to 24 and 16%, respectively; this is also shown in Table 13.[19] Simple applications of morphic features to differentiate between good and bad corn kernels[20] (Figure 18), Vitrain

$$\tan \theta = K (R_0)^X (L_0)^Y (PER)^{\bar{X}} (VAR)^U (SKE)^V$$

$K = 1.121 \times 10^3$

$X = -0.1278$

$Y = 120.97$

$\bar{X} = 0.6584$

$U = 1.795$

$V = 0.01269$

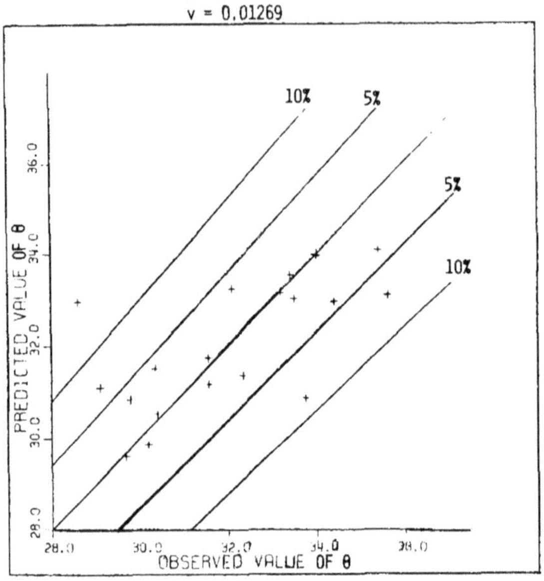


19 SAMPLES

TOTAL SQUARE ERROR = 47.10

FIGURE 15. Effect of morphology of polyethylene particles on angle of internal friction.

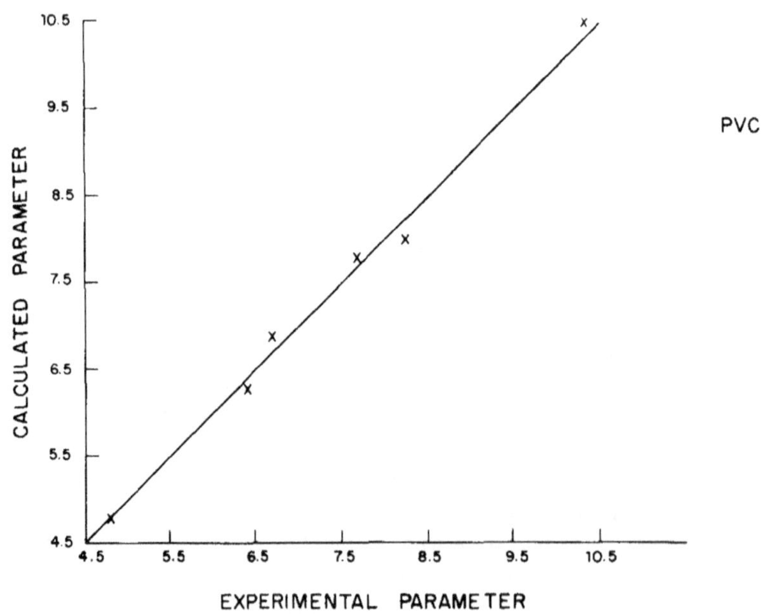


FIGURE 16. Effect of PVC particle morphology on manufacturer's processing parameter.

**Table 12**
**CLASSIFICATION STUDIES**

| Materials | Procedure |
|---|---|
| Unknown samples O and N | Discriminant analysis |
| Corn quality | Simple classification |
| Coal macerals | Simple classification |
| Corn and wheat dust | Simple classification |
| Oil shale | Morphic terms only reported |

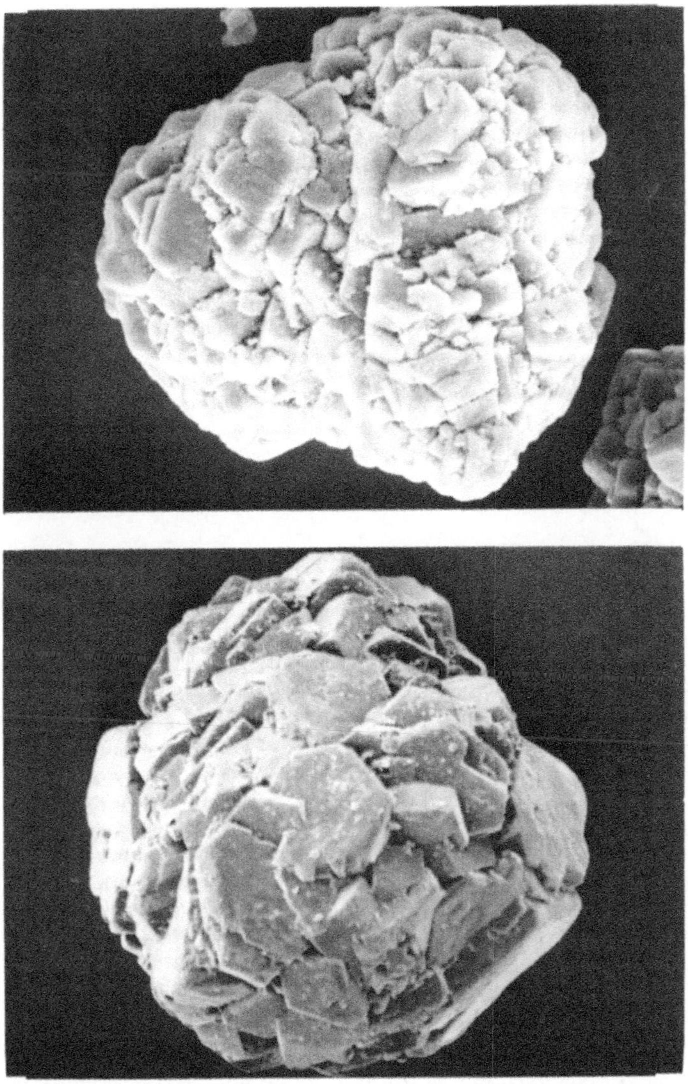

FIGURE 17. Samples O and N, scanning electron micrographs.

**Table 13**
**DISCRIMINANT SCORES FOR SAMPLES N AND O**

| | $R_0$ | 51 Variables | $R_0$ 51 Variables |
|---|---|---|---|
| ALN and ALO (I = 1) (I = 2) | $\begin{bmatrix} 0.73 & 0.27 \\ 0.34 & 0.66 \end{bmatrix}$ | $\begin{bmatrix} 0.70 & 0.30 \\ 0.30 & 0.70 \end{bmatrix}$ | $\begin{bmatrix} 0.84 & 0.16 \\ 0.24 & 0.76 \end{bmatrix}$ |

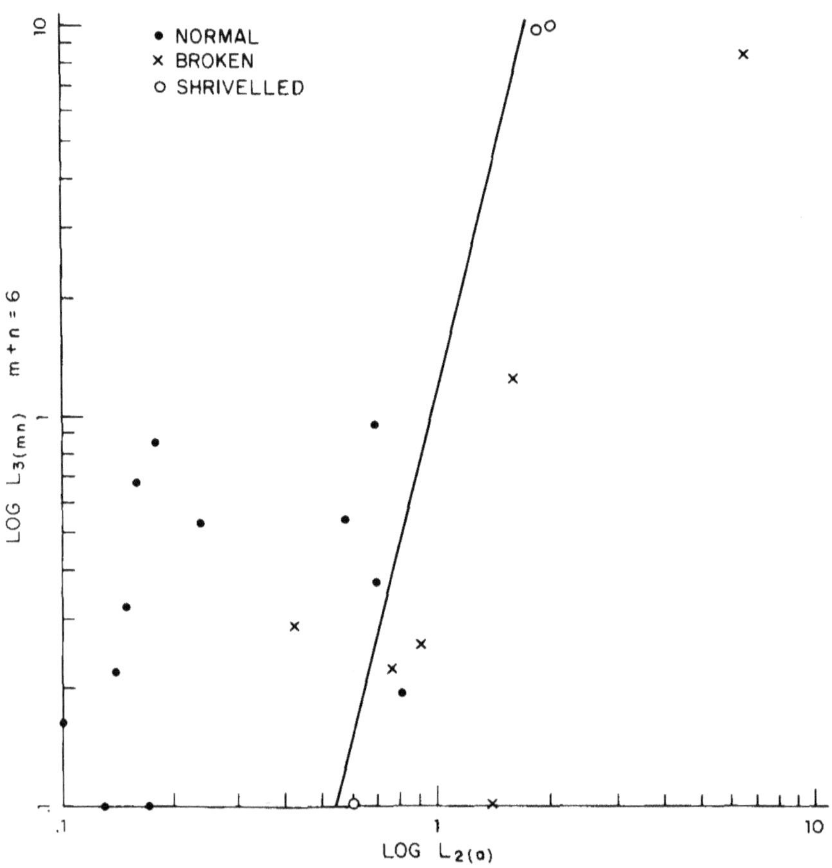


FIGURE 18. Good and bad corn kernels.

and fusain coal macerals[21] (Figure 19), and differentiating between corn and wheat dust[22] (Figure 20) illustrate the potential application of morphological analysis in classification applications. Other applications of this would be in quality control.

It should be noted that it is now possible to specify the morphological features of processed particles very precisely and therefore the manufacturer and processer have available a very precise tool for improving the accuracy of their material specifications and for controlling processing variables.

A sample of oil shale was subject to morphological analysis[23] and it is interesting to note

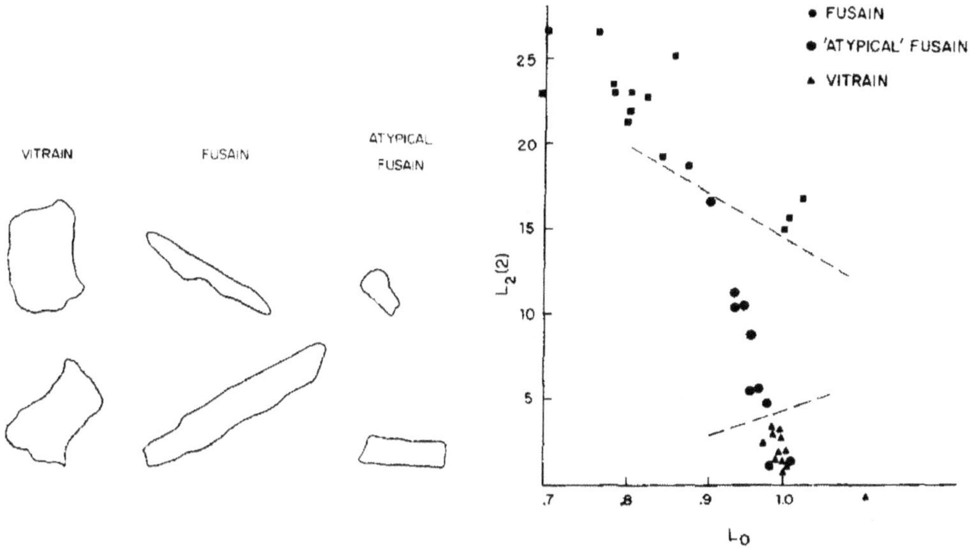


FIGURE 19. Vitrain and fusain discrimination.

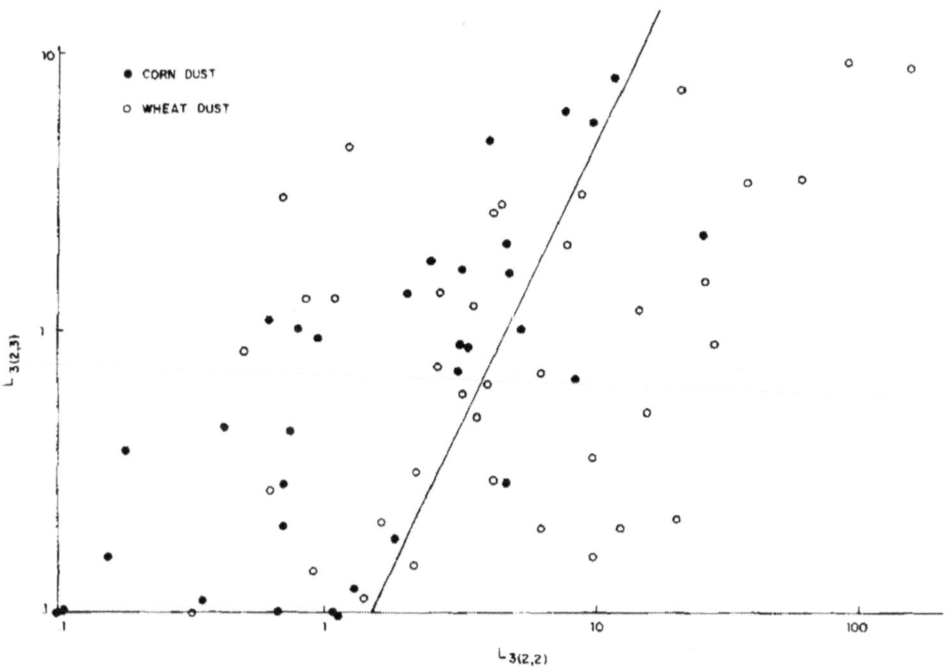


FIGURE 20. Corn and wheat dust differentiation.

that the $\mu_2$, $\mu_3$ terms locate the sample exactly in the middle of the comminution section of Figure 10.

## XI. FUTURE WORK

One of the main difficulties in trying to do morphological analysis research experiments with naturally occurring materials is that their morphology is very complex. One way of reducing the complexity of this problem is to carry out experiments using monomorphic particle sets. These have to be manufactured, and an examination of the methods that might be used to manufacture monomorphic particles led to the conclusion the most feasible way is to extrude a plastic stock of a given shape, for example, circular, triangular, square, octagonal, etc. and then chop the extrudate into particles of the required length. Once this equipment is in place, it is intended to produce large quantities of monomorphic particles in order to further investigate the structure and properties of packed beds, fluidized beds, and also slurry transportation. Other future developments in morphological analysis will include a more basic understanding of particle formation and of the behavior of particulate systems. We can also expect application of morphological analysis in process control, quality control, and systems design problems. It is also to be expected that the number of features being used will be amplified to include more dimensions than size, shape, and color. One would also expect to see in the future the development of morphological analysis instrumentation.

## ALPHABETICAL SYMBOLS

$a_0$ = Mean radius
$a_n$ = Fourier coefficient
$b_n$ = Fourier coefficient
$L_0$ = Shape term
$L_{2(n)}$ = Shape term
$L_{3(m,n)}$ = Shape term
n = Order of the coefficient
n,m = Integers
PER = Perimeter
r = Rugosity coefficient
$R_0$ = Equivalent radius

## GREEK SYMBOLS

$\mu_0$ = Mean radius
$\mu_2$ = Radial standard deviation (the radance)
$\mu_3$ = Radial skewness

## REFERENCES


1. **Beddow, J. K., Hua, L., Vetter, A. F., Chang, C.-R., and Carmichael, G. R.,** *Principles and Applications of Morphological Analysis,* Japanese Society of Powder Technicians, Kyoto, 1982, 10.
2. **Beddow, J. K. and Philip, G.,** Fourier analysis — synthesis method of particle shape analysis, *Planseeber. Pulvermetall.,* 23(1), 3, 1975.
3. **Heywood, H.,** Symposium on particle size analysis, *Trans. Inst. Chem. Eng.,* 25, 14, 1947.
4. **Bird, K. E.,** The Assessment of Particle Shape and Its Influence on Sizing Analysis, thesis, University of London, 1966.
5. **Wadel, H.,** *J. Geol.,* 40, 443, 1932; 43, 250, 1935.
6. **Heywood, H. H.,** *Proceedings of Second Lunar Science Conference,* Vol. 13, 1989, 1971.
7. **Luerkens D. W., Beddow, J. K., and Vetter, A. F.,** Invariant Fourier descriptors, *Powder Technol.,* 31(2), 209, 1982.
8. **Lenth, R.,** Unpublished work.

9. **Fong, S. T., Beddow, J. K., and Vetter, A. F.,** A refined method for particle shape representation, *Powder Technol.*, 32, 17, 1979.
10. **Luerkens, D. W., Beddow, J. K., and Vetter, A. F.,** Generalized (RS) method of morphological analysis, *Powder Technol.*, 31(2), 217, 1982.
11. **Guo, A., Beddow, J. K., and Ramakrishnan, P.,** Relationships between manufacturing method and particle morphology, *Powder Metallurgy Int.*, in press.
12. **Beddow, J. K., Fong, S. T., and Vetter, LA. LF.,** Morphological analysis of metallic wear debris, *Wear*, 58, 201, 1980.
13. **Luerkens, D. W.,** Morphological Analysis, Ph.D. thesis, University of Iowa, Iowa City, 1980.
14. **Hua, L., Beddow, J. K., and Vetter, A. F.,** Application of morphological analysis in kinetic modeling, Intl. Powder and Bulk Solids Conf., Rosemont, Ill., May 1981.
15. **Vetter, A. F. and Swanson, P.,** Size and shape characteristerics of abrasive wear particles, submitted.
16. **Carmichael, G. R.,** Estimation of drag coefficient of regularly shaped particles in slow flows from morphological descriptors, *I & EC Process Design Devel.*, 21, 401, 1982.
17. **Chang, C.-R. et al.,** Particle morphology and bulk properties in *Particle Characterization in Technology*, Vol. 2, Beddow, J. K., Ed., CRC Press, Boca Raton, Fla., 1984.
18. **Leong, P.-L. et al.,** Analysis of color as a morphic feature, Proc. Int. Powder Bulk Solids Handling Processing Conf., Rosemont, Ill., 1982, 191.
19. **Hua, L. et al.,** Morphological analysis of unknown samples, in *Particle Characterization in Technology*, Vol. 2, Beddow, J. K., Ed., CRC Press, Boca Raton, Fla., 1984.
20. **Vetter, A. F., Beddow, J. K., and Rosheim, J.,** Morphological analysis of good and bad corn kernels, unpublished data.
21. **Beddow, J. K., Vetter, A. F., and Hansen, P.,** Morphological analysis of selected coal macerals, CIC Conf., Halifax, Nova Scotia, June 1981, 87.
22. **Beddow, J. K., Vetter, A. F., and Musgrove, D.,** Morphic features of grain dust, *Part. Sci. Technol.*, in press.
23. **Goren, S.,** Personal communication.

Chapter 13

# PARTICLE MORPHOLOGY AND ABRASIVE WEAR

**Arthur F. Vetter and Paul Swanson**

## TABLE OF CONTENTS

# I. INTRODUCTION

It has long been recognized that abrasive wear is affected by both size and shape of the abrasive particles in addition to relative hardness and other factors.

It has been found[1-5] that the volume loss due to abrasion increases rapidly as the mean abrasive particle diameter increases up to a critical value of approximately 100 μm. For abrasive particles above 100 μm in diameter, abrasive wear continues to increase with increasing particle size, but at a much lower rate.

The lack of a quantitative scale for shape has, in the past, precluded correlating wear rate to particle shape. Many attempts have been made to describe shape ranging from the qualitative descriptors round, subangular, and angular, to use of selected parameters such as Feret diameter, or of ratios such as the ratio of the major axis to the perpendicular axis.

# II. MORPHOLOGICAL ANALYSIS

Beddow and Vetter[6] have developed a set of morphological parameters which have been applied to the analysis of the shape of abrasive particles. Their technique analyzes the shape of the two-dimensional profile of a particle using the well-known Fourier analysis procedure.

This yields a result

$$R(\theta) = A_0 + \sum_{n=1}^{\infty} (a_n \cos n\theta + b_n \sin \mu\theta) \tag{1}$$

where $R(\theta)$ is measured from the centroid of the projected area to the perimeter of the profile at the angle $\theta$. The reference axis from which $\theta$ is measured is arbitrary. $A_0$ is the mean value of $R(\theta)$ and $a_n + b_n$ are the coefficients of the nth Fourier terms.

The problem with this series is that it is not invariant with respect to the choice of reference axis. It has proven more convenient to use the following form of the Fourier series:

$$R(\theta) = A_0 + \sum_{n=1}^{\infty} A_n \cos (n\theta + \alpha_n) \tag{2}$$

where $A_n = \sqrt{A_n^2 + b_n^2}$, and $\alpha_n = \tan^{-1} b/a$ and is the phase angle of the nth Fourier term. In this form the $A_n$s are invariant but the $\alpha_n$s are not.

Luerkins et al.[7] have developed a set of morphological descriptors as follows:

$A_0$ = mean value of $R(\theta)$
$R_0$ = radius of a circle of the same area as the particle profile
$L_0 = A_0/R_0$
$L_1(n) = 0$ for all n.

If $L_1(n) \neq 0$ (actually small compared to $L_2(n)$ values for n smaller than 8), then the centroid has been incorrectly chosen.

$$L_2(n) = \frac{1}{2R_0^2} A_n^2 \tag{4}$$

$$L_3(m,n) = \frac{3}{4R_0^3} (A_m A_n a_{m+n} \cos (\alpha_{m+n} - \alpha_m - \alpha_n)) \tag{5}$$

$$\mu_0 = L_0 R_0 = A_0 \tag{6}$$

$$\mu_1 = R_0 \sum_{n=1}^{\infty} L_1(n) = 0 \tag{7}$$

$$\mu_2 = R_0^2 \sum_{n=1}^{\infty} L_2(n) = \frac{1}{2} \sum_{n=1}^{\infty} A_n^2 \tag{8}$$

$$\mu_3 = R_0^3 \sum_{n=1}^{\infty} \sum_{m=1}^{\infty} L_3(m,n) = \frac{3}{4} \sum_{n=1}^{\infty} \sum_{m=1}^{\infty}$$

$$(A_m A_n A_{m+n} \cos (\alpha_{m+n} - \alpha_m - \alpha_n)) \tag{9}$$

where $\mu_1$, $\mu_2$ and $\mu_3$ are the first, second, and third moments of the radial distribution about the mean, respectively.

It should be noted that while $L_3(m,n)$ is not invariant due to the included cosine term, when summed over all values of m and n it becomes invariant. At first glance it would appear that $\mu_3$ should be zero. However, the larger values of $A_m$ and $A_n$ from the Fourier series are not randomly associated with the $\alpha_n$s so that $\mu_3$ normally has a nonzero value.

## III. PHYSICAL INTERPRETATION OF THE MORPHOLOGICAL PARAMETERS

It can easily be shown that

$$R_0^2 - A_0^2 = \frac{1}{2} \sum_{n=1}^{\infty} A_n^2 \tag{10}$$

For a circle $R_0 = A_0$ since all $A_n$ terms are 0.

If one considers a circle to be a profile having no shape features, then shape data are contained in the $A_n{}^2$ terms:

$$1 - \frac{A_0^2}{R_0^2} = \frac{1}{2R_0^2} \sum_{n=1}^{\infty} A_n^2 = \sum_{n=1}^{\infty} L_2(n) \tag{11}$$

Since the left hand side of Equation 11 becomes zero when $A_0 = R_0$ (a circle), either of the other two expressions define the deviation from a circle. This might be called "noncircularity", or since

$$\mu_2/R_0^2 = {}^1/_2 R_0^2 \sum_{n=1}^{\infty} A_n^2$$

and $\mu_2$ is the second moment of the radial distribution about the mean. The term "radance" has been coined for $\mu_2$. Radance can be calculated from the mean profile radius and the area of the profile and does not require a full Fourier analysis; for a typical profile the difference between $R_0$ and $A_0$ is shown in Figure 1.

## IV. APPLICATION OF MORPHOLOGICAL ANALYSIS TO ABRASIVE WEAR

Abrasive wear is normally considered to be either a two-body or a three-body phenomenon. In two-body abrasive wear, two surfaces in contact move with respect to each other. The first features of each surface to contact the other surface are protuberances, i.e., parts of the surface extending outward from the main surface. As these slide over like protuberances from the other surface they are progressively worn off by friction and the surface becomes more smooth. This is normally referred to as "breaking-in" or "wearing-in".

This discussion is concerned with the third-body process in which a third body is introduced between the two surfaces. If the third body is harder than the material of the surfaces it will remove material from the surfaces by cutting or gouging. The abrasive particle may be introduced between the surfaces by the lubricant or it may be introduced by an air or gas

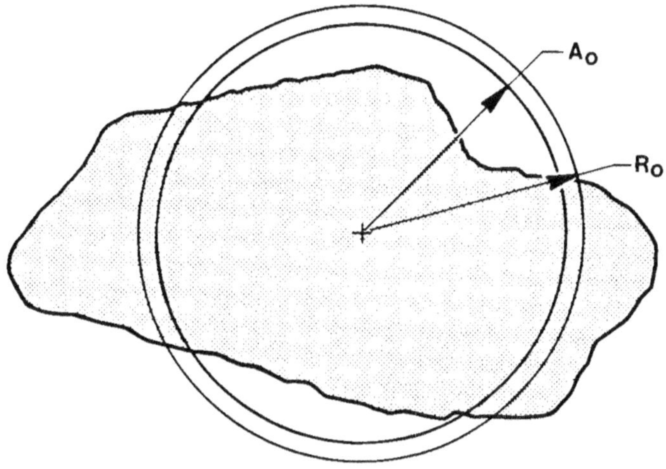


FIGURE 1. Definition of parameters ($A_0$ = mean radius; $R_0$ = radius of equal areas circle: $L_0$ = $A_0/R_0$).

stream, or by other means. The abrasive particle may be introduced from the environment, such as a sand grain, it may be introduced in the fuel or the lubricant used, or it may be a particle earlier eroded from the system and carried to the site by the lubricant.

Although corrosive action is certainly important in material surface degradation, it will not be considered here.

The literature contains many reports of abrasive studies, particularly three-body abrasion. However, in most cases it was not feasible to retroactively analyze the shapes of the abrasive particles used, since the abrasive material had been discarded and no photographs had been taken. A notable exception was the work of Avery.[8] Avery, in a private communication with Swanson, provided photographs of 13 abrasive samples he had used in abrasion studies.[9]

The objectives of this study were to attempt to classify a selection of abrasive samples using the Beddow and Vetter technique[6] and to relate the size and shape parameters to laboratory abrasion data.

In order to determine the morphological parameters of the various abrasives, a number of samples have been analyzed in the PIAS (Particle Image Analyzing System).[10] In addition to the photographs provided by Avery,[8] samples of abrasives used in the Deere & Co. abrasion research program and some samples from natural sources were analyzed. One sample of alumina has been analyzed. From 20 to 100 profiles of each sample of abrasive were analyzed and the parameters $R_0$, $1\text{-}L_0^2$ (radance) $L_2(2)$, $L_2(3)$, $L_2(4)$, and $\mu_3$ (skewness) were calculated. In most of the abrasive samples, $R_0$ was not useful as an identification tool since these samples were sieved prior to use in the abrasion experiment.

## V. ANALYSIS AND DISCUSSION

In addition to size ($R_0$ or $A_0$), radance ($1 - L_0^2$), and $\mu_3$ (skewness of the radial distribution), individual $L_2(n)$ terms carry shape information. $L_2(2)$ for example relates to ellipticity of the profile, $L_2(3)$ to equilateral triangularity, and so on. Although in principle $L_2(5)$ to $L_2(8)$ or even higher terms can be associated with pentagon, hexagon, etc. shapes, in real particles they are usually associated with surface roughness. In general, the first few $L_2(n)$ terms describe gross profile shape, higher $L_2(n)$ terms give roughness or texture.

Figure 2 is a plot of the sum of the mean values of $L_2(2) + L_2(3) + L_2(4)$ vs. $1 - L_0^2$ for each of the abrasive samples. The significance of this plot is that the deviation from the line $y = x$ represents the contribution to the radance of the higher-order terms of the Fourier series.

The importance of this figure is that, although the deviation of the sum of all the $L_2(n)$ terms (or $1 - L_0^2$) from the sum of $L_2(2) + L_2(3) + L_2(4)$ is quite small, it is this difference which accounts for surface texture, i.e., the small, sharp protuberances which are responsible for most abrasive wear. This is demonstrated by the fact that the individual scratches observed on the surface of a test specimen are much narrower than the abrasive particle diameter.

Figure 3 is a plot of the skewness vs. radance parameters for the abrasive samples. The physical significance of skewness in abrasive wear is not yet understood.

It is of interest that Avery[8] subjectively classified Ottawa sand as "round", New Jersey silica as "sub-angular", and the Wausau and Joplin samples as angular. The shape analysis described here, using the $1 - L_0^2$ and skewness parameters, agrees with Avery's classification except in the case of New Jersey silica, which falls within the range of the Ottawa samples in both radance and skewness.

Avery stated that one of the Joplin grit samples had been used in abrasion tests three times before being photographed. Although the size difference between the two Joplin samples is relatively small, the change in both the radance parameter and skewness are substantial. The behavior of the two Joplin grit samples, if they were to be used in abrasion tests, would yield quite different results, other test parameters being held constant. We will return to this point later.

The results from the Ottawa sands suggest that both skewness and "not-round" may vary with sieve fraction. Work is in progress at the University of Iowa to test this hypothesis. Very early results seem to give a similar response.

## VI. ABRASIVE WEAR MODEL

Three-body abrasion is a complex process that does not readily yield to well-defined relationships. The existence of a strong interaction between the abrasive, the test specimen, and the test environment makes establishing quantitative relations difficult. On the test specimen side, the composition of the alloy, the type of heat treatment during its manufacture, and the condition of the specimen surface all contribute to the material response. Although it is possible to control the material composition and heat treatment, the condition of the specimen surface changes during the progress of the test — and this change is related to both the material and the test conditions. The test environment includes such factors as the pressure of the abrasive against the specimen surface, the temperature of the specimen surface (raised from ambient by the friction heating and deformation caused by the abrasive) and the duration of the test. The size, shape, and hardness of the abrasive particles are also important and size and shape may change during the passage of the abrasive particle across the specimen surface. The following discussion will be limited to the effects of particle size and shape on steel test specimens.

An empirical equation of the form

$$V = aR_0^2 + b\,e^{c(1 - L_0^2)} \tag{12}$$

is proposed where V is the wear rate of the test specimen in a standard laboratory abrasion test; a, b, and c are empirical parameters which depend on the material of the test specimen, hardness of the test specimen, conditions of the test, and possibly other parameters; $R_0$ is

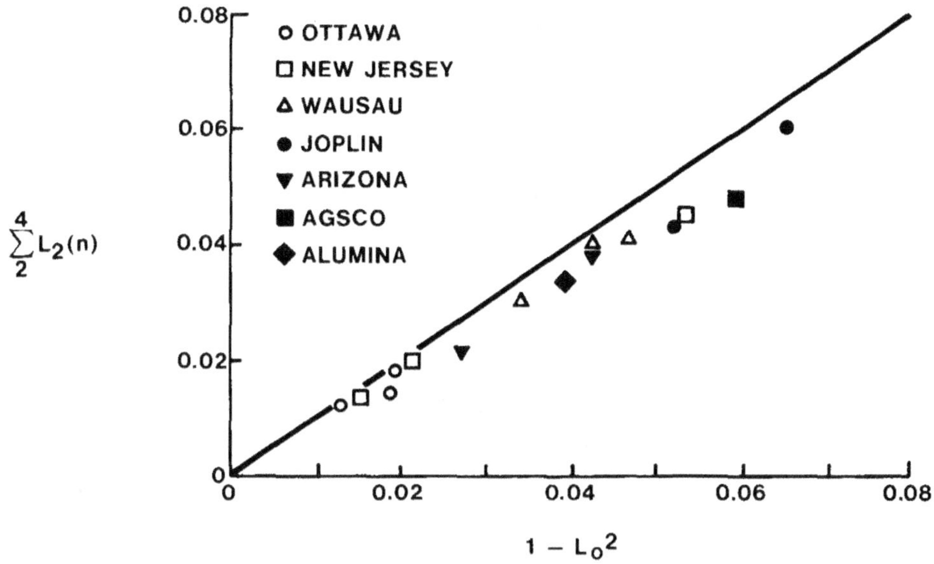


FIGURE 2. $\sum_{2}^{4} L_2(n)$ vs. $1 - L_0^2$.

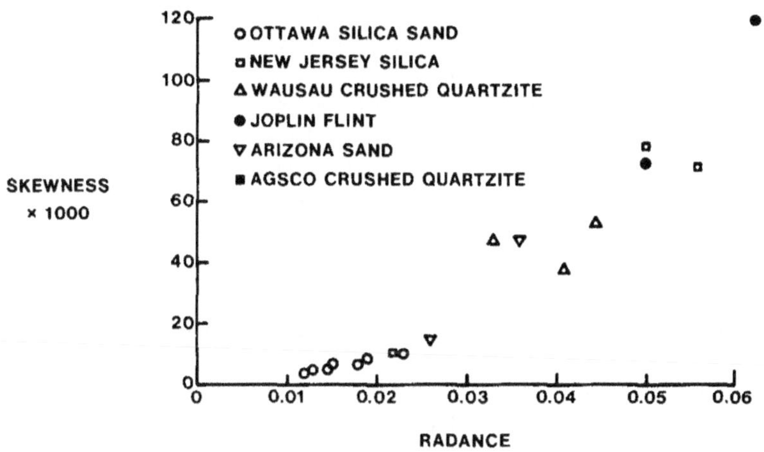


FIGURE 3. Skewness vs. radance for various sand abrasives.

the mean of the diameters of the profile areas of the abrasive particles; and $1 - L_0^2$ is the morphological parameter defining the shape parameter of the abrasive being used.

This equation separates for the first time the size and shape contributions of the abrasive in abrasive wear. The selection of $L_0^2 = A_0^2/R_0^2$ and that both $A_0$ and $R_0$ of a profile can be determined without going through the complete Fourier analysis. However, more detailed information about particle shape, either for an individual particle or an aggregate, requires a knowledge of the individual $A_n$s and the Fourier analysis is necessary.

Figure 4 shows the application of the empirical wear rate equation to data provided by Avery.[8,9] The parameters a, b, and c are determined from the data, but sophisticated methods such as least squares have not been used. This is believed to be justified by the relative inaccuracy of abrasion data in the 1950s, when Avery performed his experiments. Each datum point represents the average of several experiments, according to Avery. Where two

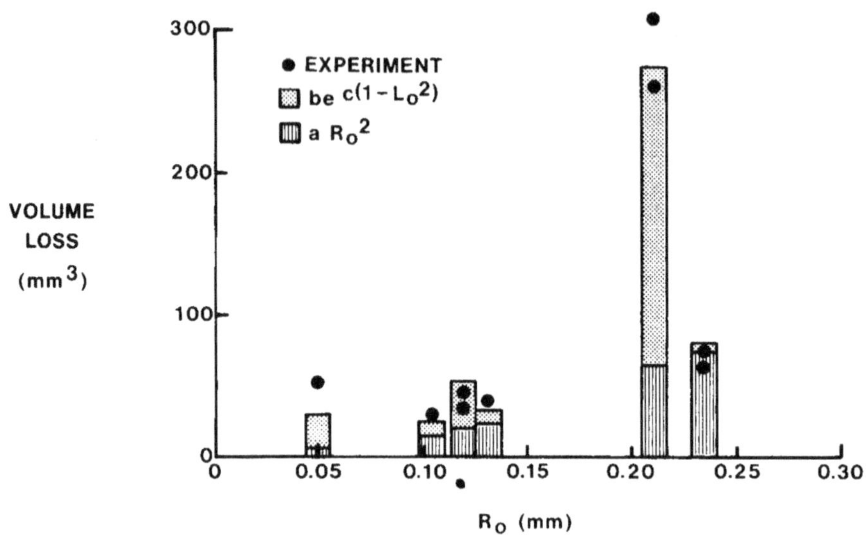


FIGURE 4. Wear vs. abrasive size and shape (Avery).

data points are shown at the same $R_0$, these represent the results obtained by using rubber wheels of different Durometer hardness in the abrasion tester. The very high abrasion rate for the Wausau crushed quartzite at $R_0 = 0.21$ mm is evident, even compared to the slightly larger Ottawa sample at $R_0 = 0.23$ mm.

Figures 5 and 6 are from later experiments by Kosel and Fiore.[11] Again, the relatively large abrasion from the crushed quartz abrasive is plainly evident. The larger shape contribution to abrasion rate from the fine silica sample compared to the 50/70 silica agrees well with the shape data using the morphological analysis technique. Experimental points are not shown on these figures, since they coincide with the top bar in each case. This is not proof of fit, but is the result of determining three empirical constants from three pieces of data.

## VII. ABRASIVE CHANGES DURING ABRASION TESTING

The propriety of using the same abrasive in more than one experiment has been debated. Clearly, if the abrasive is changed in the harsh environment of the test it is unwise to use this abrasive for subsequent tests if one wishes to compare results. Two experiments were performed at the Deere & Co. Technical Center, Moline, Ill. to examine the abrasive change after passing through the abrasive tester. These results are shown in shown in Figure 7. One experiment used a commercial crushed quartzite abrasive, the other used Ottawa test sand, both on 1045HR steel at standard test conditions. The sand made a single pass through the tester in each case. It can be seen that the crushed quartzite was noticeably changed in shape, particularly those fractions passing through the tester during the latter two thirds of the test. The greater effect on the abrasive during this period relative to that which passed through during the initial one third of the test is attributed to the fact that the test specimen was smooth at the beginning of the test, allowing more rolling of the abrasive particles, whereas during the latter two thirds of the test the specimen surface had become badly scarred and sliding was the predominant behavior of the abrasive grains. The interpretation of these data, compiled with the fact that particle size decreased only slightly during the test, is that the test specimen was removing the sharp protuberances on the abrasive particles — a sort of reverse polishing effect.

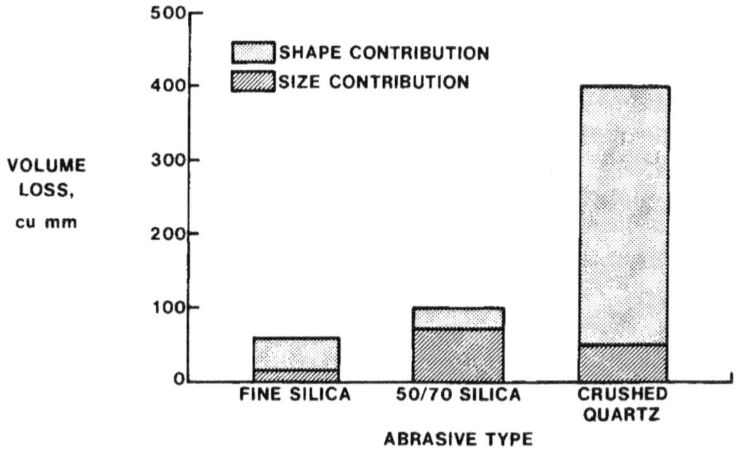


FIGURE 5. Effect of abrasive size and shape on the wear of 1020 steel.

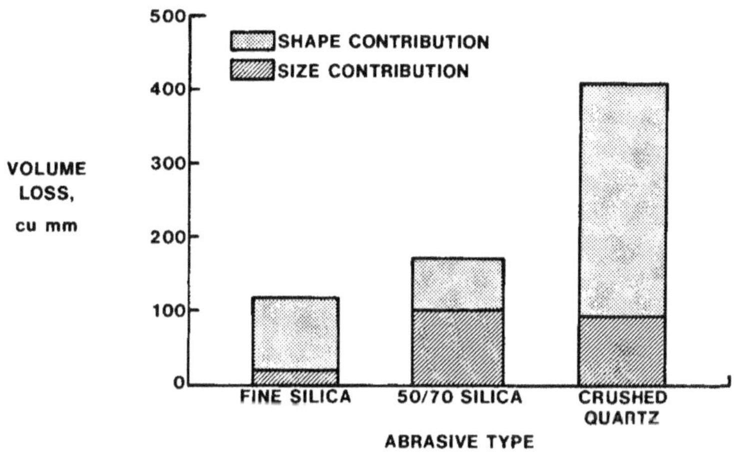


FIGURE 6. Effect of abrasive size and shape on the wear of 1045 steel.

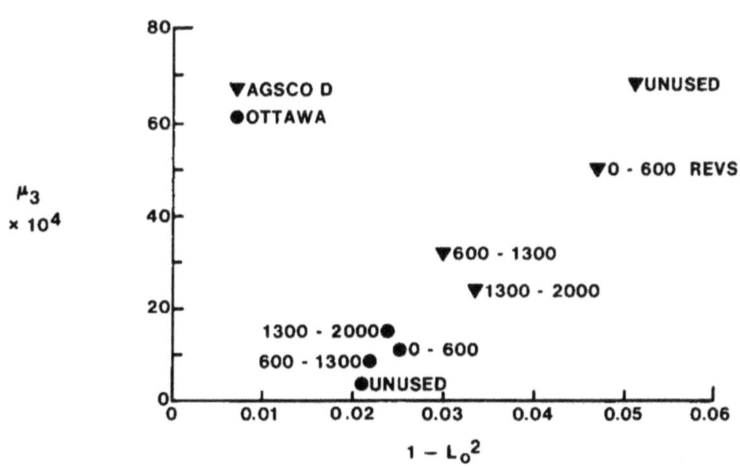


FIGURE 7. Abrasive changes in DSAT.

The Ottawa sand experiment yielded quite different results. As expected, when starting with particles already quite round in shape, the reduction in radance would be minimal. The slight increase in the radance parameter is believed to be due to fracture of a small fraction of the grains, resulting in a statistical increase in $1 - L_0^2$. This fracture effect could account for a corresponding small decrease in the mean size of the particles. The same fracture mechanism could be working in the case of the crushed quartzite abrasive, but is overshadowed by the polishing effect.

Based on these two experiments, it is recommended that sharp or angular abrasives not be reused for subsequent abrasive tests if results are to be compared. Round abrasives can probably be reused, but should be monitored to ensure that progressive changes in the abrasive have not accumulated to the point where test results will be affected. Harder test specimens are expected to show greater effects on the abrasives, but experimental data are not yet available.

## VIII. VARIABILITY IN ABRASIVES

Both natural silica abrasives and crushed material abrasives are normally provided using sieve fractions specifications only. Experiments at the Deere & Co. Technical Center have demonstrated that this may not be sufficient. Two batches of commercial crushed quartzite yielded quite different results on similar test specimens and the same test procedure. When subjected to morphological analysis, it was found that although the size of the particles from the two samples was the same within experimental error, there was a 10% difference in the shape parameter $1 - L_0^2$. The significance of this can be seen from Equation 12 and Figures 4 through 6. It is recommended that morphological analysis be incorporated to give shape as well as size parameters for abrasives used in all three-body abrasion test programs.

## IX. SUMMARY

A powerful new tool in particle morphology and its application to three-body abrasion tests has been described. The effects of size and shape of abrasive particles on metal surfaces have been separated.

## REFERENCES


1. **Mulhearn, T. O. and Samuel, L. E.,** The abrasion of metals: a model of the process, *Wear,* 5, 478, 1962.
2. **Nathan, G. K. and Jones, W. J. D.,** The empirical relationships between abrasive wear and the applied conditions, *Wear,* 9, 300, 1966.
3. **Richardson, R. C. D.,** The wear of metals by relatively soft abrasives, *Wear,* 11, 245, 1968.
4. **Larson-Badse, J.,** Influence of grit size on groove formation during sliding abrasion, *Wear,* 11, 213, 1968.
5. **Suh, N. P., Saka, N., and Sin, H.,** Effect of Abrasive Grit Size on Abrasive Wear, Progress Report to ARPA, DOD, June 1978.
6. **Beddow, J. K. and Vetter, A. F.,** A note on the use of classifiers in morphological analysis of particulates, *J. Powder Bulk Solids Technol.,* 1, 42, 1977.
7. **Luerkins, D. W., Beddow, J. K., and Vetter, A. F.,** Morphological Fourier descriptors, *Powder Technol.,* 31(2), 209, 1981.

8. **Avery, H. S.,** The nature of abrasive wear, SAE Pap. 750822, Society of Automotive Engineers, Warrendale, Pa., 1975.
9. **Avery, H. S.,** private communication to Paul Swanson, Deere & Co., February 1979.
10. **Lenth, R., Beddow, J. K., Chang, C. R., and Vetter, A. F.,** Particle image analyzing system, *Advances in Particulate Science and Technology*, Hemisphere Publishing, New York.
11. **Kosel, T. H. and Firore, N. G.,** Abrasion in Wear Resistant Steels, final report to Deere & Company, Moline, Ill., 1980.

Chapter 14

# RELATIONSHIPS BETWEEN PARTICLE MORPHOLOGY AND PRODUCTION PROCESSES

**Andy Guo, P. Ramakrishnan, and John Keith Beddow***

## TABLE OF CONTENTS



* The authors are grateful to the National Science Foundation Particulate Multiphase Processing Program for partial support under grant CPE-80-23868 and the international program of the National Science Foundation for support under grant INT-80-06619.

## I. INTRODUCTION

A study of two different materials made by several different processes was conducted: (1) adipic acid produced by atomization, comminution, and crystallization, and (2) iron powders made by atomization, chemical reduction, and electrolytic processes.

The objectives of the experimental study here include:

1. To study the morphological properties of fine particles produced by different production processes.
2. To study the relationships between morphic descriptors and production processes.

Figures 1, 2, and 3 show the flow chart of the production of iron powders D, E, and F, respectively.

From these charts, it is clear that the history of formation of each type of particle is quite complex. However, we can make certain observations with the data available:

1. Different particle formation mechanisms produce particles of different shapes. They can be differentiated from each other on the basis of their morphic properties.
2. Some of the energy consumed in the production and formation of the particles is stored internally and is observed in the morphology of the particles.
3. It is now possible to introduce precise descriptors for particle shapes in the specifications of manufactured powders.

## II. EXPERIMENTAL

The samples studied are three adipic acid powders A, B, and C:

A. Adipic acid produced by atomization

B. Adipic acid produced by comminution

C. Adipic acid produced by crystallization

and iron powders D, E, and F:

D. Produced by atomization process (RZ-365)

E. Produced by chemical reduction process (IR, produced at Bombay IIT under grant INT-80-06619)

F. Produced by electrolytic process (SINTREX-D)

The three iron powders, D, E, and F, were sieve separated with ASTM standard sieves into small (S), medium (M), and large (L) size ranges. The range of sizes are as follows:

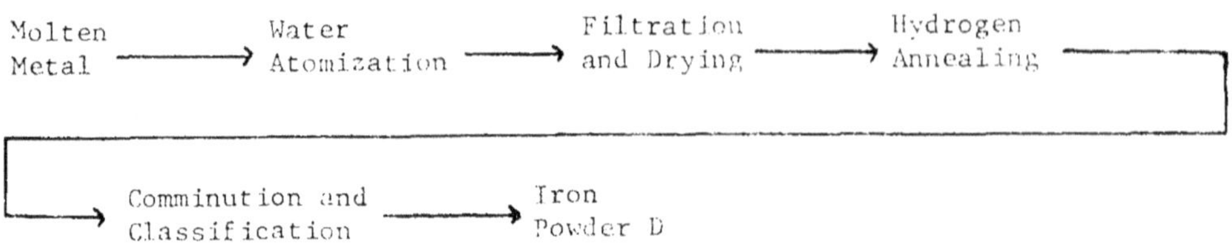


FIGURE 1. Flow chart of atomization process.

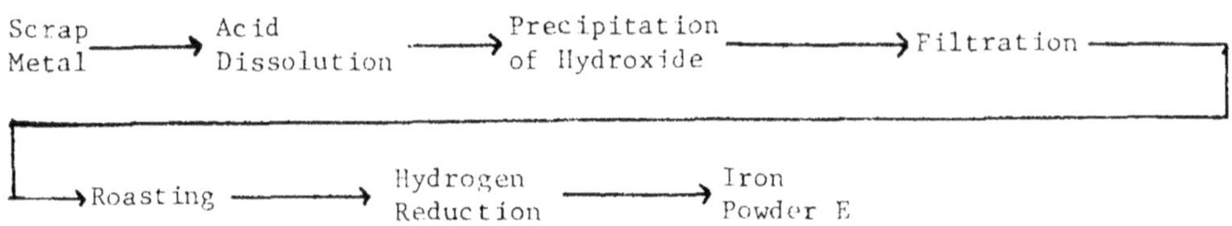


FIGURE 2. Flow chart of chemical reduction.

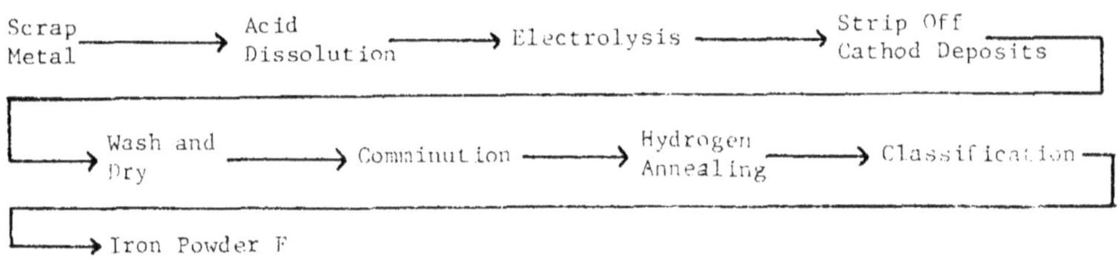


FIGURE 3. Flow chart of electrolytic process.

| | Mesh | µm |
|---|---|---|
| Small | 200—325 | 74—44 |
| Medium | 170—200 | 88—74 |
| Large | 100—150 | 149—105 |

The adipic acid was analyzed in one size range.

## III. RESULTS

A list of the morphic descriptors of adipic acid A, B, and C are shown in Table 1. From the list, we can see that in the case of the adipic acid powders:

1. Each powder has a different corresponding range of values of morphic terms VAR and SKE.
2. Crystallized particles have the highest values and atomized particles have the smallest values of morphic terms.

The morphic descriptors of iron powders D, E, and F are shown in Tables 2 through 4. From these tables it can be seen that:

**Table 1**
**ADIPIC ACID**

| | Atomized | Comminuted | Crystallized |
|---|---|---|---|
| $R_0$ | $3.0115 \times 10^{1}$ | $3.7047 \times 10^{1}$ | $2.6428 \times 10^{1}$ |
| $L_0$ | $9.9972 \times 10^{-1}$ | $9.9579 \times 10^{-1}$ | $8.8253 \times 10^{-1}$ |
| Radance | $2.4029 \times 10^{-4}$ | $7.7573 \times 10^{-2}$ | $2.1279 \times 10^{-1}$ |
| Skewness | $0.5694 \times 10^{-5}$ | $1.2239 \times 10^{-2}$ | $2.0465 \times 10^{-1}$ |
| L2(1) | $7.7067 \times 10^{-7}$ | $1.8195 \times 10^{-3}$ | $2.8373 \times 10^{-3}$ |
| L2(2) | $2.5623 \times 10^{-4}$ | $5.8642 \times 10^{-2}$ | $1.2358 \times 10^{-1}$ |
| L2(3) | $9.3942 \times 10^{-5}$ | $6.2347 \times 10^{-3}$ | $2.1098 \times 10^{-3}$ |
| L2(4) | $4.8729 \times 10^{-5}$ | $4.5173 \times 10^{-3}$ | $4.0936 \times 10^{-2}$ |
| L2(5) | $2.2268 \times 10^{-5}$ | $2.4086 \times 10^{-3}$ | $2.6239 \times 10^{-3}$ |
| L2(6) | $1.5830 \times 10^{-5}$ | $1.3420 \times 10^{-4}$ | $1.6135 \times 10^{-2}$ |

**Table 2**
**IRON POWDER**

| | | Small | Medium | Large |
|---|---|---|---|---|
| $R_0$ | D | 3.473 | 4.856 | $6.241 \times 10^{1}$ |
| | E | 4.489 | 4.646 | 7.106 |
| | F | 4.582 | 5.951 | 6.15 |
| $L_0$ | D | 9.798 | 9.853 | $9.867 \times 10^{-1}$ |
| | E | 9.68 | 9.806 | 9.856 |
| | F | 9.382 | 9.420 | 9.451 |
| PER | D | 1.320 | 1.268 | $1.302 \times 10^{0}$ |
| | E | 1.348 | 1.221 | 1.203 |
| | F | 1.306 | 1.299 | 1.278 |

**Table 3**
**IRON POWDER**

| | | Small | Medium | Large |
|---|---|---|---|---|
| IRR | D | 1.043 | 1.030 | $1.028 \times 10^{0}$ |
| $\times 10^{0}$ | E | 1.068 | 1.040 | 1.029 |
| | F | 1.144 | 1.132 | 1.123 |
| Radance | D | 3.963 | 2.903 | $2.627 \times 10^{-2}$ |
| $\times 10^{-2}$ | E | 6.247 | 3.828 | 2.846 |
| | F | 11.785 | 11.11 | 10.58 |
| Skewness | D | 0.1218 | 0.0493 | $0.0459 \times 10^{-3}$ |
| $\times 10^{-3}$ | E | 0.4732 | 0.1252 | 0.1142 |
| | F | 4.655 | 3.169 | 2.957 |

**Table 4**
**IRON POWDER**

| | | Small | Medium | Large |
|---|---|---|---|---|
| L2(1) | D | 0.8324 | 0.3172 | $0.2831 \times 10^{-3}$ |
| $\times 10^{-3}$ | E | 2.367 | 0.5581 | 0.2828 |
| | F | 1.599 | 2.614 | 1.141 |
| L2(2) | D | 2.149 | 1.377 | $1.188 \times 10^{-2}$ |
| $\times 10^{-2}$ | E | 3.963 | 2.316 | 1.675 |
| | F | 8.858 | 8.329 | 8.348 |
| L2(3) | D | 7.408 | 6.514 | $5.981 \times 10^{-3}$ |
| $\times 10^{-3}$ | E | 8.665 | 7.368 | 5.527 |
| | F | 3.692 | 5.517 | 4.285 |
| L2(4) | D | 2.879 | 2.651 | $2.326 \times 10^{-3}$ |
| $\times 10^{-3}$ | E | 4.869 | 2.545 | 2.294 |
| | F | 13.674 | 9.622 | 9.701 |
| L2(5) | D | 2.008 | 1.775 | $1.460 \times 10^{-3}$ |
| $\times 10^{-3}$ | E | 2.367 | 1.414 | 1.343 |
| | F | 2.176 | 2.954 | 1.974 |
| L2(6) | D | 1.172 | 0.8907 | $1.103 \times 10^{-3}$ |
| $\times 10^{-3}$ | E | 1.616 | 0.9172 | 0.5606 |
| | F | 3.395 | 2.133 | 1.897 |

1. For iron powders, as the size of particle increases, the magnitude of morphic descriptor decreases.
2. The magnitudes of $R_0$, $L_0$, and PER are very close together. These descriptors cannot be used to distinguish these particles from each other.
3. The magnitudes of IRR, VAR, SKE, and the L2s are not of the same order. In general, powder F has the largest values and powder D has the smallest.
4. Among these morphic descriptors, the VAR, SKE, L2(2), and L2(4) values of powder F are much larger than those of powder D and powder E. The morphic terms of powder D and powder E are fairly close.

## IV. DISCUSSION

Smaller particles tend to possess larger morphic terms. This is consistent with the idea that the absolute magnitude of the surface irregularities is controlled by the mechanism of particle formation. Consequently, the smaller the $R_0$, the more prominent do the shape terms behave.

Figure 4 is the plot of radance VAR vs. skewness of adipic acid powders A, B, and C. Figure 5 is the plot of mean radance VAR vs. mean of skewness of iron powder D, E, and F.

From Figure 6 we can see that adipic acid particles produced by different process and formation mechanism have different morphological properties. They can be classified on the basis of their radance and skewness values.

From Figure 7 we can see that important production steps have significant influence on particle morphologies. In this case, for iron particles D, E, and F, the variation of skewness is much greater than the variation of radance for different production process.

FIGURE 4. Shapes of adipic acid.

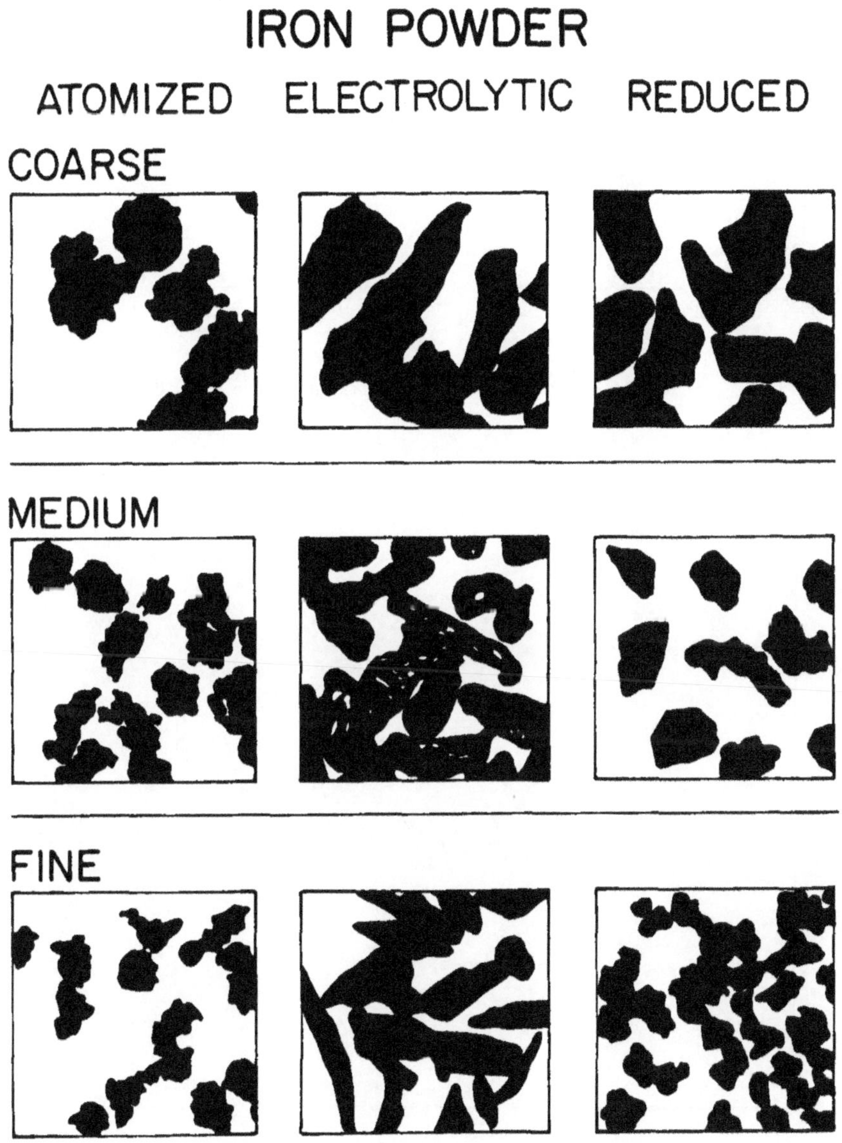


FIGURE 5. Shapes of iron powders, D, E, F (see Figures 1-3).

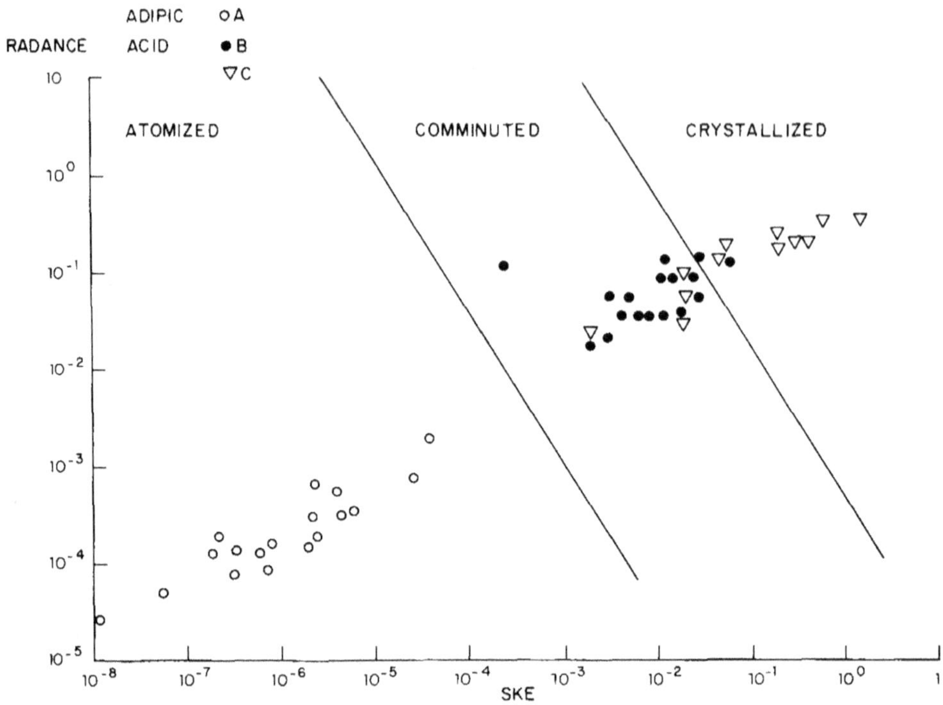


FIGURE 6. Plot of radance vs. skewness of adipic acid powders.

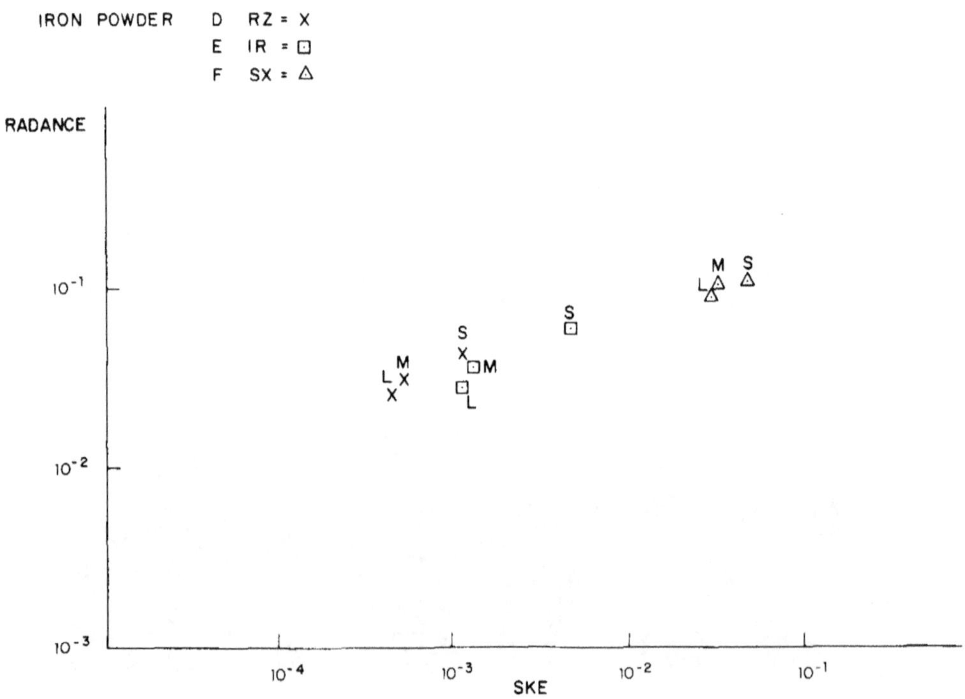


FIGURE 7. Plot of VAR vs. SKE of iron powders.

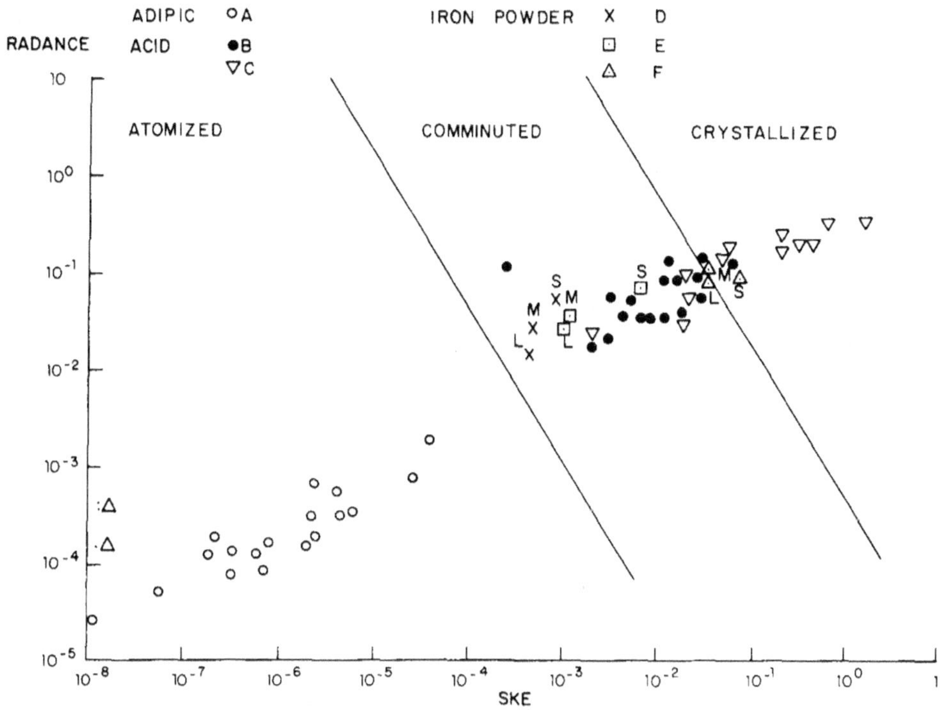


FIGURE 8. Plot of radance vs. skewness of adipic acid and iron powders.

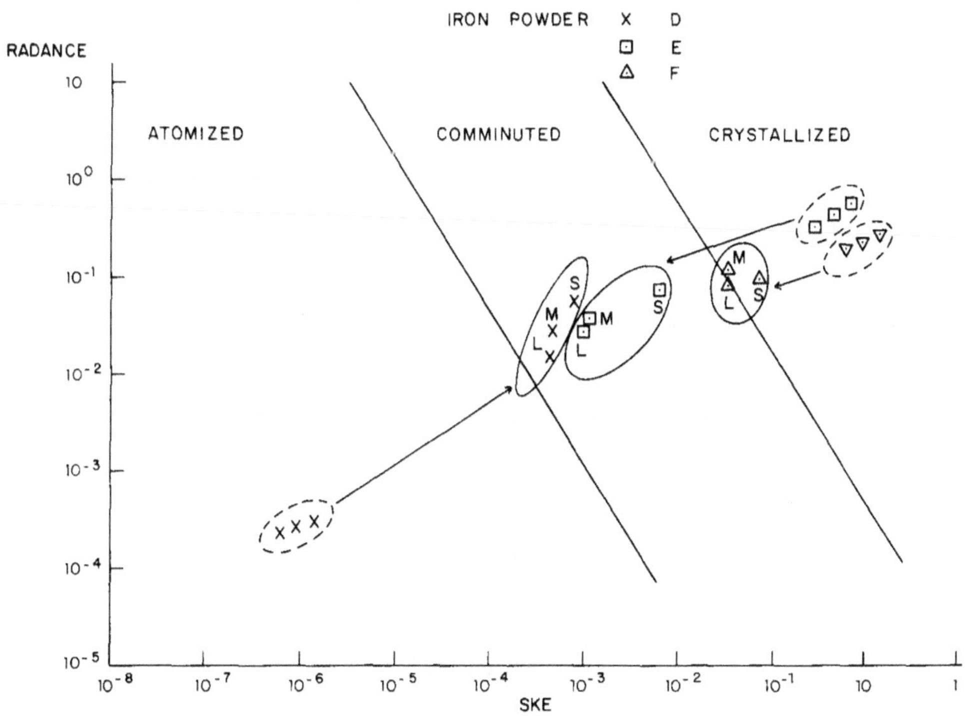


FIGURE 9. Plot of radance vs. skewness of iron powders for different processes.

The morphic properties of powders are related to their mechanism of formation which is symptomatic of the production process. The exact forms of these relationships have yet to be determined.

Figure 8 shows the iron and adipic acid plotted together. The locations of D and E are near the border of the atomized zone and comminuted zone of the adipic acid powders, the location of F is near the border of the comminuted zone and crystallized zone of adipic acid powders.

We may speculate that the locations on the morphological chart of Figure 9 of iron powders D, E, and F prior to the annealing and comminution steps of their manufacture would be within the crystallized and atomized zone of the adipic acid powders. After annealing and comminution, they shifted toward the comminuted zone. That is, particles produced by a single mechanism can be classified by their locations in the radance and skewness space. However, particles produced by a series of mechanisms will be moved from their original locations in the radance and skewness space.

## V. CONCLUSIONS

Particles produced by different mechanisms can be differentiated on the basis of their morphology.

Several production processes can be differentiated using the morphological chart (a chart of radance vs. skewness of the powders from each process).

By utilizing morphological analysis, manufacturers of powders now have the capability to quote in the specifications of powder, not only the size, but also the shape in a quantitative way.

## LIST OF SYMBOLS

$\text{IRR} = \left(\frac{R_0}{A_0}\right)^2$, defined as irregularity

$\text{Radance} = \sum_n L2(n)$, the second moment about mean of the radius distribution

$\text{Skewness} = \sum_m \sum_n L3(m,n)$, the third moment about mean of the radius distribution

$$L2(n) = \frac{1}{2R_0^2} A_n^2 \qquad m,n = 1, \ldots k$$

$$L3(m,n) = \frac{3}{4R_0^3} (A_m A_n A_{m+n} \text{COS} (\alpha_{m+n} - \alpha_m - \alpha_n)) \text{ in which}$$

$A_0$ is the mean radius

$A_n$'s and $\alpha_n$'s are Fourier coefficients

$R_0$ is the equivalent radius

## REFERENCES

1. **Lenth, R., Chang, C.-R., Beddow, J. K., and Vetter, A. F.,** Particle image analyzing system, in *Particulate Systems: Applications to Fundamentals,* Beddow, J. K., Ed., Hemisphere Publishing, New York, 1983.
2. **Luerkens, D. W., Beddow, J. K., and Vetter, A. F.,** Morphological analysis of fine particles using the Fourier method, *Powder Technol.,* 31(2), 1982.

Chapter 15

# THE EFFECTS OF PARTICLE MORPHOLOGY ON BATCH SIEVING BEHAVIOR OF FINE POWDERS

**Samir Rajpal, Louise Hua, Chi-Ren Chang, and John Keith Beddow**

## TABLE OF CONTENTS

## I. INTRODUCTION

Screening and sieving have long been the most common methods of separating particles according to size. The simplicity of these processes is complemented by their economic advantages. It is now proposed that screening and sieving may be effectively used to classify particles on the basis of both their size and shape. This is demonstrated by applying the invariant morphic descriptors to the sieving analysis. Shape and size classification by such economical methods offers tremendous advantages by way of application to continuous mass production processes. Perhaps even more important, manufacturers of powders may avail themselves of such techniques to refine their products by removing the undesired morphic fractions from the sieved powder.

This chapter presents the results of a successful attempt to show the relationship between the morphology and batch sieving behavior of three metal powders. The powders chosen were known to be of sufficiently different morphologies and represented particles formed by the atomization, electrolytic, and rotating electrode processes (REP).* The REP particles were perfectly spherical while the electrolytic particles tended to be elongated. The atomized powder contained particles more rounded in general form than those of the electrolytic powder, but not nearly spherical. Representative shapes of all three powders are shown in Figures 1 to 3.

## II. THE EXPERIMENT

The powders were sieved initially to obtain required amounts in the size range −170 to +200 mesh (74 to 88 μm). Each powder was then individually sieved through the same range on a standard Rotap, and successive batches were collected at measured time intervals. These batches were morphologically analyzed to obtain the morphic terms, which were then used to obtain graphs of the relationships between shape and sieving time. Results show a distinct pattern of shape variation during the sieving process.

## III. RESULTS AND ANALYSIS

The sieving curves obtained for the three powders are shown in Figure 4. These conform to Whitby's observations for sieving behavior and show two linear segments separated by a transition region.** Examination of the morphic descriptors suggests that the transition region represents that time interval in which the majority of deviant shapes within any given powder are sieved. The curves of Figure 4 also show a downward shift with increasing irregularity or nonsphericity. This is reasonable if one agrees with the idea that the more irregular the group of particles, the less the amount that would be sieved during any given time interval. It then follows that more irregular particles would be represented lower on the percentage-passed vs. time curve than relatively more spherical particles. This trend is quantifiably substantiated in the following data.

**Morphic Terms for Different Metal Powders**

| | Morphic term | | | | |
|---|---|---|---|---|---|
| **Metal powder** | **$R_0$** | **VAR** | **ASKE** | **IRR** | **$L_2(2)$** |
| Spherical steel | 50.09 | $4.99 \times 10^{-3}$ | $4.28 \times 10^{-4}$ | 1.005 | $4.42 \times 10^{-3}$ |
| Atomized iron | 55.80 | $4.16 \times 10^{-2}$ | $3.42 \times 10^{-3}$ | 1.045 | $2.27 \times 10^{-2}$ |
| Electrolytic iron | 70.16 | $1.35 \times 10^{-1}$ | $6.01 \times 10^{-2}$ | 1.171 | $9.79 \times 10^{-2}$ |

* Manufactured and supplied by Nuclear Metals, Inc., Concord, Mass.

** Whitby, K. J., ASTM Special Technical Publication 234, American Society for Testing and Materials, Philadelphia, 1958, 3.

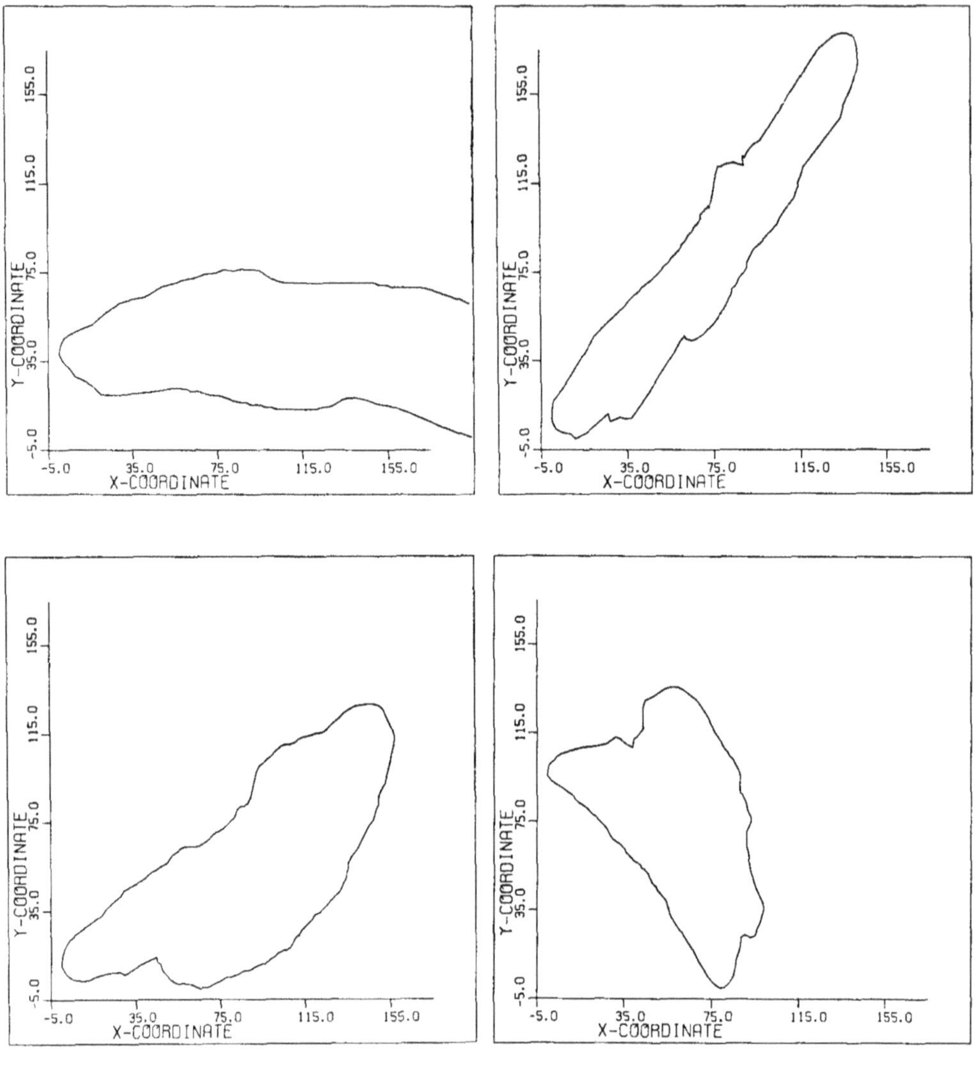


FIGURE 1. Representative electrolyte particle profiles.

Figure 5 shows the variation of size ($R_0$) with sieving time. As can be seen from the graph, the smaller particles (within a given size range) pass through the sieve during the initial sieving while the larger particles are sieved during later intervals of time. The slight discrepancy indicated by the nonuniformity of the spherical-powder curve is explained at the end of this section. Figures 6 and 7 show the graphs of $L_0$ vs. time and radance vs. time, respectively. Not only are the trends similar for all three powders, but it is apparent that the trends indicate a distinct relationship between the shape of particles and their sieving time. It would appear, then, that it is feasible to prepare sieve fractions of desired morphologies on the basis of the obtained curves. In other words, once such curves are known for a particular powder, one may sieve the powder for a predetermined time to obtain particles of the desired morphology. On a similar basis, one may manipulate the sieving process to remove undesired morphic fractions and thereby refine a given powder. Results obtained for the spherical powder support this idea. Figure 8 shows the relationship between radance and time and skewness vs. time for the spherical powder. The spherical particle profile shown earlier, while typical of the majority of particles in the spherical powder, is

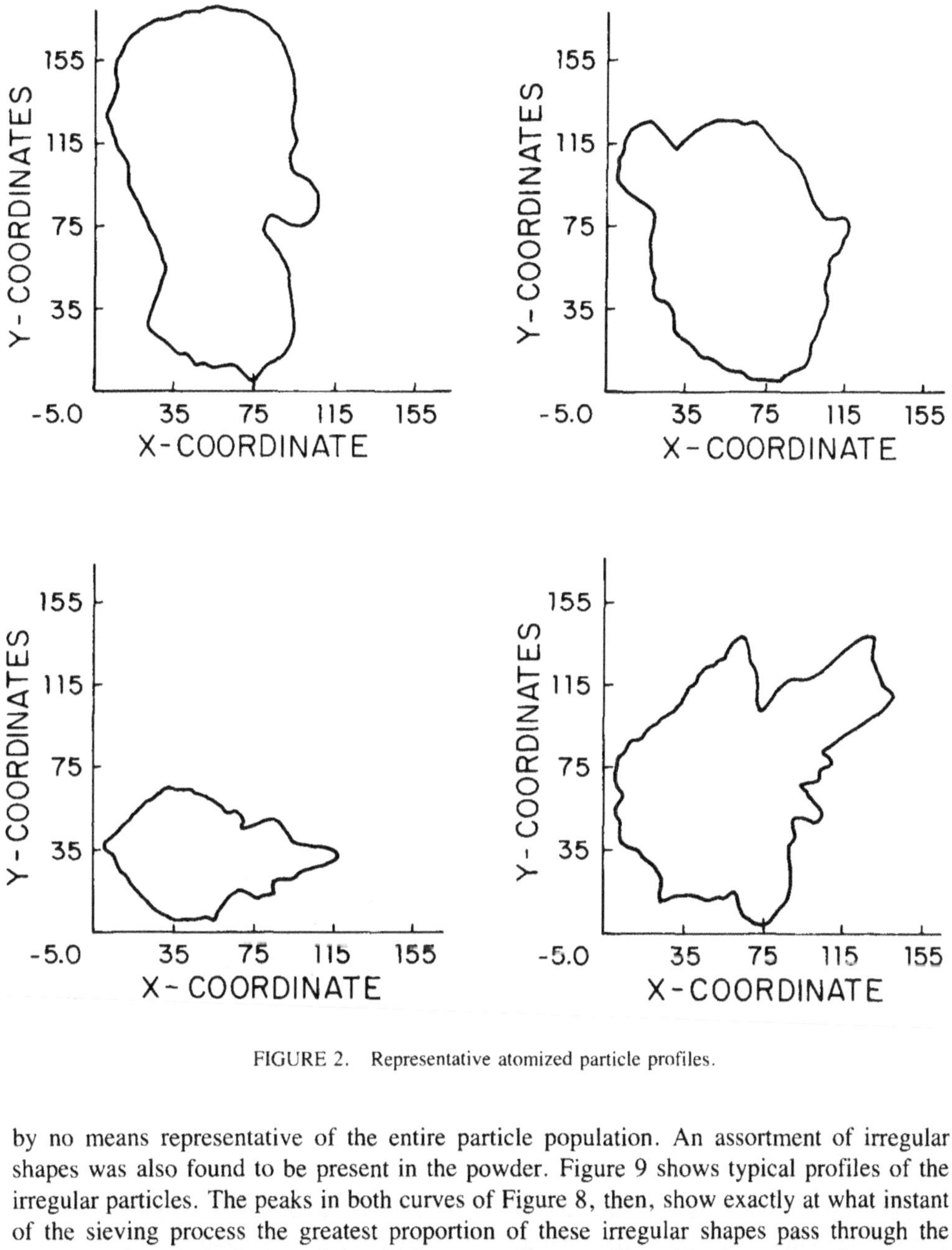


FIGURE 2. Representative atomized particle profiles.

by no means representative of the entire particle population. An assortment of irregular shapes was also found to be present in the powder. Figure 9 shows typical profiles of the irregular particles. The peaks in both curves of Figure 8, then, show exactly at what instant of the sieving process the greatest proportion of these irregular shapes pass through the sieve. Again, manipulation of the sieving procedure would enable the systematic removal of these deviant shapes. Results show that the irregular particles sieved in the 35 to 75-sec interval corresponding to the peaks in the curves represented 1.07% of the total sample.

An interesting observation is that in the case of the spherical powder the variation of $R_0$ with time is nonuniform. This is shown in Figure 10. Not only do "larger" particles appear to be passing through the sieve before many smaller ones, but the size of the particles is also larger than the mesh opening (88 μm). This can be explained by the fact that the sample contains many particles which are ellipsoidal rather than spherical. These particles contribute to a larger equivalent radius, $R_0$, while at the same time passing through the sieve in their smallest dimension.

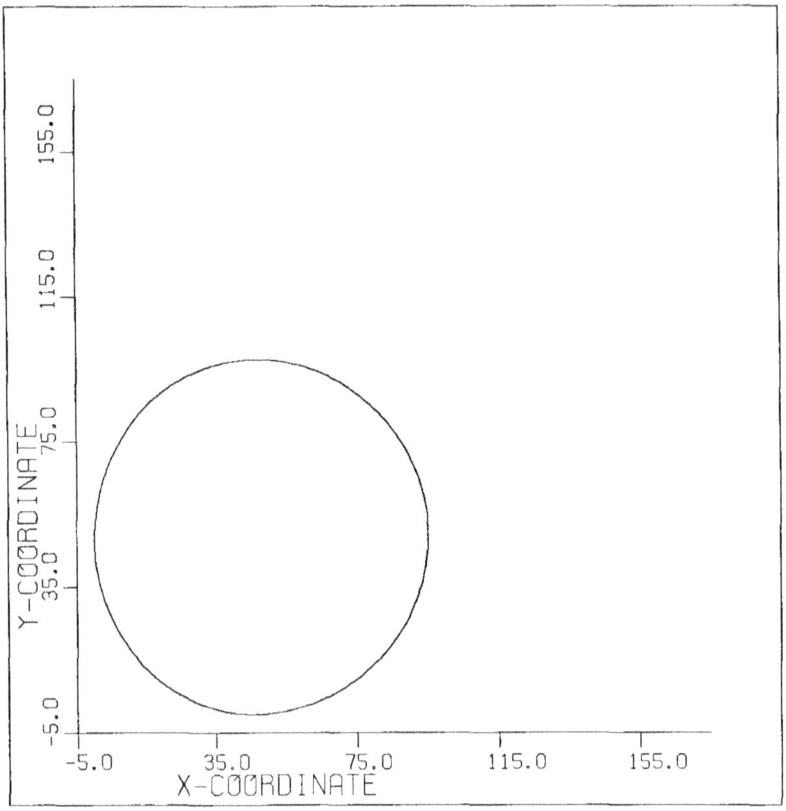


FIGURE 3. Representative REP particle profile.

## IV. PREDICTING SIEVING BEHAVIOR FROM KNOWN MORPHOLOGIES

Based on data obtained, it has been possible to propose a general correlation that relates the cumulative fraction passing through the sieve with time and morphology of the particles. The analysis is as follows.

If m = the fraction remaining on the screen, and k is the probability of a particle passing through the sieve, then

$$\frac{dm}{dt} = -km \tag{1}$$

The probability k is based on two considerations: (1) the number of times a particle strikes the screen, and (2) the probability that a strike is effective, i.e., the particle is able to pass through. The probability of a strike being effective is determined for the case of a spherical particle as follows. If D = the sieve opening, and $R_0$ = the radius of the particle, then, the particle is likely to pass through the sieve only if it strikes the shaded area. The probability of such an occurrence is

$$\frac{(D - 2R_0)}{D^2} = (1 - 2r)^2,$$

$$r = \frac{R_0}{D}$$

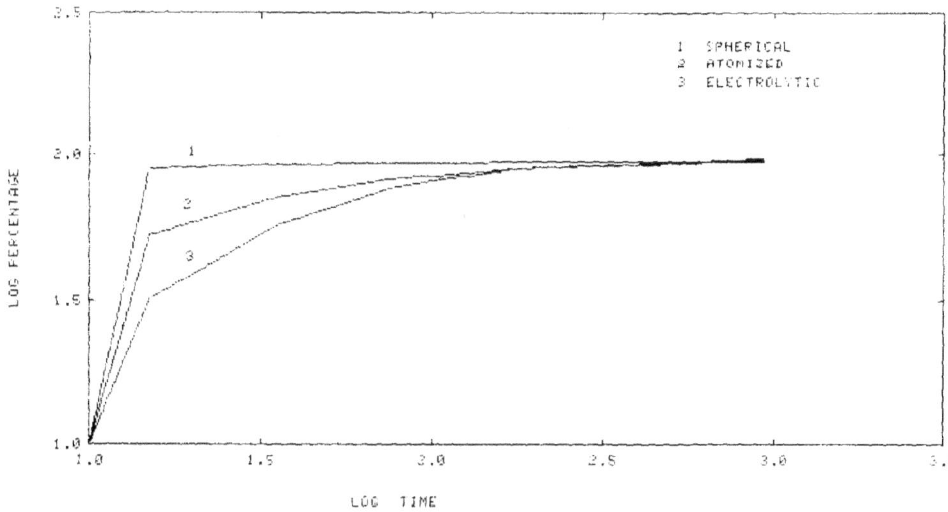


FIGURE 4. Whitby sieving curves: (1) REP spherical powder; (2) atomized powder; and (3) electrolytic powder.

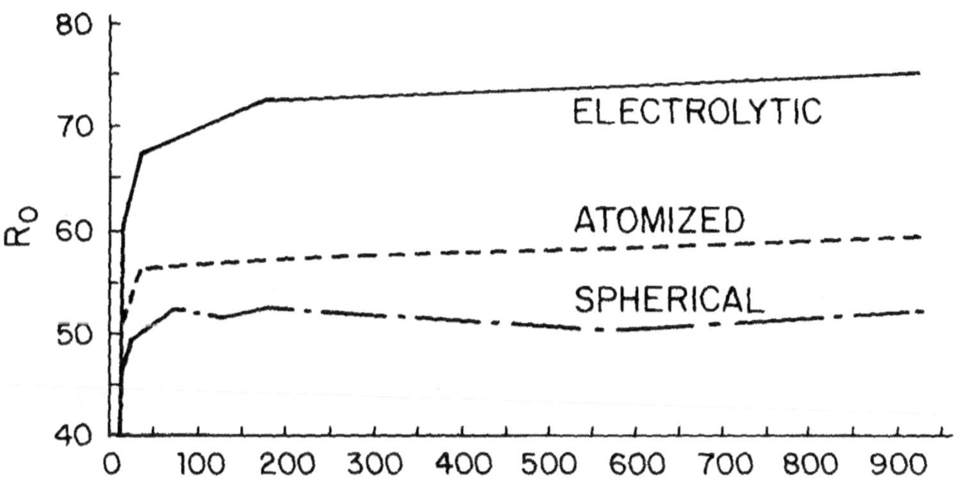


FIGURE 5. $R_0$ vs. time for different powders.

in order to generalize this for nonspherical particles, we introduce the effective radius, $R_{0eff}$. This is defined in terms of size and shape as follows:

$$R_{0eff} = R_0 (1 - \text{shape factors})$$

The shape factors used in our correlation are $L_0$, VAR, ASKE, and $L_2(2)$. Therefore,

$$R_{0eff} = R_0 (1 - x_1 L_0^{x_2}\, VAR^{x_3}\, ASKE^{x_4}\, L_2(2)^{x_5})$$

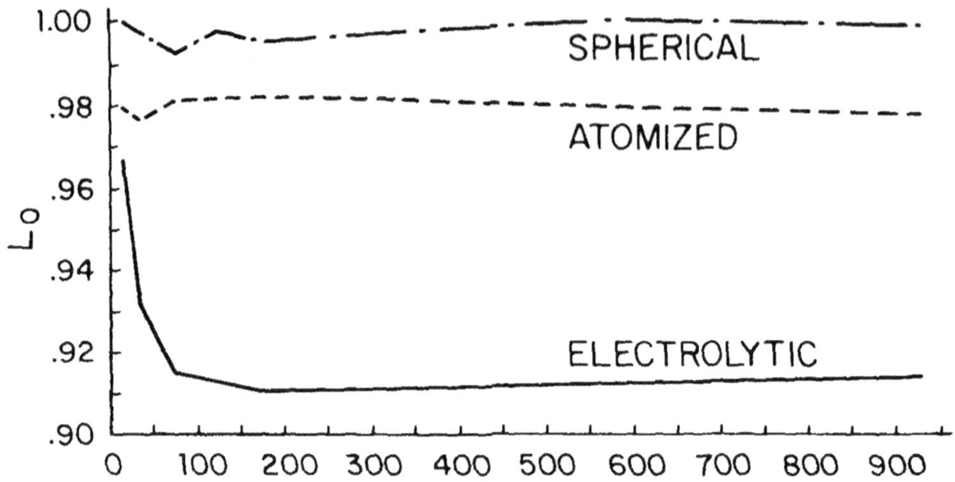


FIGURE 6. $L_0$ vs. time for different powders.

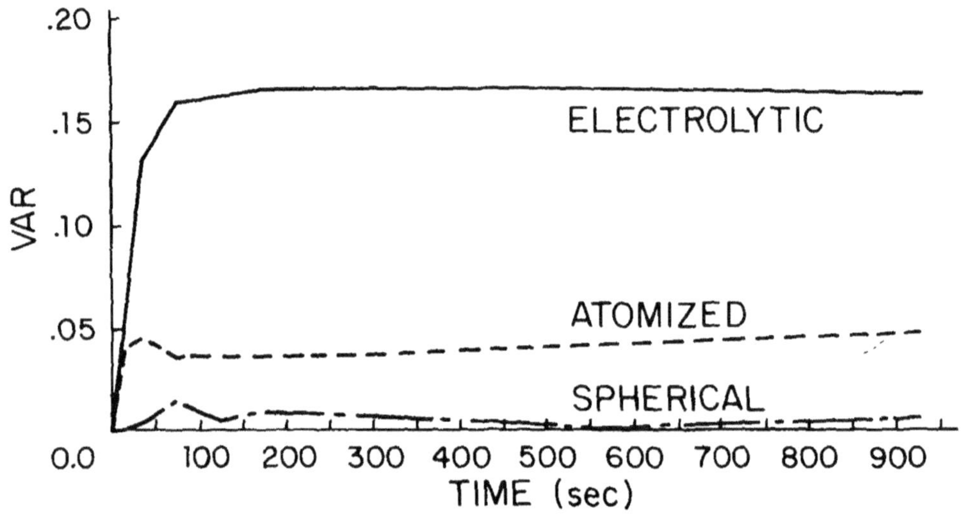


FIGURE 7. Radance (VAR) vs. time for different powders.

and

$$r = \frac{R_{0eff}}{D}$$

where $x_1$ through $x_5$ are constants to be determined.
Using these terms and combining the two effects upon which k is dependent,

$$k = A\,(1 - 2r)^2$$

where A accounts for the number of times the particle strikes the screen, and is a function of the sieving conditions such as Rotap operating specifications and the amount of material

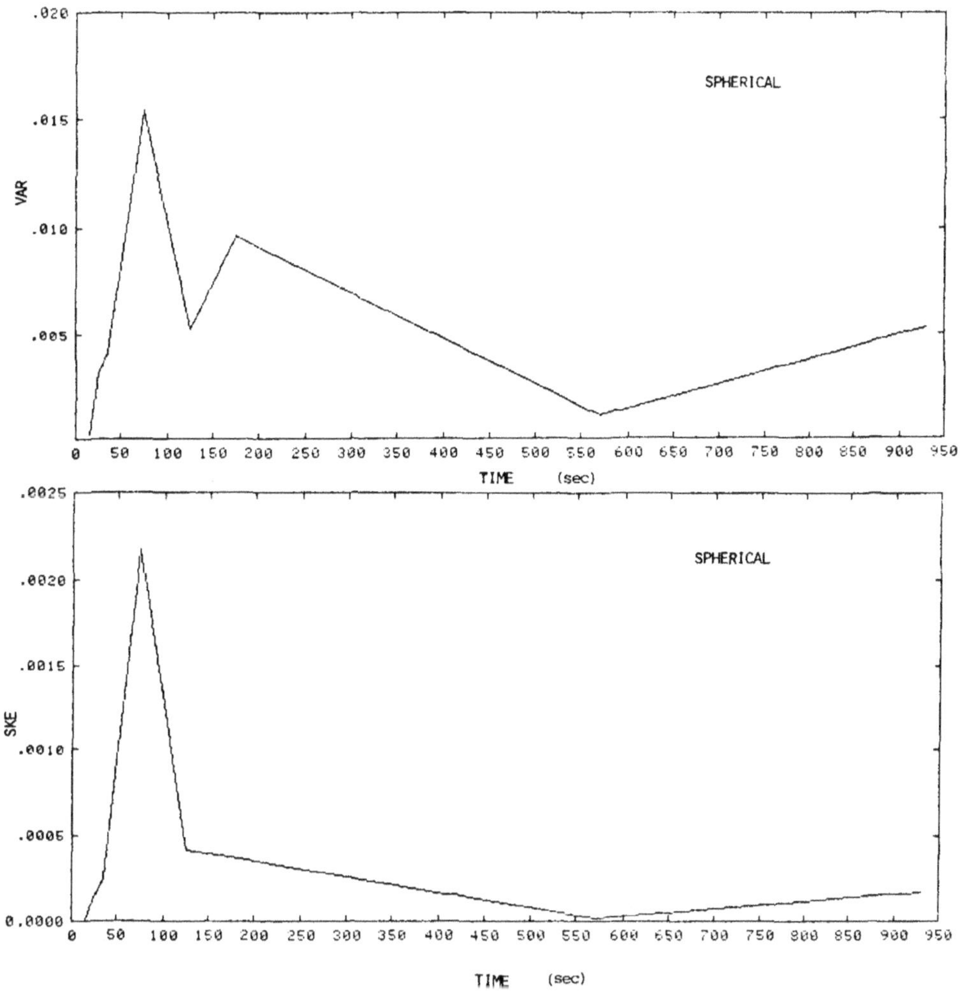


FIGURE 8. Shape terms, radance (VAR) and skewness (SKE) vs. time for REP powder.

on the screen. The amount of material is not of much significance unless the amount is either very large or very small. Then

$$k = A\left[1 - \frac{2R_0}{D}(1 - x_1 L_0^{x_2}\ VAR^{x_3}\ ASKE^{x_4}\ L_2(2)^{x_5})\right] \quad (2)$$

Since size and shape are not constant during sieving, it is clear that k is also not constant and varies with time. However, if we consider a small time interval during sieving, it is reasonable to assume k is constant within that interval. Integration of Equation 1 over the small time interval then yields

$$\int_{m_0}^{m_j} \frac{dm}{m} = -\sum_{i=1}^{j} \int_{t_{i-1}}^{t_i} k_i\, dt$$

$$t_0 = 0$$

If $m_j$ = amount remaining after jth time interval then

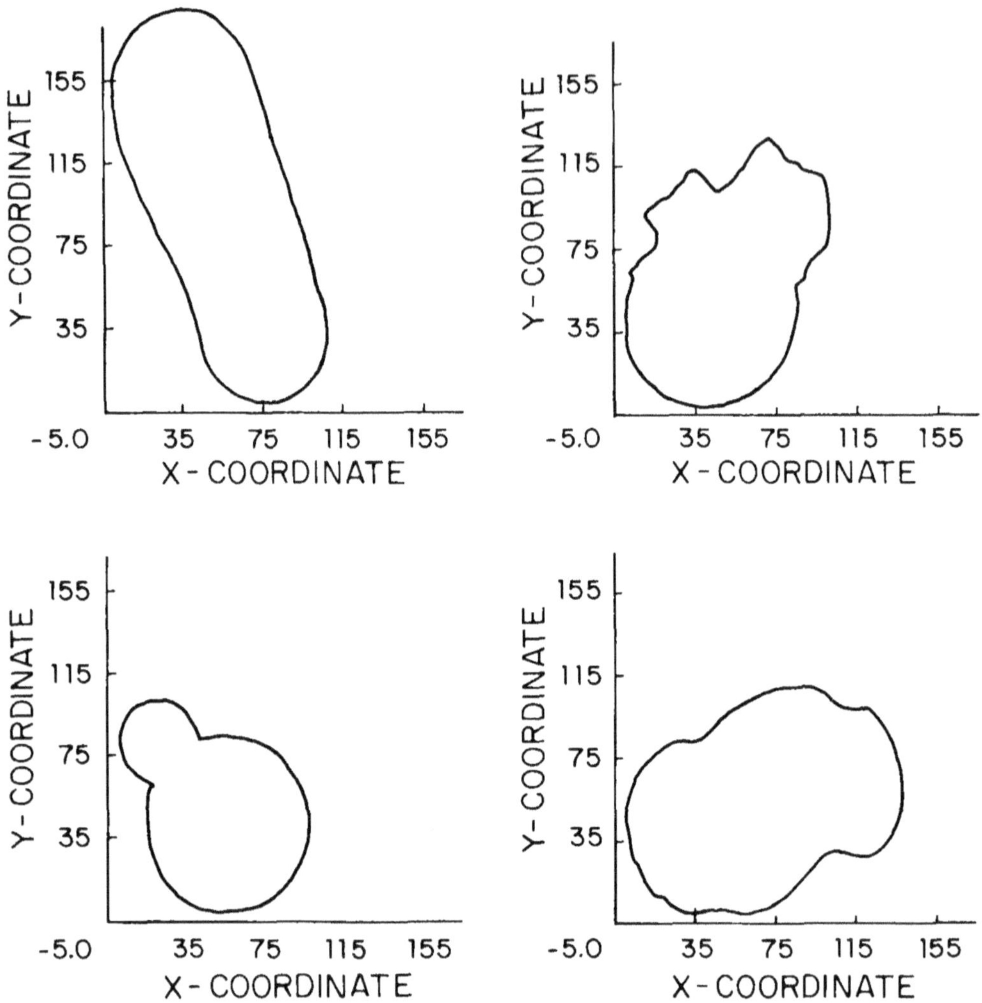


FIGURE 9. Representative shapes of irregular REP (spherical) powder.

$$\ell n \frac{m_j}{m_0} = - \sum_{i=1}^{j} k_i \Delta t_i$$

$$\Delta t_i = t_i - t_{i-1}$$

Also, $1 - m_j/m_0$ = cumulative fraction passing through sieve = $f_j$,

$$\therefore \ell n\, f_j = - \sum_{i=1}^{j} k_i \Delta t_i \quad (3)$$

From Equation 3, we may calculate $k_i$ having measured $f_i$ and $\Delta t_i$ for each powder and for each interval. Using all the calculated values of $k_i$ in Equation 2, we may determine the least square fit to find the values of A and $x_1$ through $x_5$.

Knowing the values of A and Xs, we may then estimate k from Equation 2 and use this in Equation 3 to estimate the fraction sieved. Figures 11, 12, and 13 show the agreement between the observed and the estimated values of the fraction passing vs. time for the three powders tested.

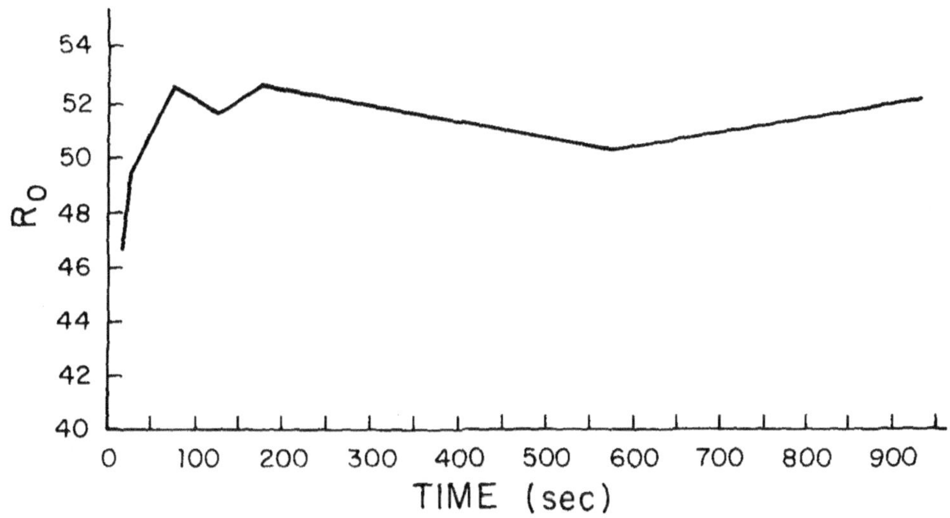


FIGURE 10. $R_0$ vs. time for REP (spherical) powder.

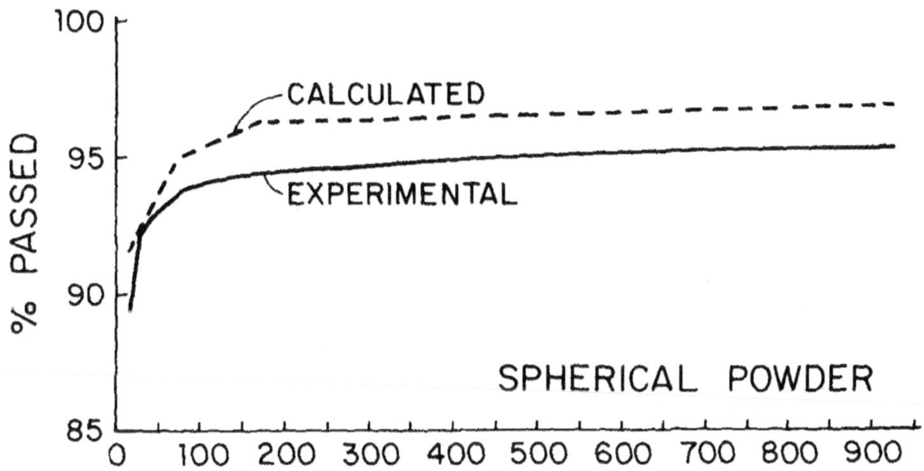


FIGURE 11. Calculated vs. experimental values of fraction sieved for REP (spherical powder).

Figure 11, for the spherical powder, shows a slightly greater discrepancy between the calculated and experimental values than expected, but this is due mainly to the fact that an extremely large fraction (90%) of the sample passed through the sieve in the first time interval. This left too small a remaining sample to provide sufficient correlation.

The following correlation is proposed for the general set of fine powders:

$$\ln f_i = -k_i \, \Delta t_i$$

where $f_i$ = fraction passing through the screen in the ith time interval and $k_i$ is expressed by Equation 2 with:

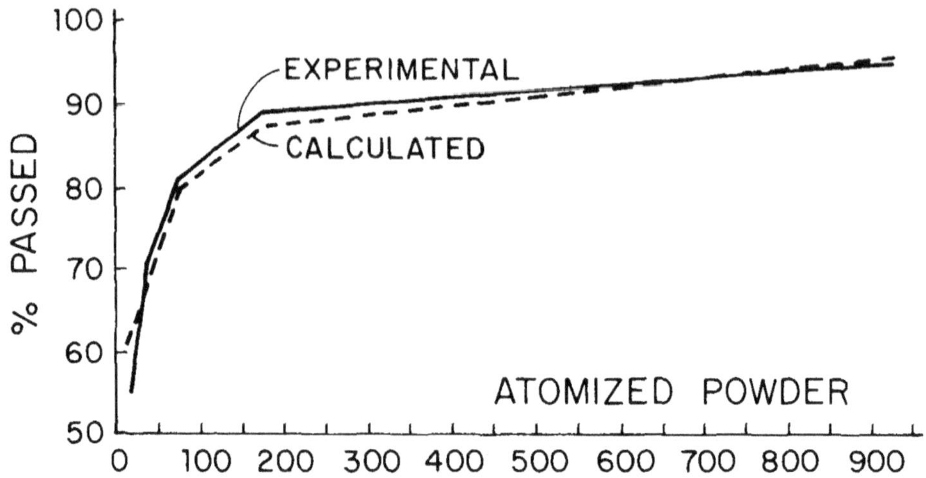


FIGURE 12. Calculated vs. experimental values of fraction sieved for atomized powder.

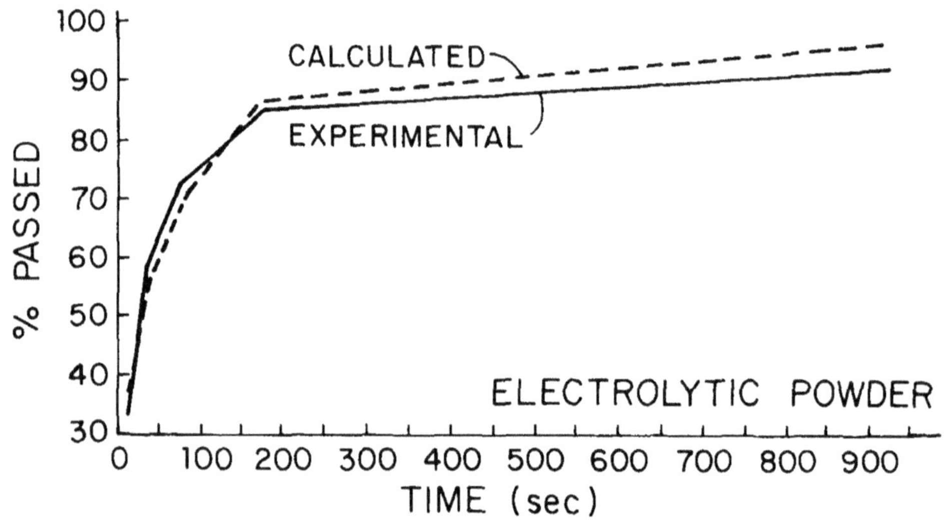


FIGURE 13. Calculated vs. experimental values of fraction sieved for electrolytic powder.

A = 1.035
X1 = 0.61
X2 = 0.20
X3 = 0.385
X4 = 0.339
X5 = −0.668

It is clear from these results that one may actually determine the sieving curve for a powder once its morphology is known. Thus, it is not necessary to actually sieve the powder. The actual relationship to predict the sieving curve from a knowledge of the powder morphology is currently under investigation. However, from this work it seems likely that *for predictive purposes,* the k values determined in the foregoing analysis may not be applicable for two reasons. First, the time intervals used were not small enough to justify the assumption of constant K (since this has a greater bearing on the predictive equations). Second, the k determined represents the overall or mean value of the particles sieved within any time interval, and thus represents an average of the various shapes present in the batch collected. However, in order to derive an expression capable of predicting the sieving curves from a knowledge of the morphology, it is necessary to determine values of k for monomorphic batches. To achieve this condition in practice requires reexperimentation with different samples, each monomorphic in nature. Work is currently underway to produce such particles, and further investigation of sieving behavior is to follow.

## V. SUMMARY

On the basis of the above analysis, it is clear that the shape of particles being sieved is a determinant function of time and that this shape-dependent sieving behavior may be characterized in terms of the morphic descriptors developed earlier. It would also seem possible to say that the trends shown may be used to separate and classify particles on the basis of both their size and shape. A direct result of this would be the capability to refine a given powder by sieving it to prepare various morphic fractions and subsequently discarding the undesired fractions. This application of sieving technology is especially appealing for its simplicity and economy, with potential applications in various processes requiring dry separation techniques.

## LIST OF SYMBOLS

| | |
|---|---|
| A = | Factor accounting for number of times particle strikes screen |
| D = | Diameter of sieve opening |
| f = | Fraction of particles passing through screen |
| IRR = | Morphic descriptor |
| k = | Probability of particle passing through screen |
| $L_0$ = | Morphic descriptor |
| $L_2(2)$ = | Morphic term |
| m = | Fraction of particles remaining on screen |
| r = | Ratio of particle radius to sieve opening, $R_0/D$ |
| $R_0$ = | Morphic descriptor, equivalent radius of particle |
| $R_{0_{eff}}$ = | Effective particle radius including shape effects |
| ASKE = | Skewness of particle radial distribution |
| VAR = | Variance of particle radial distribution, also known as radance |
| Xs = | Constants of correlation |

Chapter 16

# THE EFFECT OF SHAPE ON PARTICLE SOLIDS FLOW

**Gregory R. Carmichael**

## TABLE OF CONTENTS

## I. INTRODUCTION

Estimation of the drag force exerted by a fluid on a solid is important in the design of solids flow systems and separation equipment. The shape characteristics of the solid particles can exert a profound influence on the drag. For shapes other than spheres, the drag must be estimated from approximation theories or numerical solutions, or determined experimentally. The assessment of the direct effects of shape on the drag has been difficult, due, in part, to the inability to systematically and quantitatively describe shape itself.

In this chapter, the effects of particle shape and orientation on the drag of freely settling particles is discussed. A general theoretical background on the drag force and applications and utilizations of the drag force in design are presented. In addition, a new model for the estimation of the drag exerted by different geometric objects for flows in the Stokes region, derived using morphological analysis, is introduced.

## II. THEORETICAL CONSIDERATIONS

The resultant motion of a body falling in a fluid is determined by a balance of forces acting on the body. The total force exerted by the fluid on an immersed body is composed of a normal component, N, and a tangential component, $\tau$, i.e.,

$$F = \int_{surface} N \, d\Omega + \int_{surface} \tau \, d\Omega \tag{1}$$

The component of the resultant force parallel to the undisturbed initial velocity is the drag, D, and the component perpendicular is the lift, L. Frequently the drag and the lift are referred to in terms of the dimensionless coefficients

$$C_D = \frac{D}{{}^1/_2 \, \rho \, V^2 A}$$

$$C_L = \frac{L}{{}^1/_2 \, \rho \, V^2 A} \tag{2}$$

when ${}^1/_2 \, \rho \, V^2$ is the dynamic head and A is area. By convention, A is taken as the frontal area exposed by the body to the flow direction.

It has been shown theoretically and experimentally that the drag depends on the size, shape, projected area, and relative velocity of the body, and on the density and viscosity of the fluid. For dynamically similar flows (i.e., geometrically similar stream lines) it is required that the Reynolds numbers be equal, i.e.,

$$R_e = \frac{\text{inertial forces}}{\text{frictional forces}} \simeq \frac{\rho \, u \dfrac{\partial u}{\partial x}}{\mu \dfrac{\partial^2 u}{\partial x^2}} \simeq \frac{\rho \, V^2/d}{\mu \, V/d^2} = \frac{\rho V d}{\mu} \tag{3}$$

where d is the characteristic length, V the characteristic velocity and $\rho$ and $\mu$ are the fluid density and viscosity, respectively. For geometrically similar bodies which have the same orientation with respect to the free stream direction, the drag coefficient, $C_D$, is a function of the Reynolds number only

$$C_D = f(R_e, \text{only}) \tag{4}$$

(This equation is true when the only forces acting on the body are due to friction and inertia.)

In most cases $f(R_e)$ can only be determined experimentally. The relationship between $R_e$ and $C_D$ for spheres and cylinders is plotted in Figure 1. It is important to note that each geometric object has a unique relationship between $C_D$ and $R_e$, emphasizing the fact that particle shape has an important influence on $C_D$.

There are a few cases where $f(R_e)$ can be determined analytically. Spherical particles in slow flow is one case. The equations of motion for an incompressible fluid can be written in Cartesian coordinates

$$\frac{\partial u_i}{\partial t} + u_j \frac{\partial u_j}{\partial x_j} = -\frac{1}{\rho}\frac{\partial P}{\partial x_i} + \nu \frac{\partial^2 u_i}{\partial x_j \partial x_j} \tag{5}$$

For very slow flows ($R_e < 0.1$), the importance of the inertial terms is minimal. By neglecting the inertial terms and assuming steady state, Stokes[2] solved Equation 5 (along with the continuity equation $\partial u_i/\partial x_i = 0$). The solution is

$$u = U_\infty \left[\frac{3}{4}\frac{Rx^2}{r^3}\left(\frac{R^2}{r^2} - 1\right) - \frac{1}{4}\frac{R}{r}\left(3 + \frac{R^2}{r^2}\right) + 1\right] \tag{6}$$

$$v = U_\infty \frac{3}{4}\frac{Rxy}{r^3}\left(\frac{R^2}{r^2} - 1\right) \tag{7}$$

$$w = U_\infty \frac{3}{4}\frac{Rxz}{r^3}\left(\frac{R^2}{r^2} - 1\right) \tag{8}$$

$$P - P_\infty = -\frac{3}{2}\mu \frac{U_\infty Rx}{r^3} \tag{9}$$

where the origin is at the center of the sphere and $r^2 = x^2 + y^2 + z^2$. Note that these equations satisfy the no-slip condition at the surface. These equations can be used to determine that

$$D = 6\pi \mu R U_\infty \tag{10}$$

of which 1/3 is due to pressure and 2/3 to shear, and

$$C_D = \frac{D}{{}^1/_2\, \rho V_\infty^2 A} = \frac{6\pi \mu R U_\infty}{{}^1/_2\, \rho U_\infty^2 \cdot \pi R^2} = \frac{24}{R_e} \tag{11}$$

Oseen[3] improved Stokes' analysis by including a linearization of the inertial terms and calculated

$$C_D = \frac{24}{R_e}\left(1 + \frac{3}{16}R_e\right) \tag{12}$$

A similar analysis for cylinders was carried out by Lamb.[4] However, analytic expressions for most other geometric objects and higher Reynolds number flows do not exist. The drag coefficient for these flows must either be estimated from approximation theories (e.g., boundary layer theory) or numerical solutions, or determined experimentally.

Before returning to the discussion of the determination of $C_D = f(R_e)$ for various geometric objects, a few examples of the use of $C_D$ in practical applications will be presented.

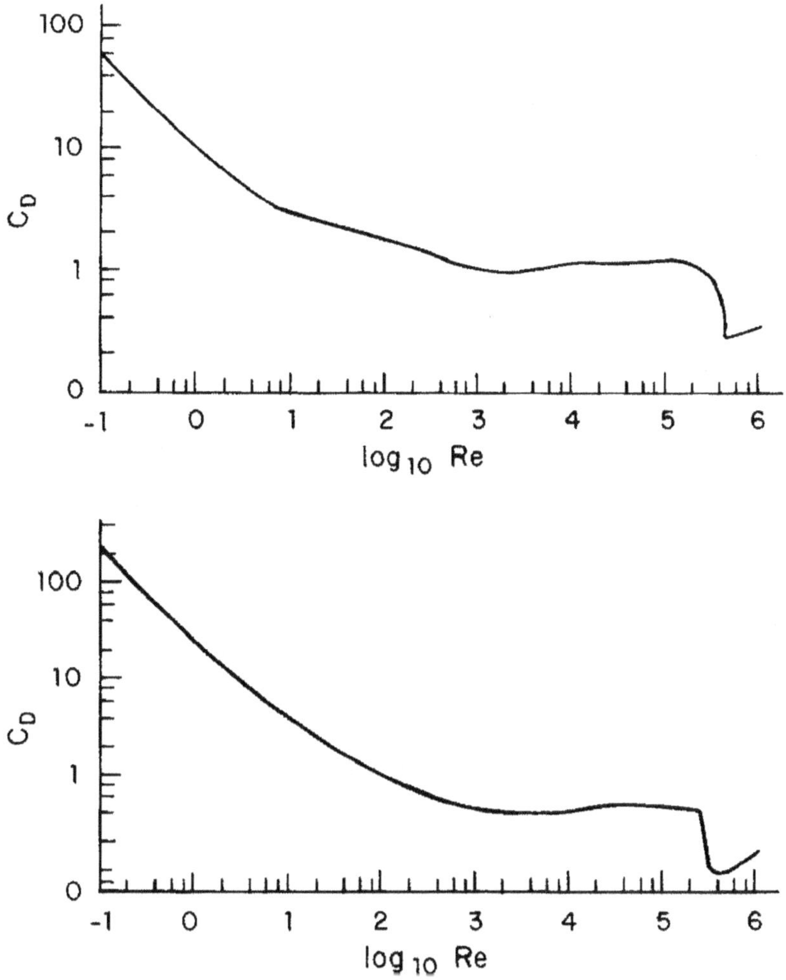


FIGURE 1. $C_D$ vs. $R_e$ for cylinders (top curve) and spheres (bottom curve).

## III. APPLICATION AND UTILIZATION OF $C_D$

### A. Settling Velocity

Consider a particle moving in a fluid as shown in Figure 2. The three forces acting on the particle are the external force, $F_e$, the drag force, $F_D$, and the buoyant force, $F_B$; where

$$F_e = m\, a_e \tag{13}$$

$$F_D = D = \frac{C_D\, V^2 \rho A}{2} \tag{14}$$

$$F_B = \left(\frac{m}{\rho_s}\right) \rho\, a_e \tag{15}$$

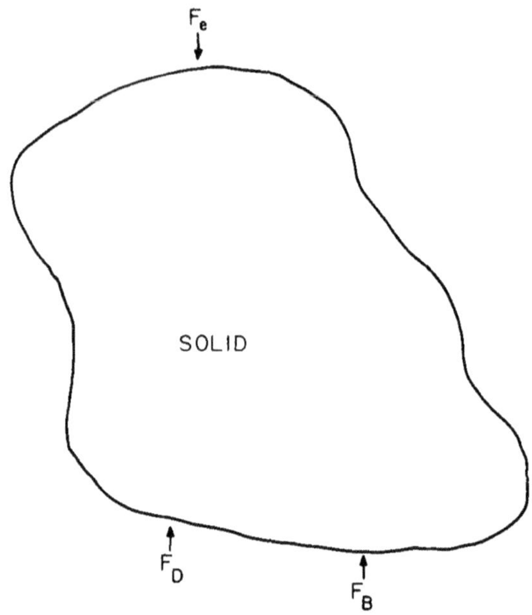


FIGURE 2. Forces acting on a settling particle.

and $a_e$ is the acceleration due to the external field and $\rho$ and $\rho_s$ are the fluid and solid densities, respectively. A force balance on the particle determines that

$$\frac{dV}{dt} = a_e - \frac{\rho\, a_e}{\rho_s} - \frac{C_D\, V^2\, \rho A}{2m} \tag{16}$$

If the external field is gravitational, then $a_e = g$ and Equation 16 becomes

$$\frac{dV}{dt} = g(1 - \rho/\rho_s) - \frac{C_D\, V^2\, \rho A}{2m} \tag{17}$$

If the field is centrifugal, then $a_e = r\,\omega^2$, where $\omega$ is the angular velocity.

Now, if the projected area of the particle is substituted for A and a value for the mass is known, then the terminal velocity can be determined from Equation 16 by setting $dV/dt = 0$ and solving for V. For example, for a spherical particle $A = \pi\, d^2{}_p/4$, $m = \rho_s \cdot \pi/6\, d^3{}_p$ and

$$V_t = \sqrt{\frac{4}{3}\,\frac{(\rho_s - \rho)g\, d_p}{C_D\, \rho}} \tag{18}$$

For very slow flows, Equation 11 can be used for $C_D$ and Equation 18 can be simplified to

$$V_t = \frac{(\rho_s - \rho)g\, d_p^2}{18\, \mu} \tag{19}$$

In practice Equation 18 is frequently used for nonspherical particles, but where $d_p$ and $C_D$ are based on equivalent sphere dimensions.

It is important to recognize that $V_t$, $d_p$, and $C_D$ are all interrelated in Equation 18 (since $C_D = f(d_p,V_t)$) and therefore $V_t$ cannot be determined directly. However, if $d_p$ is known then Equation 18 can be solved for $C_D$, i.e.,

$$\ell n\ C_D = \ell n \left[\frac{4g\ d_p(\rho_s - \rho)}{3\rho}\right] - 2\ \ell n\ V_t \qquad (20)$$

and

$$\ell n\ R_e = \ell n \left(\frac{d_p \rho}{\mu}\right) + \ell n\ V_t \qquad (21)$$

$V_t$ can be eliminated to yield

$$\ell n\ C_D = -2\ \ell n\ R_e + \ell n \left[\frac{4g\ d_p^2\ \rho\ (\rho_s - \rho)}{3\mu^2}\right] \qquad (22)$$

Now on a log-log plot of $C_D$ vs. $R_e$, Equation 22 represents a line with slope $-2$ and passes through the point ($C_D$, 1) where

$$\tilde{C}_D = \frac{4g\ d_p^2\ \rho(\rho_s - \rho)}{3\mu^2}$$

The intersection of this line with the proper $C_D$ vs. $R_e$ curve (i.e., the $C_D$ vs. $R_e$ curve for spheres if dealing with spherical particles, the curve for cylinders if dealing with cylinders, etc.) yields the actual $R_e$ value for these conditions. $V_t$ can be calculated from the definition of the Reynolds number, i.e.,

$$V_t = \frac{R_e \mu}{\rho\ d_p}$$

If $V_t$ is measured instead, then the size of the particle, $d_p$, can be determined by eliminating $d_p$ from Equations 20 and 21 and then following a similar procedure.

So far, only a single particle flowing in an infinite fluid where the size of the particle is much larger than the mean free path of the fluid has been considered. However, as the particle becomes very small (i.e., approaches the mean free path of the fluid) the resistance to motion is decreased since particle-fluid molecule collisions alter the flow and the particle may slip by the fluid molecules unscathed. One way of taking this into account is to use the Cunningham correction factor,[5] i.e.,

$$\tilde{F}_D = F_D/C \qquad (23)$$

where $F_D$ is the uncorrected drag and

$$C = 1 + \frac{2\lambda}{d_p}\,[1.26 + 0.4\ \exp(-2.2\ d_p/\lambda)] \qquad (24)$$

In air systems this correction is important for $d_p \lesssim 1\ \mu m$.

If the fluid is not infinite, then the effects of the wall must be considered. Wall effects, in general, tend to increase the drag. Correction factors are available. Those suggested by Schiller[6] are listed below:

For spheres moving || to a wall:

$$F_w = F_D \left(1 + \frac{9}{32}\frac{d_p}{x}\right) \qquad (25)$$

For spheres moving ⊥ to a wall:

$$F_w = F_D\left(1 + \frac{9}{16}\frac{d_p}{x}\right) \tag{26}$$

For spheres moving along the axis of an infinite cylinder:

$$F_w = F_D\left(1 + \beta/2\,\frac{d_p}{x}\right), \qquad R_e < 1 \tag{27}$$

$$F_w = F_D\left(\frac{1 + \beta^4}{[1 - \beta^2]^2}\right), \qquad R_e > 1000 \tag{28}$$

where x is the tank diameter and β values are as tabulated in the *Chemical Engineers' Handbook.*[7]

In slurry or suspension applications it is necessary to consider the influence of other particles on the system. Under these conditions, the settling properties of the slurry can be quite different than those of an individual particle. Typically, this "hindered settling" is treated by reevaluating the physical properties of the system. The density of the fluid phase becomes the bulk density of the suspension, and is thus increased. The viscosity of the slurry is considerably higher than that of the fluid due to the interaction of the boundary layer around neighboring particles and because of the increased drag caused by the solid particles. At present, the slurry viscosity, $\mu_s$, must be determined experimentally. For one particular slurry of nonclustering spheres, a correlation has been obtained:[8]

$$\frac{\mu_s}{\mu} = \frac{10^{1.82(1-\chi)}}{\chi} \tag{29}$$

where $\chi$ is the volume fraction of liquid.

Ideally, it would be desirable to have a correlation that would allow the calculation of hindered settling rates based only on the physical properties of the solid and liquid; i.e.,

$$V_H = \frac{(\rho_s - \rho)g\,d_p^2}{18\mu}\,C_H \tag{30}$$

where $C_H$ is the correction for hindered settling. At present there are very few correlations of this form. Hindered settling is usually important for solids fractions >0.1%.

## B. Applications

Equipment designed to take advantage of the fact that particles of different size, shape, and density settle at different rates include air pollution cleaning devices such as cyclone separators and particle classifiers such as gravity-settling tanks, elutriators, and Spitzkastens.[9]

To illustrate the fundamental operation of this type of equipment, consider the case where there is a mixture of particles of the same shape and density, but of two different sizes. These particles can be separated by placing the mixture in a rising current of water and adjusting the water velocity to a value between the terminal velocities of the particles. The larger particles will settle out the bottom and the smaller particles will move upward in the water.

Now suppose there is a mixture of two different materials, a and b, of different shapes and sizes. A given separation will then contain a certain range of sizes and shapes of both materials. The range can be calculated by recognizing that any cut in the separation contains particles with the same terminal velocities. Therefore, for the two materials

$$V_t = \sqrt{\frac{4}{3}\frac{(\rho_a - \rho)g\,d_a}{\rho\,C_{D_a}}} = \sqrt{\frac{4}{3}\frac{(\rho_b - \rho)g\,d_b}{\rho\,C_{D_b}}} \tag{31}$$

or

$$\frac{d_a}{d_b} = \left(\frac{\rho_b - \rho}{\rho_a - \rho}\right) \frac{C_{D_a}}{D_{D_b}} \tag{32}$$

where $d_a$ and $d_b$ are equivalent sphere diameters. The size ratio of separation can be controlled by varying the fluid density. For example, as $\rho \rightarrow \rho_b$, $d_a/d_b \rightarrow 0$ and all sizes can be separated. This is the basis of heavy-medium separation. The liquid density can be increased by dissolving a soluble material or by dispersing a heavy solid so finely ground that its settling velocity is negligible.

Equation 32 indicates that the drag coefficient is an important parameter in the design of particle classifiers. In fact, $C_D$ is important in all solids flow and separation processes: the design of sedimentation ponds and continuous thickeners, the calculation of pressure drop in packed beds, slurry pipelines, pneumatic conveyers and fluidized beds, and the design of filtration systems are just a few examples.

## IV. INFLUENCE OF SHAPE ON $C_D$

As stated previously, the drag coefficient is a function of Reynolds number and particle shape. The shape characteristics of a solid particle can exert a profound influence on its ability to absorb momentum from a moving fluid stream and, therefore, can greatly affect the settling velocity. For example, particles with shapes markedly different than spherical settle with velocities which are only a fraction of that for a sphere with the same density and volume.

Although hundreds of experiments measuring settling velocities of different shaped particles have been conducted, a correlation which can accurately predict the settling velocity from knowledge of the fluid and shape parameters for both regularly and irregularly shaped particles has not been found. The primary difficulty in obtaining such a correlation and in assessing the effect of shape on the terminal velocity has been the inability to systematically and quantitatively describe shape itself.

At least 13 different shape parameters have been used in analyzing the effect of shape on settling velocity (see Table 1). These quantities include sphericity, circularity, flatness factor, and the roundness factor. These parameters have been useful in obtaining first-order correlations for the settling velocity for regularly shaped particles of the same shape class (e.g., spheroidal, nonisometric-nonspheroidal, etc.), but are of limited applicability for particles of a different shape class or orientation or irregularly shaped particles (see Table 2).

However, recent advances in particle characterization now make it possible to describe particle shape uniquely and quantitatively. These developments should facilitate the development of improved models for predicting the shape dependency of the drag coefficient. Work along these lines is ongoing. Reported here are results for predicting $C_D$ for slow flows (i.e., $R_e < 0.1$).

### A. Data

Data from three experimental studies were selected for analysis.[11,13,14] In these experiments, the settling velocities of isometric particles consisting of tetrahedrons, cubes, cube octahedrons, and spheres, and nonisometric particles of cylinders, square prisms, double cones, and spheroids of different aspect ratios and settling orientation positions were measured, and the drag coefficients and Reynolds numbers calculated. Different particle material, fluids, and experimental techniques were utilized.

The experimental data were reported and analyzed according to the model

$$C_D = \frac{24}{R_e \cdot K} \tag{33}$$

**Table 1**
**SHAPE FACTORS USED IN CORRELATING THE DRAG COEFFICIENT AND THE REYNOLDS NUMBER**

| | |
|---|---|
| $\alpha = \frac{A_n}{A_t}$ | $= \frac{\text{normal projected area of particle}}{\text{normal projected area of equivalent sphere}}$ |
| $\beta = \frac{A_t}{A_n}$ $A_t$ | $=$ surface area tangential to the flow |
| $K = \frac{U}{U_s}$ | $= \frac{\text{settling velocity of particle}}{\text{settling velocity of sphere of equal volume}}$ |
| $\frac{\ell}{s}$ | $= \frac{\text{length taken} \perp \text{to its max. circular or square section}}{\text{dia. of max. circular or square section}}$ |
| $\lambda$, $\phi$ (circularity) | $= \frac{\text{circumference of circle with same projected area}}{\text{actual circumference of projected area}}$ |
| $k = \frac{\pi/6\ d_p^3}{d_n^3}$ | $= \frac{\text{volume of particle}}{\text{volume of sphere}}$ |
| $d_c$ | $=$ diameter of circle having same $A_n$ |
| $d_s$ | $=$ diameter of sphere of equal volume |
| $\psi$ (sphericity) | $= \frac{\text{surface area of sphere of equal volume}}{\text{surface area of actual particle}}$ |
| m (mean hydraulic radius) | $\frac{A_n}{\text{perimeter of projected area}}$ |
| area ratio | $= \frac{\text{projected area}}{\text{area of circumscribing circle}}$ |
| prismoidal ratio | $= \frac{\text{mean thickness}}{\text{max. thickness}}$ |
| F (flatness factor) | $= \frac{b + c}{2a}$, a, b, c shortest → longest axis |
| $R_c$ (roundness factor) | $= \frac{\text{radius of sharpest corner}}{\text{longest diameter}}$ |
| SF (Corey's shape factor) | $= \frac{a}{(b \cdot c)^{2.5}}$ |

where K is the shape correction term and is equal to $V_t/V_{ts}$. For spherical particles, $K = 1$ and Equation 32 reduces to the Stokes equation (Equation 11). The data from these studies are presented in Table 3. In all instances each point represents the average response of a large number of repeated experiments. The reported variations within a given experiment were quite small (on the order of a few percent) and all reported K values were corrected for wall effects. The drag coefficient was evaluated based on the particle projected area normal to the flow direction and the Reynolds number based on the equivalent sphere diameter.

### B. Analysis

The model represented by Equation 33 is appealing since the influence of shape is lumped into a single parameter, K, and geometrically similar particles have the same K value. This is shown in Figure 3. Geometrically similar particles of different size and density are shown to have similar K values. (Each datum point represents a different particle with particles of

**Table 2**
**EMPIRICAL EQUATIONS FOR USE IN PREDICTING THE DRAG COEFFICIENT IN THE STOKES REGION[a]**

| Expression | Comments | Ref. |
|---|---|---|
| $C_D = AR_e^{-1} + BRe^{-0.5} + C$ | $Re \leq 10$ | 10 |
| $A = 24.66\ SF^{-0.157}$ | | |
| $B = 4.07\ SF^{-0.0162}$ | | |
| $C = 0.49\ SF^{-1.399}$ | | |
| $C_D = \dfrac{24}{K \cdot Re}$ | $Re \leq 1$ | 11 |
| $K = 0.843\ \ln\{\psi/0.065\}$ | $0.4 \leq SF \leq 0.8$ | 12 |
| $K = 0.946\ SF^{-0.378}$ | $SF < 0.4$ | 12 |
| $K = 2.18 - 2.09\ SF$ | | |
| $K = \dfrac{\alpha \cdot \beta}{\{0.811/\lambda^{1.56}\} + 0.089}$ | spheroids | 13 |
| $\log_{10}(K) = 0.25\sqrt{\phi d_c/d_s}\,(d_c/d_s - 1) + \log_{10}(d_c/d_s\sqrt{\phi})$ | $\lambda$ varies with orientation | 14 |
| $\log_{10}(K) = \dfrac{-0.270}{\sqrt{\psi}\,(d_c/d_s)^{0.345}}(d_c/d_s - 1) + \log_{10}(d_c/d_s\sqrt{\phi})$ | $\lambda$ constant | 14 |

[a] See Table 1 for definition of shape parameters.

the same number having the same shape.) Note that there is no apparent dependency of K on $R_e$.

Some general observations can be made from these data. From Malaika's data[13] it is seen that, for particles with aspect ratio 4, the spheroid has the lowest drag (i.e., the highest K value), followed by the cylinder, prism, and double cone, respectively. For aspect ratio 1, the ranking followed the series sphere, cylinder, double cone, and prism, while for aspect ratio 1/4 the progression was cylinder, sphere, prism, and double cone.

For all orientations, the drag increases as the aspect ratio deviates further from 1. For example, K values for cylinders settling roundside decrease as the length of the cylinder increases for a fixed cross-sectional area. In addition, for all represented geometries the drag decreases when the particle is rotated to the edgewise position (e.g., a cylinder falling flat-end first). It should be noted that in very slow flows, when particles are symmetric about three mutually perpendicular planes (as are those in Table 3), the particles are stable in any settling position. However, for higher Reynolds numbers ($0.1 < R_e$) particles tend to orient themselves with the largest cross section in a position normal to the direction of motion.

The importance of tangential area can be seen in the fact that for constant normal cross-section-area objects, the drag increases as the characteristic length in the direction of motion increases.

The observed values of K are plotted against some commonly used shape parameters in Figures 4 to 7. It is apparent from these figures that these shape factors alone cannot provide adequate correlations for K. However, useful correlations based on a combination of these parameters have been developed (see Table 2). In most cases, these correlations were developed based on limited data sets and the accuracy of the predicted values usually decrease when the orientation of the particle is changed or if applied to different types of particles.

**Table 3**
**DATA SET DESCRIPTION**

| Particle # | Description[a] | K-measured | Aspect ratio |
|---|---|---|---|
| 1 | Circular cylinder | 0.78 | 4 |
| 2 | Circular cylinder | 0.96 | 1 |
| 3 | Circular cylinder | 0.76 | 0.25 |
| 4 | Square prism | 0.76 | 4 |
| 5 | Square prism | 0.92 | 1 |
| 6 | Square prism | 0.71 | 0.25 |
| 7 | Double cone | 0.75 | 4 |
| 8 | Double cone | 0.94 | 1 |
| 9 | Double cone | 0.67 | 0.25 |
| 10 | Spheroid | 0.79 | 4 |
| 11 | Spheroid | 1.00 | 1 |
| 12 | Spheroid | 0.73 | 0.25 |
| 13 | Rectangular parallelopiped | 0.72 | 0.25 |
| 14 | Square projected area | 0.84 | 0.50 |
| 15 | Square projected area | 0.93 | 1 |
| 16 | Square projected area | 0.96 | 2 |
| 17 | Square projected area | 0.97 | 3 |
| 18 | Square projected area | 0.92 | 4 |
| 19 | Rectangular parallelopiped | 0.87 | 0.25 |
| 20 | Rectangular projected area | 0.93 | 0.5 |
| 21 | Rectangular projected area | 0.93 | 1 |
| 22 | Rectangular projected area | 0.86 | 2 |
| 23 | Rectangular projected area | 0.82 | 3 |
| 24 | Rectangular projected area | 0.76 | 4 |
| 25 | Circular cylinders | 0.91 | 0.25 |
| 26 | Roundwise fall | 0.96 | 0.5 |
| 27 | Roundwise fall | 0.94 | 1 |
| 28 | Roundwise fall | 0.88 | 2 |
| 29 | Roundwise fall | 0.81 | 3 |
| 30 | Roundwise fall | 0.76 | 4 |
| 31 | Circular cylinders | 0.76 | 0.25 |
| 32 | Edgewise fall | 0.86 | 0.5 |
| 33 | Edgewise fall | 0.96 | 1 |
| 34 | Edgewise fall | 0.97 | 2 |
| 35 | Edgewise fall | 0.96 | 3 |
| 36 | Edgewise fall | 0.93 | 4 |
| 37 | Spheroids | 0.91 | 0.25 |
| 38 | Edgewise fall | 0.96 | 4 |
| 39 | Cube octahedron | 0.97 | 1 |
| 40 | Octahedron | 1.10 | 1 |
| 41 | Tetrahedron | 1.22 | 1 |

*Note:* Description is given as shape and orientation. For Malaika's data all orientations are with the maximum projected area perpendicular to the flow.

[a] Particles 1 to 12 and 37 and 38 are from Malaika[13]; 13 to 36 from Heiss and Coull;[11] and 39 to 41 from Pettyjohn and Christiansen.[14]

### C. A New Correlation

The data given in Table 3 represent many different types of particles as well as orientation effects, and therefore provide a meaningful test set. Based on this data set, a model for K in terms of the new morphological descriptors developed by Luerkens (this volume) was determined.

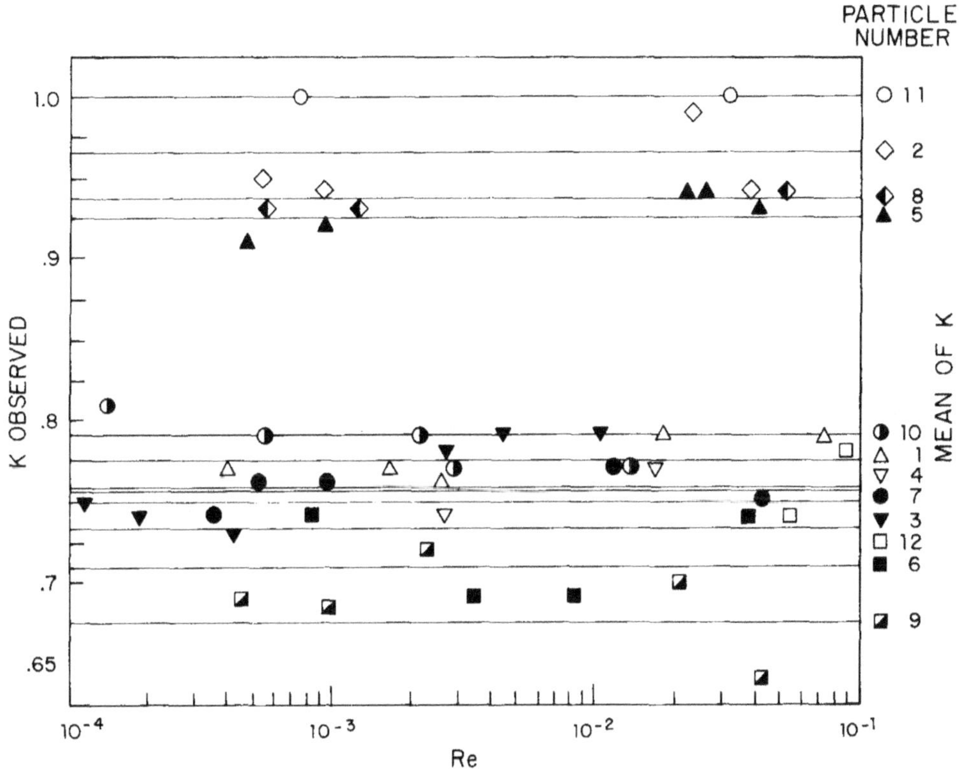


FIGURE 3. Observed K values plotted against Reynolds number (data from Reference 13).

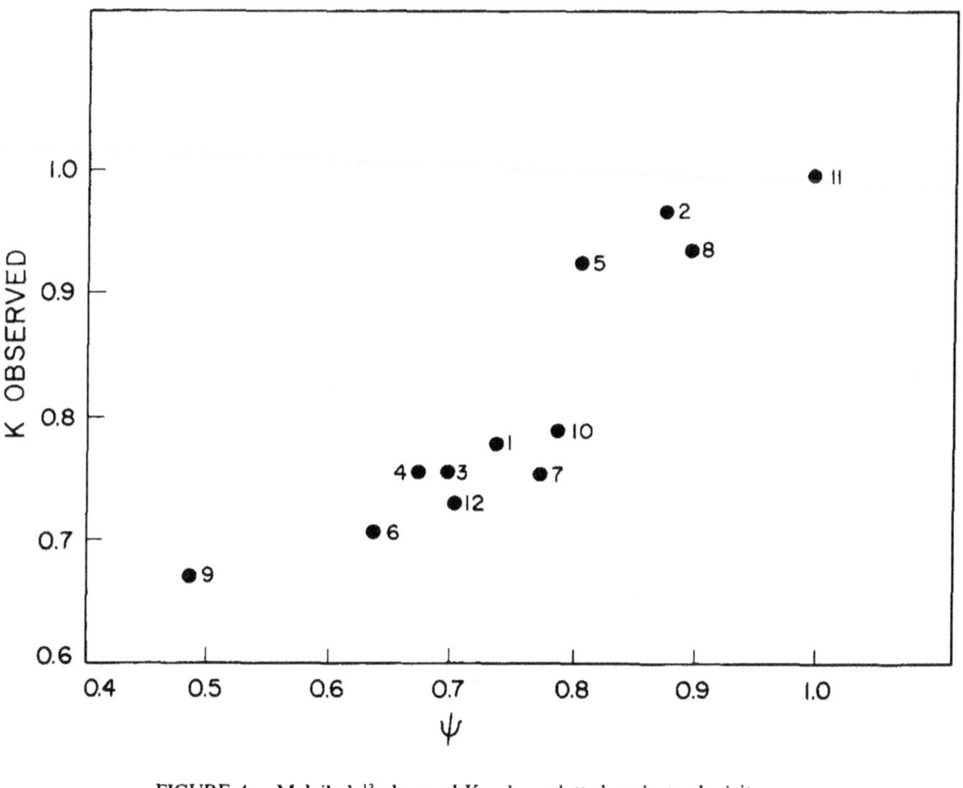


FIGURE 4. Malaika's[13] observed K values plotted against sphericity.

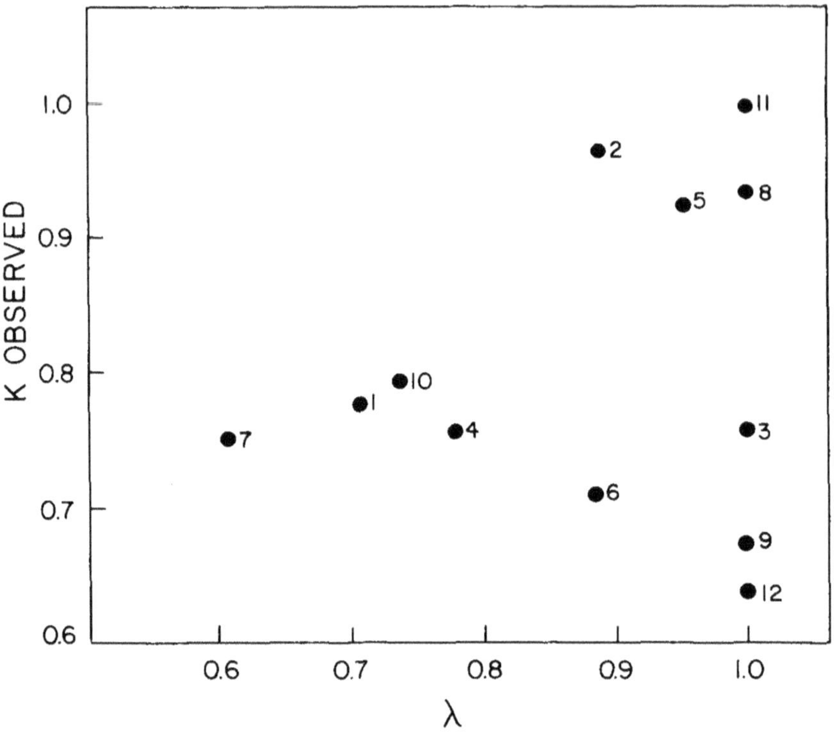


FIGURE 5. Heiss and Coull's[11] observed K values plotted against sphericity.

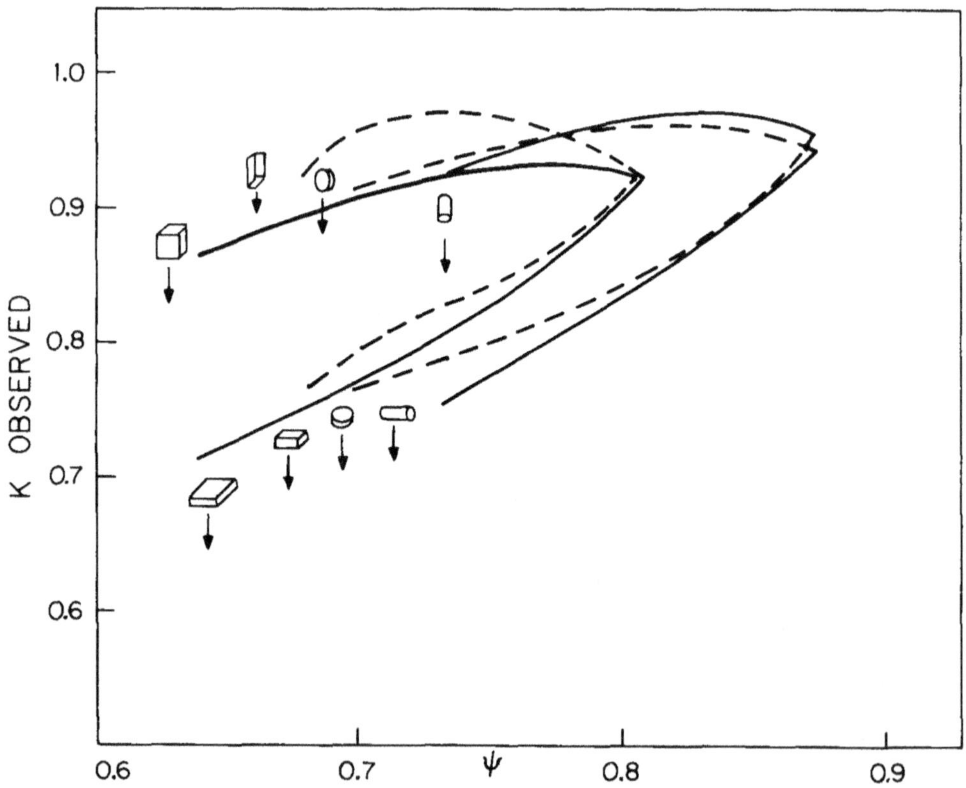


FIGURE 6. Malaika's[13] observed K values plotted against circularity.

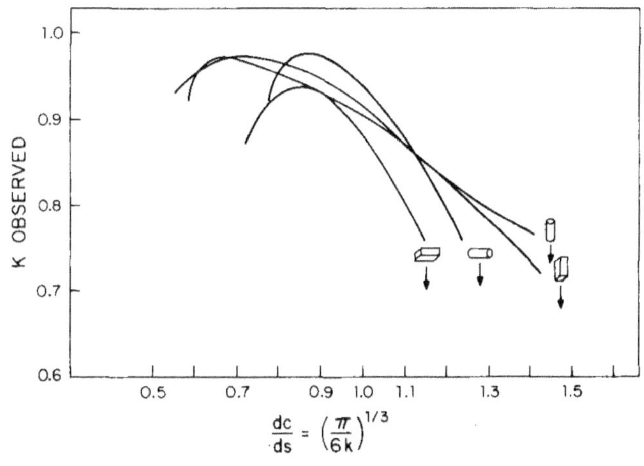


FIGURE 7. Observed K values plotted against k.

## Table 4
## MORPHOLOGICAL DESCRIPTORS OF THE NORMAL AND TANGENTIAL PROJECTED AREAS FOR THE PARTICLES DESCRIBED IN TABLE 3 AND THE PREDICTED K VALUES

| Particle | $d_c/d_s$ | $L_0^N/L_0^T$ | $\Sigma L_2^N$ [a] | $\Sigma L_2^T$ | $\Sigma L_3^T$ | K predicted |
|---|---|---|---|---|---|---|
| 1 | 1.24 | 0.86 | 0.248 | 0.000 | 0.0000 | 0.79 |
| 2 | 0.99 | 0.99 | 0.011 | 0.000 | 0.0000 | 0.92 |
| 3 | 1.39 | 1.17 | 0.000 | 0.270 | 0.1520 | 0.77 |
| 4 | 1.36 | 0.92 | 0.171 | 0.011 | 0.0009 | 0.76 |
| 5 | 1.20 | 1.03 | 0.001 | 0.057 | 0.0008 | 0.95 |
| 6 | 1.44 | 1.16 | 0.011 | 0.260 | 0.1340 | 0.73 |
| 7 | 1.27 | 0.90 | 0.184 | 0.000 | 0.0000 | 0.81 |
| 8 | 1.26 | 1.01 | 0.000 | 0.011 | 0.0009 | 0.86 |
| 9 | 2.25 | 1.10 | 0.000 | 0.176 | 0.1400 | 0.64 |
| 10 | 1.26 | 0.88 | 0.221 | 0.000 | 0.0000 | 0.80 |
| 11 | 1.00 | 1.00 | 0.000 | 0.000 | 0.0000 | 0.95 |
| 12 | 1.59 | 1.13 | 0.000 | 0.220 | 0.1240 | 0.73 |
| 13 | 1.44 | 1.15 | 0.011 | 0.249 | 0.1300 | 0.73 |
| 14 | 1.15 | 1.04 | 0.011 | 0.085 | 0.0058 | 0.86 |
| 15 | 0.92 | 1.00 | 0.011 | 0.011 | 0.0009 | 0.90 |
| 16 | 0.72 | 1.04 | 0.011 | 0.085 | 0.0058 | 0.97 |
| 17 | 0.63 | 1.09 | 0.011 | 0.170 | 0.0478 | 0.93 |
| 18 | 0.57 | 1.15 | 0.011 | 0.249 | 0.1300 | 0.93 |
| 19 | 0.72 | 0.87 | 0.249 | 0.011 | 0.0009 | 0.88 |
| 20 | 0.81 | 0.96 | 0.085 | 0.011 | 0.0009 | 0.90 |
| 21 | 0.92 | 1.00 | 0.011 | 0.011 | 0.0009 | 0.90 |
| 22 | 1.02 | 0.96 | 0.085 | 0.011 | 0.0009 | 0.85 |
| 23 | 1.09 | 0.91 | 0.170 | 0.011 | 0.0009 | 0.81 |
| 24 | 1.14 | 0.87 | 0.249 | 0.011 | 0.0009 | 0.77 |
| 25 | 0.78 | 0.86 | 0.249 | 0.000 | 0.0000 | 0.89 |
| 26 | 0.80 | 0.95 | 0.085 | 0.000 | 0.0000 | 0.94 |
| 27 | 0.99 | 0.99 | 0.011 | 0.000 | 0.000 | 0.92 |
| 28 | 1.11 | 0.95 | 0.085 | 0.000 | 0.0000 | 0.87 |
| 29 | 1.18 | 0.91 | 0.170 | 0.000 | 0.0000 | 0.83 |
| 30 | 1.24 | 0.86 | 0.249 | 0.000 | 0.0000 | 0.79 |
| 31 | 1.39 | 1.15 | 0.000 | 0.249 | 0.1300 | 0.77 |
| 32 | 1.10 | 1.05 | 0.000 | 0.085 | 0.0058 | 0.91 |
| 33 | 0.87 | 1.00 | 0.000 | 0.011 | 0.0009 | 0.94 |
| 34 | 0.69 | 1.05 | 0.000 | 0.085 | 0.0058 | 1.03 |
| 35 | 0.61 | 1.10 | 0.000 | 0.170 | 0.0478 | 0.98 |
| 36 | 0.55 | 1.15 | 0.000 | 0.249 | 0.1300 | 0.98 |
| 37 | 0.80 | 0.88 | 0.220 | 0.000 | 0.0000 | 0.90 |
| 38 | 0.63 | 1.13 | 0.000 | 0.221 | 0.1240 | 0.93 |
| 39 | 0.97 | 1.00 | 0.011 | 0.011 | 0.0009 | 0.88 |
| 40 | 1.08 | 0.98 | 0.043 | 0.016 | 0.0003 | 0.94 |
| 41 | 1.22 | 1.00 | 0.043 | 0.043 | 0.0097 | 0.76 |

[a] Zero values are actually $1 \times 10^{-10}$, which is the resolution limit of image analyzing instrument.

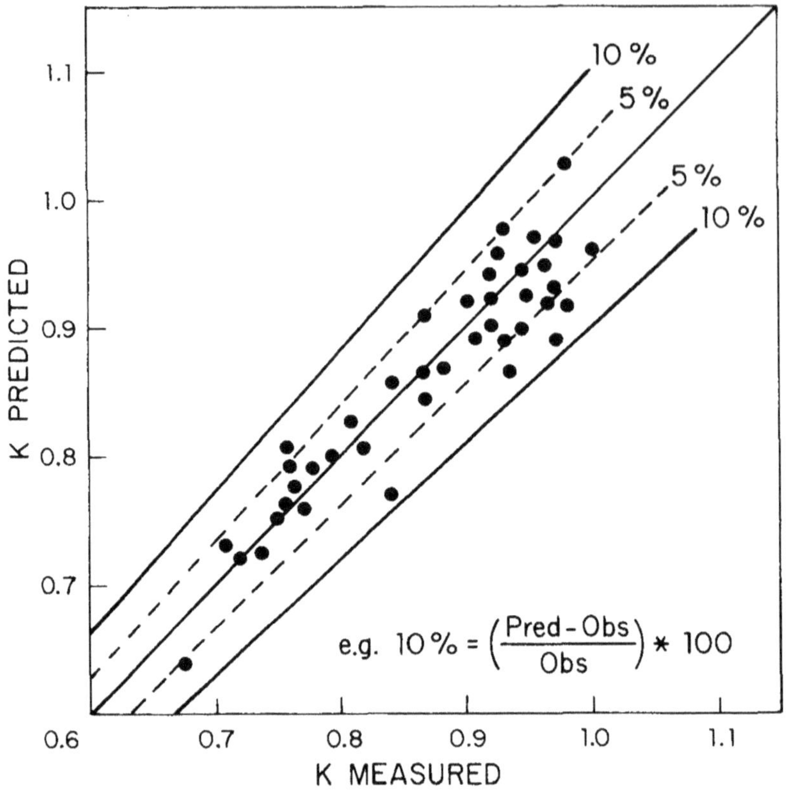


FIGURE 8. K values predicted by Equation 34 against observed K values.

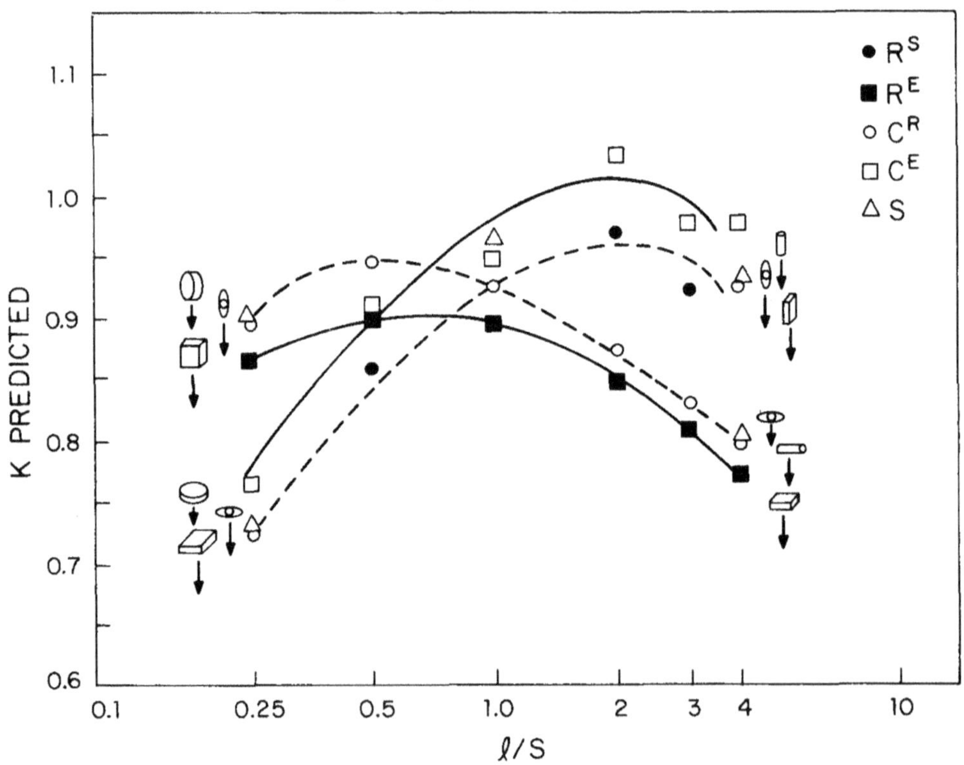


FIGURE 9. K values predicted by Equation 34 for various particle shapes and orientations as a function of aspect ratio.

The normal and tangential projected profiles of each particle were analyzed and the corresponding morphological descriptors are tabulated in Table 4. It was found that K can be estimated from the expression

$$K = 0.69 \left(\frac{d_c}{d_s}\right)^{-0.264} \left(\frac{L_0^N}{L_0^T}\right)^{0.581} \left(\Sigma L_2^N\right)^{-0.002} \left(\Sigma L_2^T\right)^{0.060} \left(\Sigma L_3^T\right)^{-0.072} \tag{34}$$

All the parameters in this equation except $d_s$ are determined directly from the particle profiles.

Values of K predicted from Equation 34 are plotted against observed values in Figure 8. The predicted K values, as a function of aspect ratio for different orientations, are presented in Figure 9. These trends are also present in the measured data. This figure demonstrates that not only can this model predict K values accurately, but it can also model the influence of particle orientation.

## ALPHABETICAL SYMBOLS*

| | |
|---|---|
| A | Frontal area of body exposed to the flow direction |
| $a_e$ | Acceleration due to external field |
| C | Cunningham correction factor |
| $C_D$ | Coefficient of drag |
| $C_H$ | Correction factor for hindered settling |
| $C_L$ | Coefficient of lift |
| d | Characteristic diameter |
| $d_c$ | Diameter of circle having some normal area |
| $d_s,d_p$ | Diameter of sphere of equal volume |
| D | Drag force |
| F | Force exerted by fluid |
| $F_B,F_D,F_e$ | External, drag, and buoyant forces, respectively |
| $\tilde{F}_D$ | Corrected drag force |
| $F_w$ | Drag force corrected for wall effects |
| g | Acceleration due to gravity |
| K | Shape correction term |
| L | Lift force |
| m | Particle mass |
| N | Normal component of F |
| $P,P_\infty$ | Pressure |
| r | Radial distance |
| R | Spherical particle radius |
| $R_e$ | Reynolds number |
| u,v,w | Velocities in the x, y, and z directions, respectively |
| $U_i$, $U_\infty$ | ith Component and free stream velocities, respectively |
| V | Fall velocity |
| $V_H$ | Terminal velocity in hindered settling |
| $V_t$ | Terminal settling velocity |
| X | Tank diameter |

* Shape factors are defined in Table 1.

## GREEK SYMBOLS

| | |
|---|---|
| $\beta$ | Correction constants for wall effects |
| $\lambda$ | Mean free path of fluid, also used as the circularity shape factor |
| $\mu$ | Fluid dynamic viscosity |
| $\mu_s$ | Slurry viscosity |
| $\nu$ | Fluid kinematic viscosity |
| $\rho$ | Fluid density |
| $\rho_s$ | Particle density |
| $\tau$ | Tangential component of F |
| $\chi$ | Volume fraction of liquid |
| $\omega$ | Angular velocity |

## REFERENCES


1. **Schlichting, H.,** *Boundary Layer Theory,* 6th ed., McGraw-Hill, New York, 1968, 17.
2. **Stokes, G. G.,** On the effect of internal friction of solids on the motion of pendulums, *Trans. Cambridge Philos. Soc.,* 9(II), 8, 1951.
3. **Oseen, C. W.,** Über die Stokessche Formel und über die verwandtle Aufgabe in der Hydrodynamics, *Ark. Mat. Astron. Fys.,* 6, #29, 1910.
4. **Lamb, B. H.,** *Hydrodynamics,* 6th ed., Cambridge University Press, Cambridge, 1945.
5. **Cunningham, E.,** *Proc. R. Soc. London Ser. A,* 83, 357, 1910.
6. **Schiller, L.,** *Handbook of Experimental Physics,* Leipzig, 4, 342, 1932.
7. **Perry, R. H. and Chilton, C. H.,** *Chemical Engineers' Handbook,* 5th ed., McGraw-Hill, New York, 1973, 5.
8. **Brown, G. G.,** *Principles of Unit Operations,* Foust, A. S., Ed., John Wiley & Sons, New York, 1950, chap. 18.
9. **Foust, A. S., Wenzel, L. A., Clump, C. W., Maus, L., and Anderson, L. B., Eds.,** *Principles of Unit Operations,* John Wiley & Sons, New York, 1980, chap. 22.
10. **Brezina, J.,** Granometric, D-6903 Neckargemund, 1979.
11. **Heiss, J. F. and Coull, J.,** The effect of orientation and shape on the settling velocity of non-isometric particles in a viscous medium, *Chem. Eng. Progr.,* 48, 133, 1952.
12. **Komar, P. D. and Reimers, C. E.,** Grain shape effects on settling rates, *J. Geol.,* 86, 193, 1978.
13. **Malaika, J. W.,** Effect of Shape of Particles on Their Settling Velocity, Ph.D. thesis, State University of Iowa, Iowa City, 1949.
14. **Pettyjohn, E. S. and Christiansen, E. B.,** Effect of particle shape on free settling rates of isometric particles, *Chem. Eng. Prog.,* 44, 157, 1948.

Chapter 17

# ON RELATING BULK PROPERTIES TO PARTICLE MORPHOLOGY OF SELECTED GRAINS*

**Chi-Ren Chang, John Keith Beddow, Ming-Jau Yin, Arthur F. Vetter, Gordon Butters, and David L. O. Smith**

TABLE OF CONTENTS



* The authors are grateful to the National Science Foundation Particulate Multiphase Processing Program for partial support under grant CPE-80-23868. The authors are also grateful to the International Fine Particle Research Institute.

# I. INTRODUCTION

Experimental results relating bulk properties to newly developed shape terms are presented in this chapter: Hausner ratio (H.R.), processing parameter, flow time, angle of internal friction, and void ratio of several materials.

Although it is well known that particle shape affects the properties of bulk material, reports on the relationship between bulk properties and particle shape are not extensive. The major problem has been the lack of an appropriate quantity to represent "shape". The theory of morphological analysis and some of its applications are dealt with in these volumes.

In the studies described here, attempts were made to use Fourier descriptors to relate particle shape to the bulk properties (angle of internal friction, H.R., flow time, void ratio). The results are very encouraging.

# II. RELATIONSHIPS BETWEEN THE BULK PROPERTIES OF SELECTED AGRICULTURAL GRAINS AND THEIR SHAPE

In designing storage bins, two major problems are present: ensuring a satisfactory flow of the material through the bin and removing the material successfully from the bottom of the bin. Thus, grain morphology is of primary importance in bin behavior. There is an excellent reference work available on this subject.[1]

Corn, wheat, barley, oats, and soybean samples were subjected to morphological analysis using the Particle Image Analyzing System (PIAS). Empirical relationships were obtained which correlate grain morphic feature and bulk properties, such as void ratio and angle of internal friction.

## A. Void Ratio of Five Different Grains (Corn, Wheat, Barley, Oat, Soybean)

Table 1 shows the void ratio and morphic data. It was assumed that the correlation will take the form of Equation 1:

$$V.R. = B + M(IRR) \qquad (1)$$

In Equation 1, the correlation obtained based on linear least square was found to have a B value of −2.6513 and an M value of 3.0770. Figure 1 shows a plot of the calculated void ratio from Equation 1 vs. experimental values.

## B. Angle of Internal Friction of Selected Agricultural Grains

Table 2 lists the morphic data and angle of internal friction of grains. A correlation equation of the form of Equation 2 was assumed

$$\theta = A + B \times EXP(C \times IRR) \qquad (2)$$

and the values obtained for A, B, C are

$A = -1.35 \times 10^3$
$B = 1.34 \times 10^3$
$C = 0.23 \times 10^{-1}$

Table 3 lists the experimental data, calculated data, and square error; the range of error is between 0.4 and 7.8%.

From Equations 1 and 2 we find that high irregularity results not only in high void ratio but also in high angle of internal friction. This is contrary to expectation and is believed to

**Table 1**
**VOID RATIO AND MORPHIC DATA OF SELECTED AGRICULTURAL GRAINS**

| Grain | Void ratio | $R_0$ (mm) | $A_0$ (mm) | IRR |
|---|---|---|---|---|
| Corn | 0.575 | 5.11 | 5.01 | 1.04 |
| Wheat | 0.560 | 2.24 | 2.16 | 1.08 |
| Barley | 0.858 | 2.63 | 2.50 | 1.11 |
| Oat | 1.260 | 2.51 | 2.23 | 1.27 |
| Soybean | 0.455 | 3.22 | 3.22 | 1.00 |

*Note:* $\text{Void ratio} = \dfrac{\text{Particle density} - \text{bulk density}}{\text{bulk density}}$

$R_0$: equivalent radius.
$A_0$: mean radius.
$IRR = (R_0/A_0)^2$, defined as irregularity.

be due to the fact that oat and barley are much rougher than wheat and corn (see Figure 2). Because of this roughness, even in low density packing, oat and barley have a high value of angle of internal friction.

## III. RELATIONSHIPS BETWEEN THE BULK PROPERTIES OF PLASTIC POWDERS AND THEIR SHAPE

The selected agricultural grains listed in the previous section are not the same material, which might affect the relationships being found. So, it was decided to use plastic powders made from different processes.*

### A. Processing Parameter of PVC Powders

A processing parameter of six PVC powders was studied. Figure 3 shows two of the samples used, and Table 4 lists the morphic data and the value of the processing parameter of the samples. A correlation equation of the form of Equation 3 was assumed

$$P = k(R_0)^x(PER)^y(RD\ C)^z(SKE)^u \tag{3}$$

and the values obtained for k, x, y, z, and u are

$k = 2.058 \times 10^4$
$x = -0.871$
$y = -7.54$
$z = -0.553$
$u = 0.666$

Figure 4 is a plot of calculated values against observed values of this processing parameter.

### B. Angle of Internal Friction of Polyethylene Powders

The angle of internal friction of 19 polyethylene powders (PE) from different processes were studied. Figure 5 shows some of these powders. Table 5 summarizes the morphic data

* The processing parameter is not ready to be fully disclosed at this time.

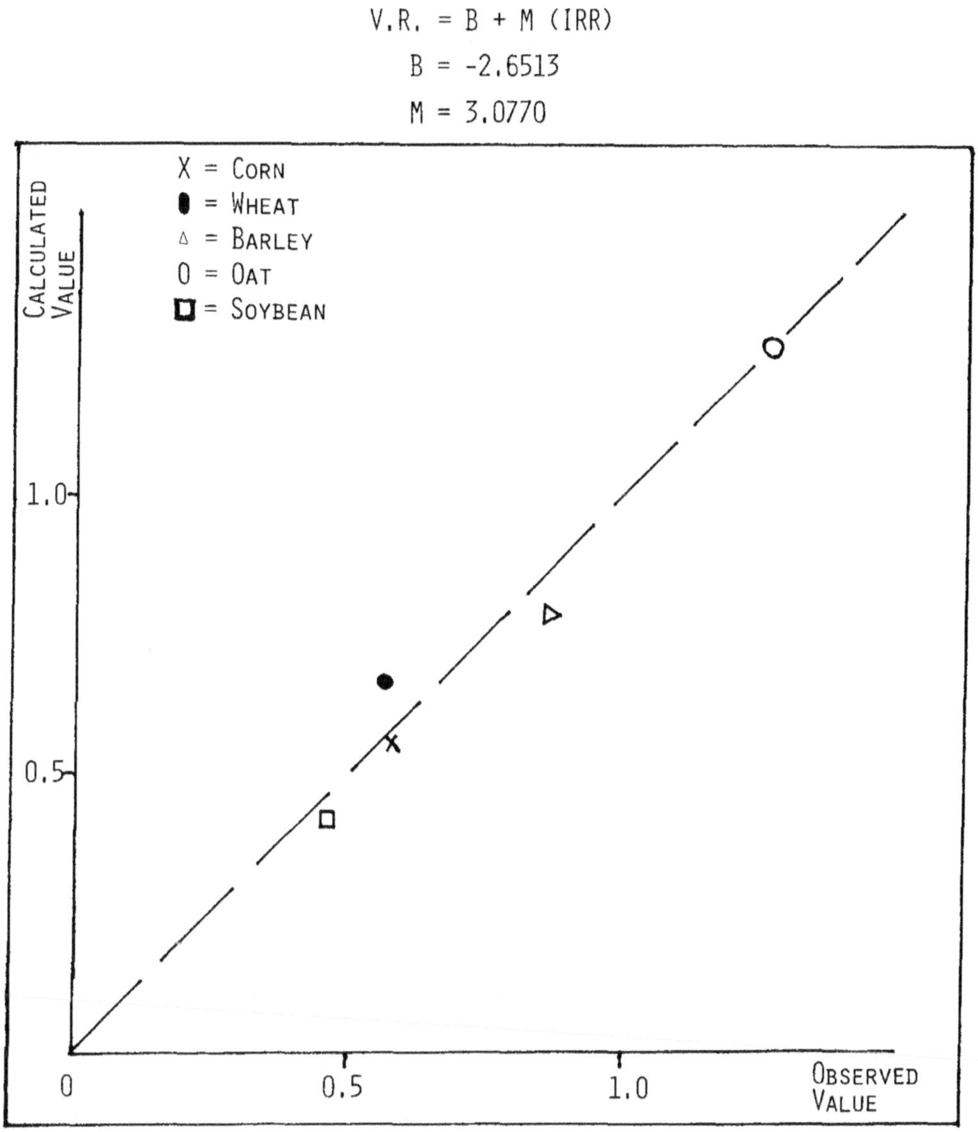


FIGURE 1. Void ratio of grains.

and angle of internal friction of the 19 PE powders. It was assumed that the correlation will take the form of Equation 4:

$$\tan \theta = k(R_0)^x(L_0)^y(PER)^z(RD\ C)^u(SKE)^v \tag{4}$$

where $R_0$ is in μm and θ is in degrees.

In Equation 4, k is considered to be a variable of the condition of experiment and the nature of the material only, and since the samples represent different sizes and shapes of PE powders, k in this case is considered to be a constant. The best correlation obtained based on linear least square was found to have the following k, x, y, z, u, and v values.

**Table 2**
**ANGLE OF INTERNAL FRICTION AND MORPHIC DATA OF SELECTED AGRICULTURAL GRAINS**

| Grain | Friction angle (degree) | IRR |
|---|---|---|
| Corn | 26.0 | 1.04 |
| Wheat | 25.8 | 1.08 |
| Barley | 30.9 | 1.11 |
| Oat | 32.3 | 1.27 |

**Table 3**
**EXPERIMENTAL AND CALCULATED VALUES OF ANGLE OF INTERNAL FRICTION OF AGRICULTURAL GRAINS**

| Grain | Exp. data ($\times 10^2$) | Cal. data ($\times 10^2$) | Square error |
|---|---|---|---|
| Corn | 0.260 | 0.261 | $0.212 \times 10^{-1}$ |
| Wheat | 0.258 | 0.275 | $0.287 \times 10^{1}$ |
| Barley | 0.309 | 0.285 | $0.554 \times 10^{1}$ |
| Oat | 0.323 | 0.328 | $0.254 \times 10^{0}$ |

$k = 3.485 \times 10^3$
$x = -0.0619$
$y = 126.7$
$z = 0.7877$
$u = 1.905$
$v = -0.03951$

Figure 6 shows a plot of calculated angle of internal friction from Equation 4 vs. experimental values. Among the 19 samples, 1 has an error of more than 10% and 6 are off by more than 5%. It was noted that the PE powders have a rather broad size distribution and this might affect the property. Therefore, it was decided to take size distribution into account, and fewer variables were used in correlation. Table 6 summarizes the morphic data used and angle of internal friction of the 19 PE powders. It was assumed that the correlation will take the form of Equation 5:

$$\tan \theta = (k_1 R_0^{1/3} + k_2 L_0^x)(1 + k_3 \sigma) \quad (5)$$

where $k_1 \simeq 0$, $k_2 = 0.50$, $k_3 = 1.88$, $x = 6.9$, and $\sigma$ = the standard deviation of the size distribution.

The correlation constant for $R_0$, $k_1$ being zero implies that, within the tested range, the angle of internal friction is not dependent on size. Figure 7 shows the result.

## IV. BULK PROPERTIES OF SANDS

The plastic powders in the previous section all have quite a broad size distribution which might complicate the relationships being sought and distort the results. In order to reduce the number of variables which might affect the bulk properties, it was decided to use different sources of silica sands (to give different shapes), sieved to a very narrow size range. The

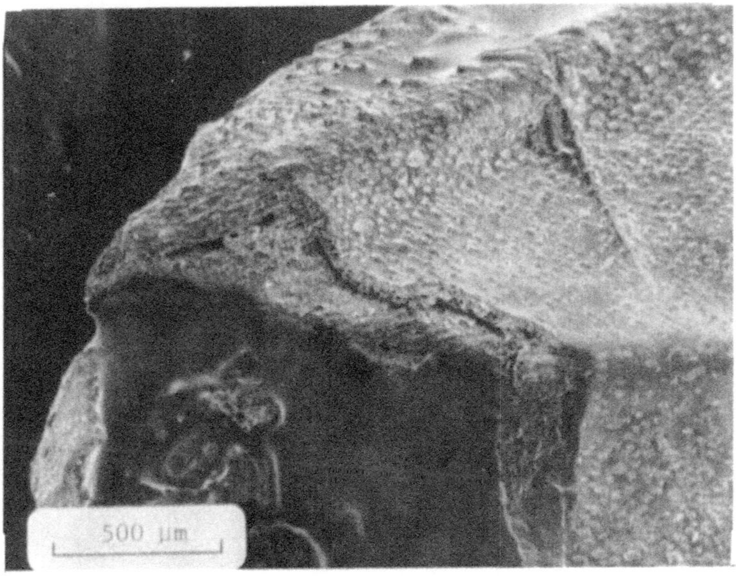


A

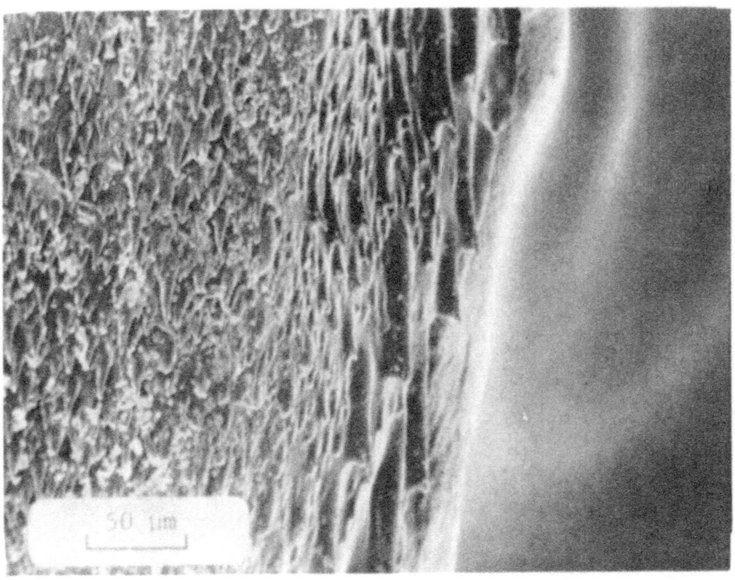


B

(A) Barley. (Magnification × 50.) (B) Barley. (Magnification × 300.)

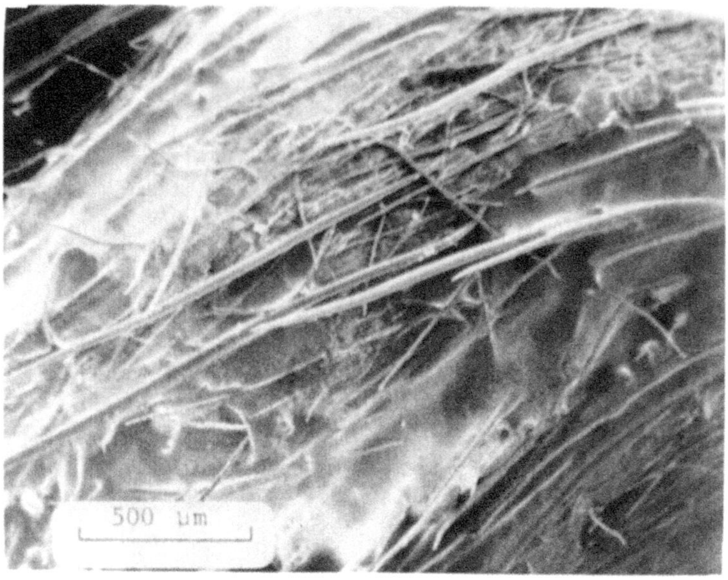


C

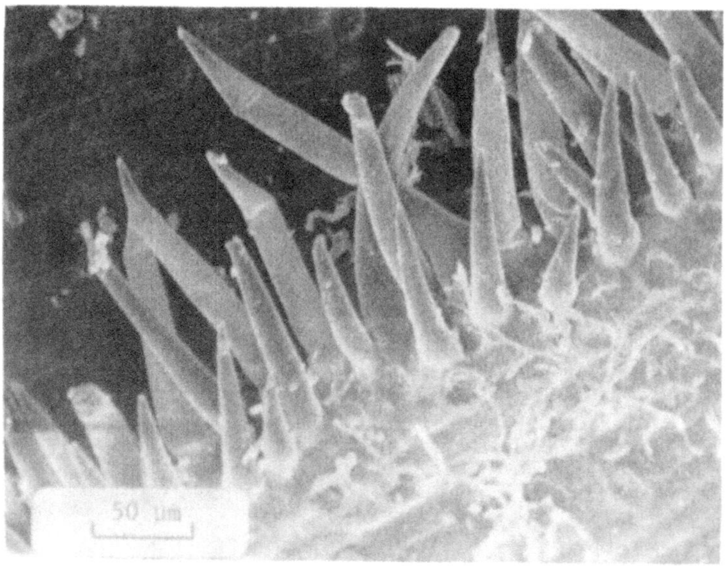


D

FIGURE 2. (C) Oat. (Magnification × 50.) (D) Oat. (Magnification × 300.)

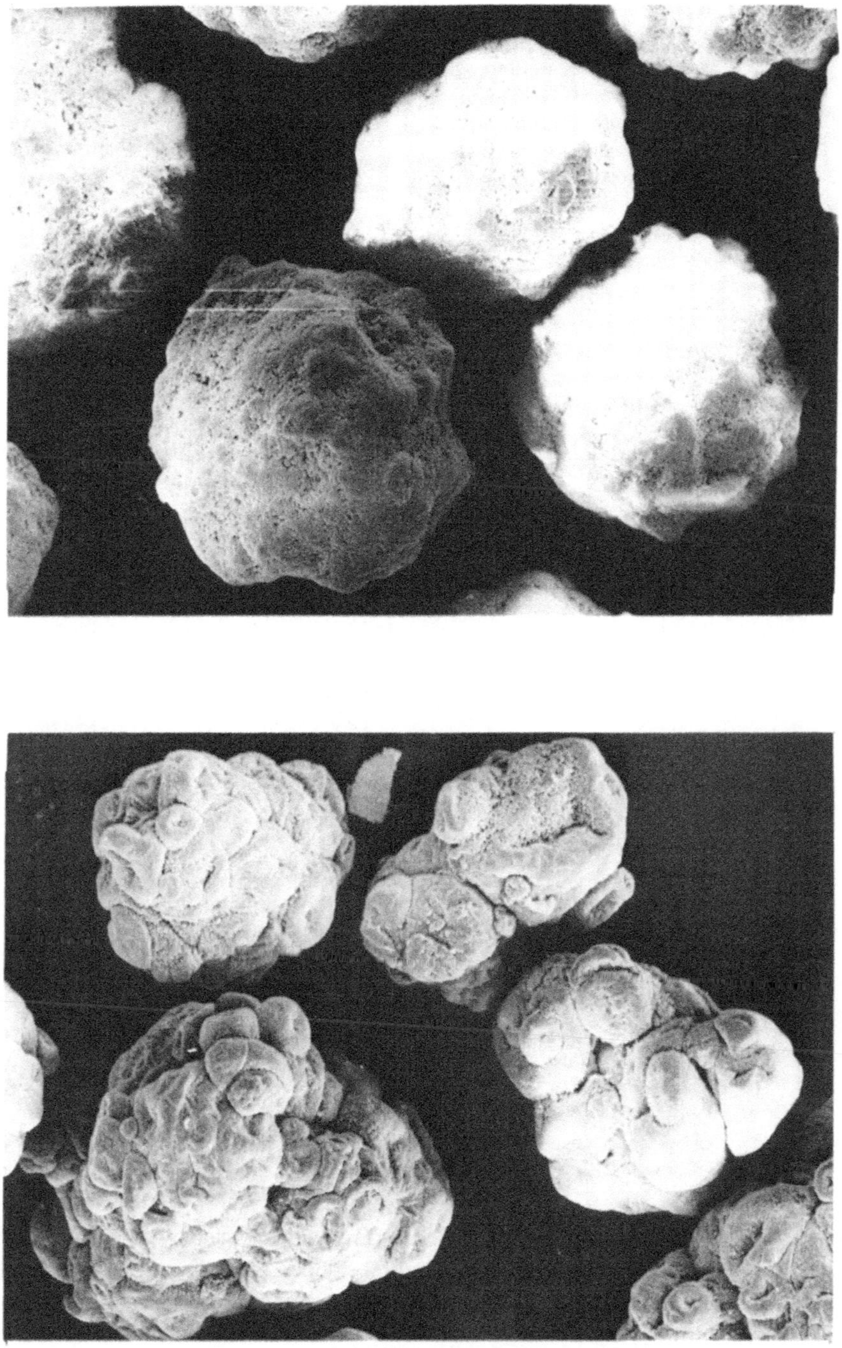

FIGURE 3. Representative photomicrographs of PVC powders.

**Table 4**
**PROCESSING PARAMETER (P) AND MORPHIC DATA OF PVC POWDERS**

| Sample | P | $R_0$ (μm) | PER | RDC ($\times 10^{-2}$) | SKE ($\times 10^{-3}$) |
|---|---|---|---|---|---|
| PVC1 | 7.7 | 144.4 | 1.154 | 1.737 | 0.8372 |
| PVC2 | 6.7 | 203.6 | 1.137 | 1.856 | 0.9841 |
| PVC3 | 10.1 | 118.9 | 1.129 | 1.564 | 0.6941 |
| PVC4 | 8.2 | 171.4 | 1.122 | 1.437 | 0.6844 |
| PVC5 | 4.8 | 207.9 | 1.095 | 0.8667 | 0.2038 |
| PVC6 | 6.4 | 194.0 | 1.154 | 2.094 | 1.046 |

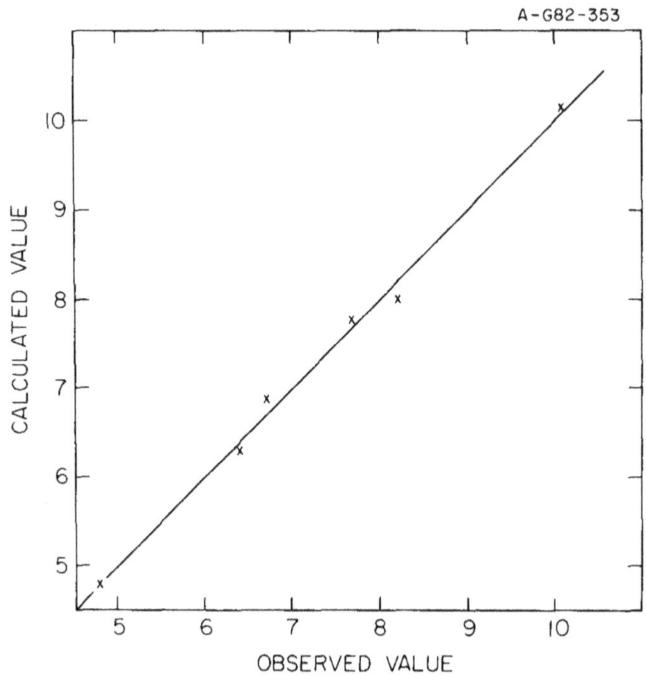


FIGURE 4. Processing parameter of PVC powders.

sands used were from Rehoboth Beach, AGSCO, Ottawa, New Jersey, Whitehead Brothers, and Pennsylvania Glass Sand companies. Figure 8 shows some of the sands and Table 7 lists the sands. The Hausner ratio (H.R.) (ratio of tap density to apparent density, i.e., TD/AD) and the flow time of the sands were measured. Tables 8 and 9 summarize the morphic data, H.R., and flow time.

### A. The Hausner Ratio (H.R.) of the Sands

The apparent densities of the sands were determined by using the ASTM standard method for the apparent densities of metal powders.[2] Sands were allowed to run through a 60° cone funnel to fill up a 25-$cm^3$ cup which sits 1 in. below the cone. The weights of the sands were then measured and the densities calculated. The tap densities were determined by using the Ro-Tap method suggested by Beddow and Kostelnik.[3] Each sand was tapped for 20 min.

Figure 9 is a plot of H.R. against the size of sand. It was observed that Pennsylvania Glass sand (PGB), Whitehead Brothers sand (WB), and Ottawa sand (OTW) roughly form

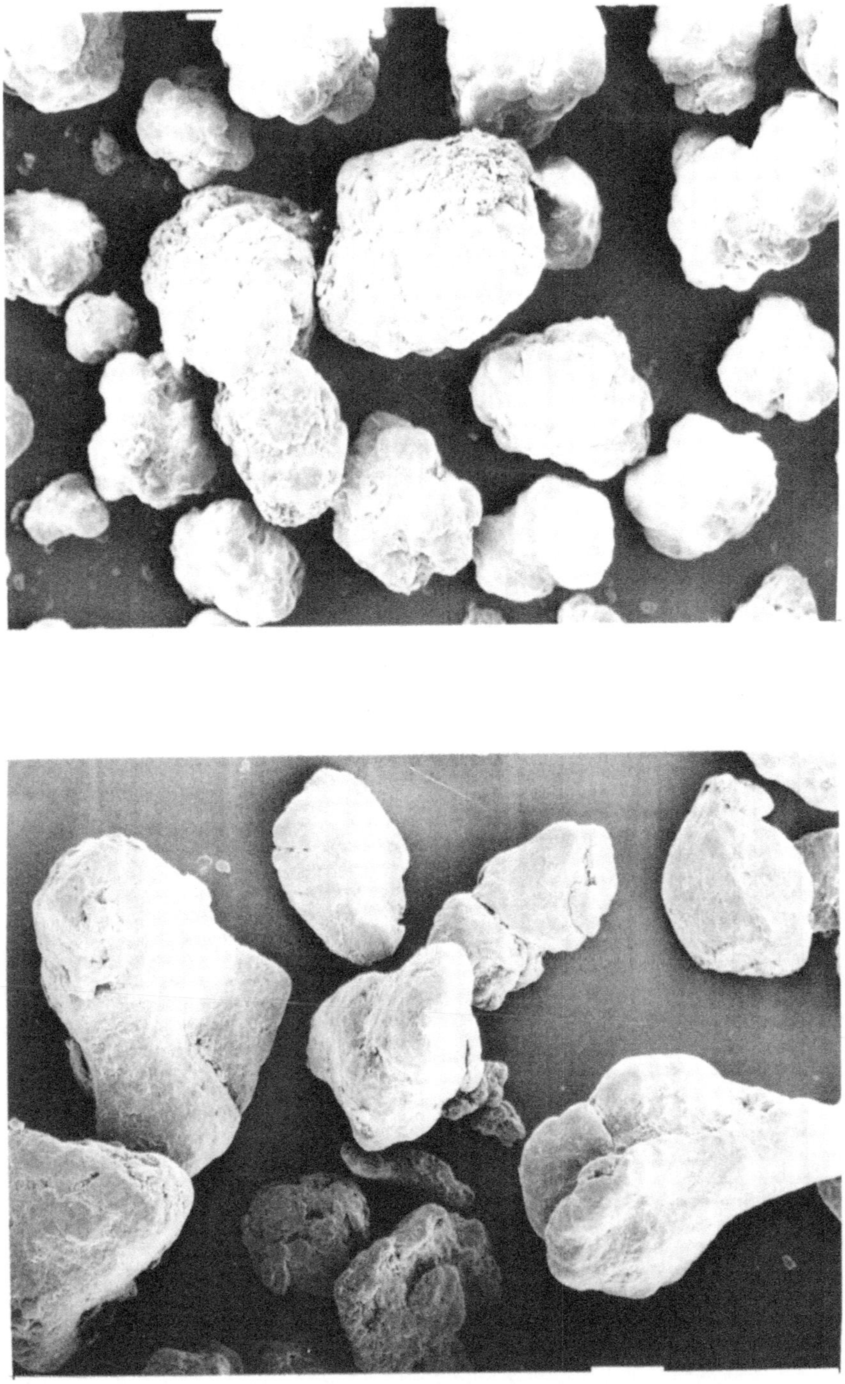

FIGURE 5. Representative photomicrographs of PE powders.

## Table 5
## ANGLE OF INTERNAL FRICTION θ AND MORPHIC DATA OF PE POWDERS

| Sample | θ (degrees) | $R_0$ (μm) | $L_0$ | PER | RDC (× $10^{-2}$) | SKE (× $10^{-3}$) |
|---|---|---|---|---|---|---|
| PEA | 32.4 | 466 | 0.9842 | 1.165 | 3.105 | 1.996 |
| PEB | 35.4 | 199 | 0.9826 | 1.210 | 3.428 | 2.551 |
| PEC | 33.4 | 401 | 0.9839 | 1.217 | 3.164 | 2.697 |
| PED | 30.2 | 457 | 0.9822 | 1.170 | 3.499 | 2.419 |
| PEE | 34.0 | 561 | 0.989 | 1.325 | 2.167 | 0.9567 |
| PEF | 33.2 | 562 | 0.9871 | 1.296 | 2.553 | 1.655 |
| PEG | 31.6 | 914 | 0.9873 | 1.256 | 2.498 | 1.628 |
| PEH | 28.6 | 387 | 0.9868 | 1.214 | 2.595 | 1.272 |
| PEI | 30.4 | 433 | 0.9893 | 1.290 | 2.116 | 1.312 |
| PEJ | 32.1 | 496 | 0.9857 | 1.262 | 2.813 | 1.815 |
| PEK | 29.8 | 530 | 0.9833 | 1.196 | 3.260 | 3.733 |
| PEL | 30.4 | 552 | 0.9828 | 1.194 | 3.383 | 2.943 |
| PEM | 29.1 | 529 | 0.9827 | 1.209 | 3.400 | 2.773 |
| PEN | 33.5 | 710 | 0.9849 | 1.300 | 2.986 | 2.083 |
| PEO | 35.6 | 609 | 0.9878 | 1.284 | 2.409 | 1.198 |
| PEP | 33.8 | 661 | 0.9876 | 1.164 | 2.455 | 1.713 |
| PEQ | 29.7 | 285 | 0.9909 | 1.214 | 1.806 | 0.6672 |
| PER | 34.4 | 511 | 0.9835 | 1.301 | 3.259 | 2.189 |
| PES | 31.6 | 552 | 0.9839 | 1.195 | 3.168 | 2.530 |

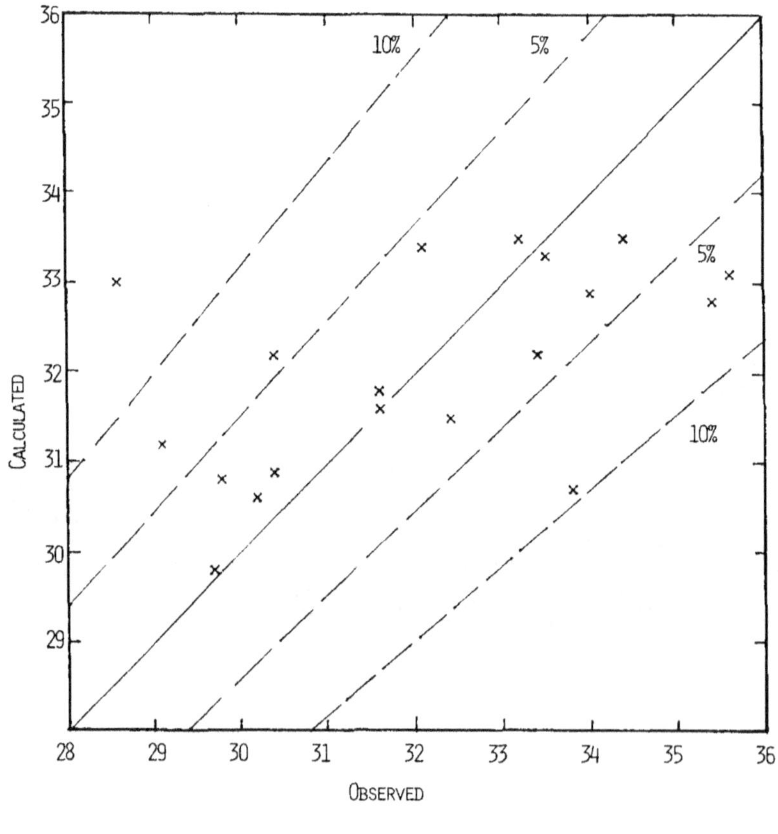


FIGURE 6. Angle of internal friction of PE powders.

**Table 6**
**ANGLE OF INTERNAL FRICTION θ AND MORPHIC DATA OF PE POWDERS**

| Sample | θ (degree) | $R_0$ (μm) | $L_\theta$ | σ |
|---|---|---|---|---|
| PEA | 32.4 | 466 | 0.9842 | 0.207 |
| PEB | 35.4 | 199 | 0.9826 | 0.278 |
| PEC | 33.4 | 401 | 0.9839 | 0.218 |
| PED | 30.2 | 457 | 0.9822 | 0.181 |
| PEE | 34.0 | 561 | 0.9890 | 0.258 |
| PEF | 33.2 | 562 | 0.9891 | 0.241 |
| PEG | 31.6 | 914 | 0.9873 | 0.238 |
| PEH | 28.6 | 387 | 0.9868 | 0.168 |
| PEI | 30.4 | 433 | 0.9893 | 0.160 |
| PEJ | 32.1 | 496 | 0.9857 | 0.225 |
| PEK | 29.8 | 530 | 0.9833 | 0.194 |
| PEL | 30.4 | 552 | 0.9828 | 0.189 |
| PEM | 29.1 | 529 | 0.9827 | 0.197 |
| PEN | 33.5 | 710 | 0.9849 | 0.265 |
| PEO | 35.6 | 609 | 0.9878 | 0.239 |
| PEP | 33.8 | 661 | 0.9876 | 0.148 |
| PEQ | 29.7 | 285 | 0.9909 | 0.125 |
| PER | 34.4 | 511 | 0.9835 | 0.193 |
| PES | 31.6 | 552 | 0.9839 | 0.185 |

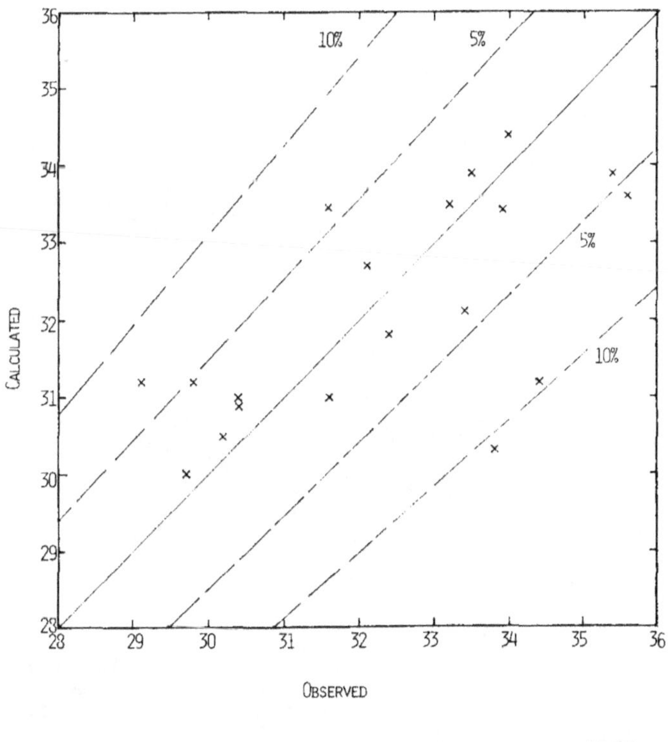


FIGURE 7. Friction angle of 19 samples (total square error = 38.98).

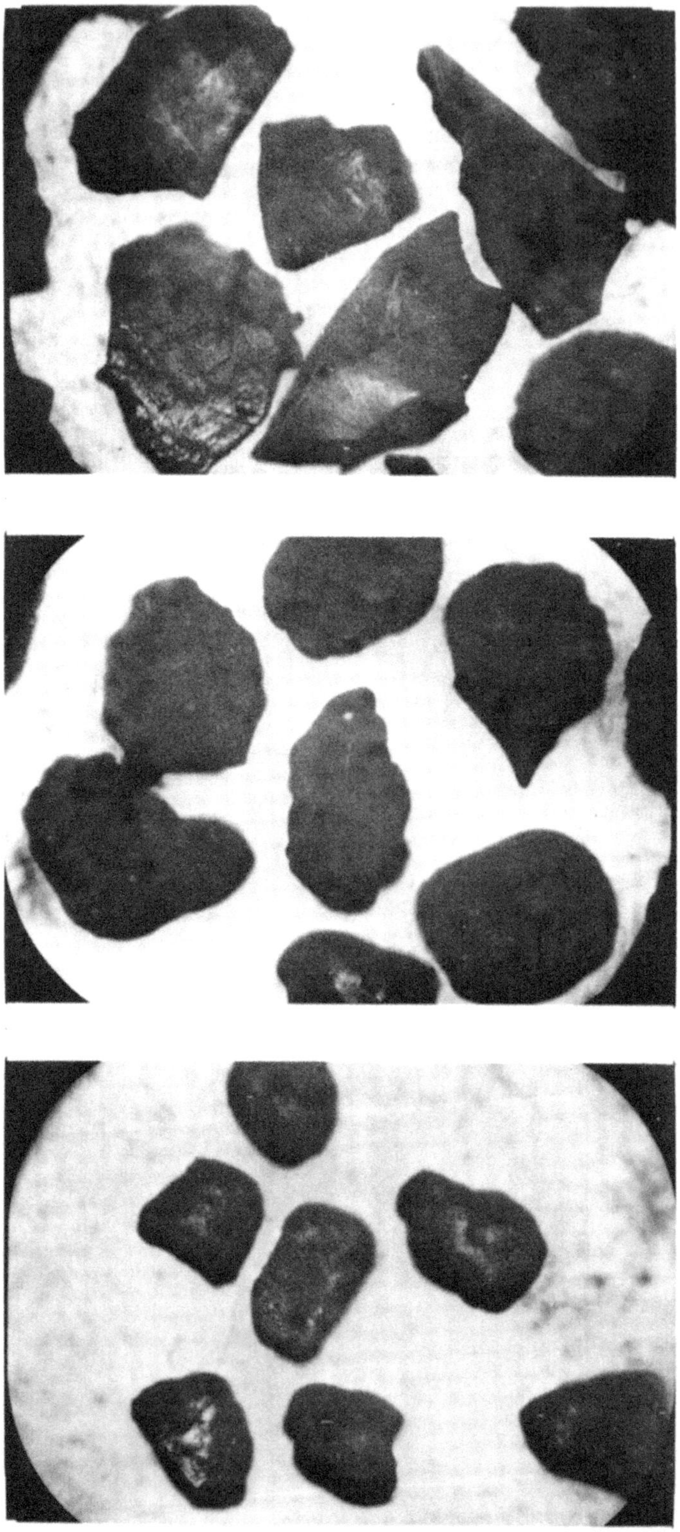

FIGURE 8. Representative photomicrographs of sands.

**Table 7**
**LIST OF SANDS**

| Sand | Prefix | Shape |
|---|---|---|
| AGSCO | AG | Angular |
| Ottawa | OTW | Subrounded |
| New Jersey | NJ | Subangular |
| Penn. Glass Sand | PG | Rounded |
| Rehoboth Beach | DC | Subrounded |
| Whitehead Brothers | WB | Subrounded |

**Table 8**
**MORPHIC DATA AND H.R. OF SANDS**

| | $R_0$ (μm) | $L_0$ | PER | RDC ($\times 10^{-2}$) | SKE ($\times 10^{-3}$) | H.R. |
|---|---|---|---|---|---|---|
| AG1A | 181.39 | 0.9873 | 1.138 | 2.507 | 1.932 | 1.208 |
| AG2B | 216.54 | 0.9776 | 1.169 | 4.398 | 5.601 | 1.246 |
| AG2A | 325.50 | 0.9847 | 1.152 | 3.028 | 2.991 | 1.187 |
| AG3A | 348.99 | 0.9782 | 1.176 | 4.292 | 6.105 | 1.205 |
| AG3C | 410.70 | 0.9807 | 1.179 | 3.804 | 4.241 | 1.194 |
| PGBD | 241.16 | 0.9882 | 1.165 | 2.327 | 1.984 | 1.185 |
| PGBB | 263.44 | 0.9881 | 1.137 | 2.349 | 1.452 | 1.171 |
| PGBA | 302.35 | 0.9899 | 1.121 | 2.007 | 1.357 | 1.166 |
| PGBC | 386.84 | 0.9885 | 1.161 | 2.273 | 1.772 | 1.153 |
| WB1A | 184.54 | 0.9887 | 1.115 | 2.232 | 1.214 | 1.182 |
| WB1E | 238.06 | 0.9893 | 1.142 | 2.111 | 1.438 | 1.171 |
| WB1C | 252.65 | 0.9886 | 1.118 | 2.249 | 1.630 | 1.163 |
| WB1B | 304.96 | 0.9890 | 1.099 | 2.175 | 1.125 | 1.149 |
| WB1D | 382.28 | 0.9901 | 1.123 | 1.957 | 1.039 | 1.144 |
| DC1B | 263.20 | 0.9888 | 1.102 | 2.209 | 1.236 | 1.133 |
| DC1A | 304.67 | 0.9918 | 1.083 | 1.620 | 0.7625 | 1.137 |
| DC1D | 384.31 | 0.9915 | 1.112 | 1.682 | 0.7731 | 1.112 |
| OTW9 | 119.25 | 0.9904 | 1.097 | 1.909 | 0.9181 | 1.211 |
| OTW8 | 137.94 | 0.9915 | 1.089 | 1.689 | 0.6754 | 1.205 |
| OTW2 | 153.03 | 0.9904 | 1.106 | 1.893 | 0.8778 | 1.196 |
| OTW1 | 169.24 | 0.9938 | 1.075 | 1.221 | 0.3722 | 1.177 |
| OTW7 | 233.84 | 0.9917 | 1.119 | 1.643 | 0.7653 | 1.160 |
| OTW4 | 283.78 | 0.9927 | 1.082 | 1.450 | 0.4395 | 1.142 |
| OTW3 | 301.95 | 0.9940 | 1.057 | 1.197 | 0.3999 | 1.140 |
| OTW6 | 370.54 | 0.9945 | 1.096 | 1.089 | 0.3223 | 1.129 |

three parallel curves. Furthermore, if plotted against ($R_0^{-1}$), three roughly parallel straight lines would be obtained. These observations suggested that the relationship between H.R. and shape of particle may take the form of Equation 6:

$$\text{H.R.} = 1 + k_1R_0^{-1} + k_2f(\text{shape}) \qquad (6)$$

in which $k_1$, $k_2$ are constants and f(shape) means function of shape.

The second term of Equation 6 is contributed by the size of particles and the third term by their shape. The constant "1" is introduced into the equation to account for the fact that the H.R. is by nature always greater than 1. From Equation 6 a plot of H.R. vs. ($R_0^{-1}$) of different sizes of particles of the same shape should give a straight line as shown in Figure 10. Note that the shape data of different sized sands within one sand group, although close

**Table 9**
**MORPHIC DATA AND FLOW TIME OF SANDS**

| | $R_0$ (μm) | $L_0$ | PER | RDC ($\times 10^{-2}$) | SKE ($\times 10^{-3}$) | Flow time (sec) |
|---|---|---|---|---|---|---|
| AG5A | 122.57 | 0.9805 | 1.177 | 3.839 | 5.497 | 43.9 |
| AG1B | 144.28 | 0.9668 | 1.212 | 6.464 | 11.69 | 46.0 |
| AG1A | 181.39 | 0.9873 | 1.138 | 2.507 | 1.932 | 49.8 |
| AG2B | 216.54 | 0.9776 | 1.169 | 4.398 | 5.601 | 55.6 |
| NJ0B | 157.38 | 0.9878 | 1.135 | 2.415 | 1.268 | 50.8 |
| NJ0A | 180.62 | 0.9889 | 1.123 | 2.199 | 1.301 | 54.1 |
| NJ0E | 240.24 | 0.9894 | 1.158 | 2.094 | 1.044 | 58.6 |
| WB1A | 184.54 | 0.9887 | 1.115 | 2.232 | 1.214 | 53.7 |
| WB1E | 238.06 | 0.9893 | 1.142 | 2.111 | 1.438 | 57.6 |
| DC1B | 263.20 | 0.9888 | 1.102 | 2.209 | 1.236 | 58.3 |
| DC1A | 304.67 | 0.9918 | 1.083 | 1.620 | 0.7625 | 65.1 |
| OTW9 | 119.25 | 0.9904 | 1.097 | 1.909 | 0.9181 | 45.0 |
| OTW8 | 137.94 | 0.9915 | 1.089 | 1.689 | 0.6754 | 47.9 |
| OTW2 | 153.03 | 0.9904 | 1.106 | 1.893 | 0.8778 | 50.9 |
| OTW1 | 169.24 | 0.9938 | 1.075 | 1.221 | 0.3722 | 55.3 |
| OTW7 | 233.84 | 0.9917 | 1.119 | 1.643 | 0.7653 | 60.6 |
| OTW4 | 283.78 | 0.9927 | 1.082 | 1.450 | 0.4395 | 64.3 |
| PG0C | 116.03 | 0.9909 | 1.079 | 1.805 | 0.6182 | 45.1 |
| PG0B | 132.95 | 0.9939 | 1.055 | 1.218 | 0.4011 | 47.0 |
| PG0A | 150.24 | 0.9940 | 1.079 | 1.192 | 0.2524 | 48.9 |

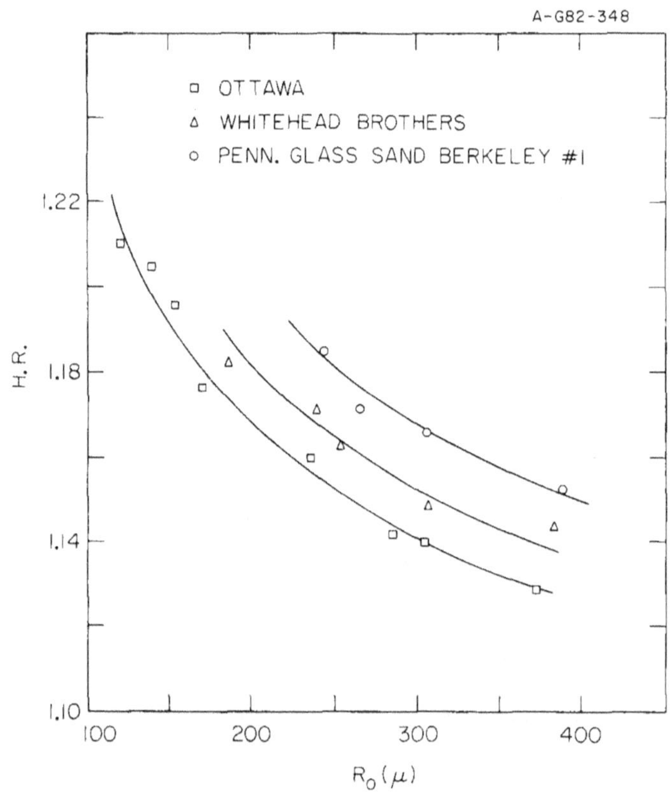


FIGURE 9. Plot of H.R. vs. $R_0$.

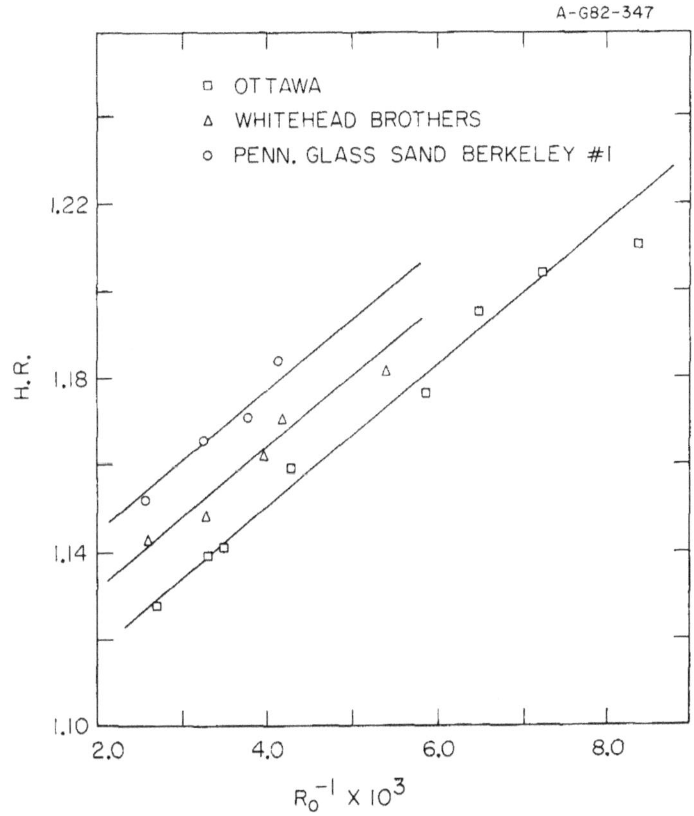


FIGURE 10. Plot of H.R. vs. $R_0^{-1}$.

to one another, are not exactly the same and deviations from the straight lines are to be expected. The linear least square fit gave the following correlation:

$$H.R. = 1 + 15.57R_0^{-1} + 0.01015L_0^{-59.34}PER^{0.8374}RD\ C^{-1.078}SKE^{0.3899} \quad (7)$$

where $R_0$ is in μm.

Figure 11 shows a plot of H.R. calculated from Equation 7 vs. observed H.R.

## B. The Flow Time of the Sands

The ASTM standard method for the determination of the free flow time of metal powders[4] was adopted to measure the flow time of the sands. The time required for 25 $cm^3$ of sand to freely flow through a 60° cone funnel which has an opening of 0.1 in. was measured. When plotting flow time vs. size, parallel curves were observed (Figure 12) as in the case of H.R. Plot of flow time vs. ($R_0^{1/2}$) gave two roughly parallel straight lines (Figure 13). The correlation equation was thus assumed to have the form of Equation 8.

$$F.T. = k_1R_0^{1/2} + k_2(L_0)^x(PER)^y(RD\ C)^z(SKE)^u \quad (8)$$

where F.T. is in seconds and $R_0$ is in μm.

The values obtained are $k_1 = 2.98$, $k_2 = 3.03$, $x = -3.53$, $y = 1.25$, $z = -0.263$, and $u = 0.0338$.

Figure 14 plots flow time calculated from Equation 8 vs. observed flow time.

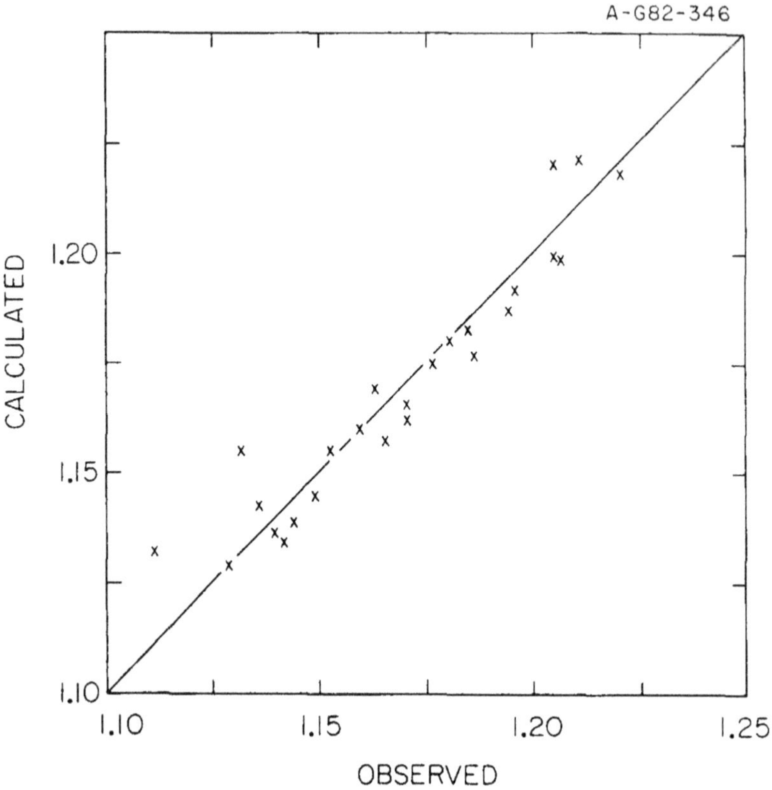


FIGURE 11. Plot of calculated value of H.R. vs. observed.

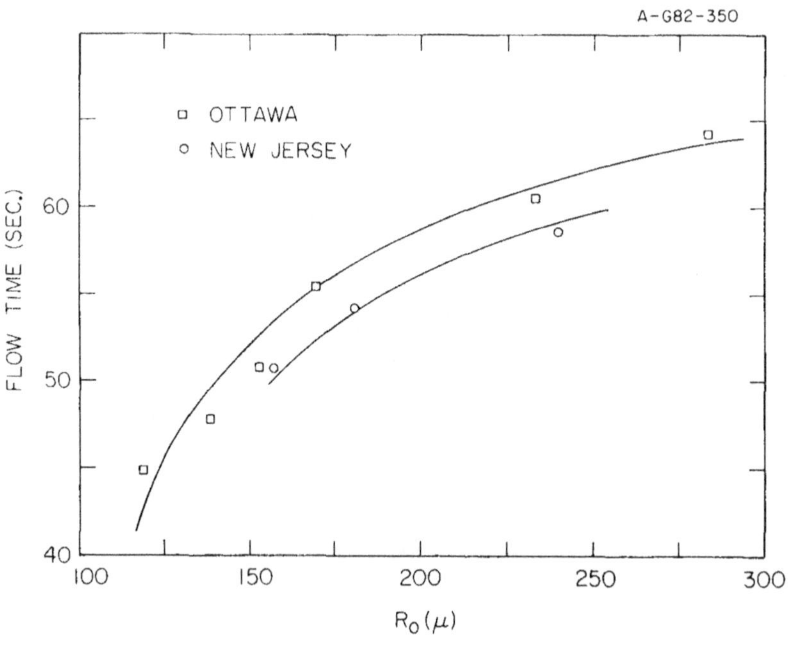


FIGURE 12. Plot of flow time vs. $R_0$.

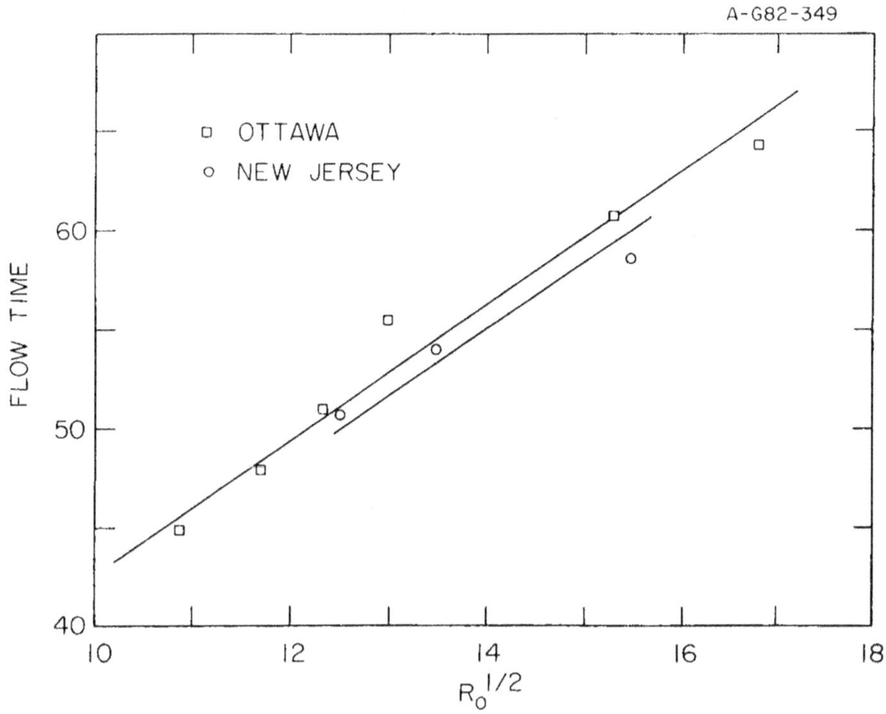


FIGURE 13. Plot of flow time vs. $R_0^{1/2}$.

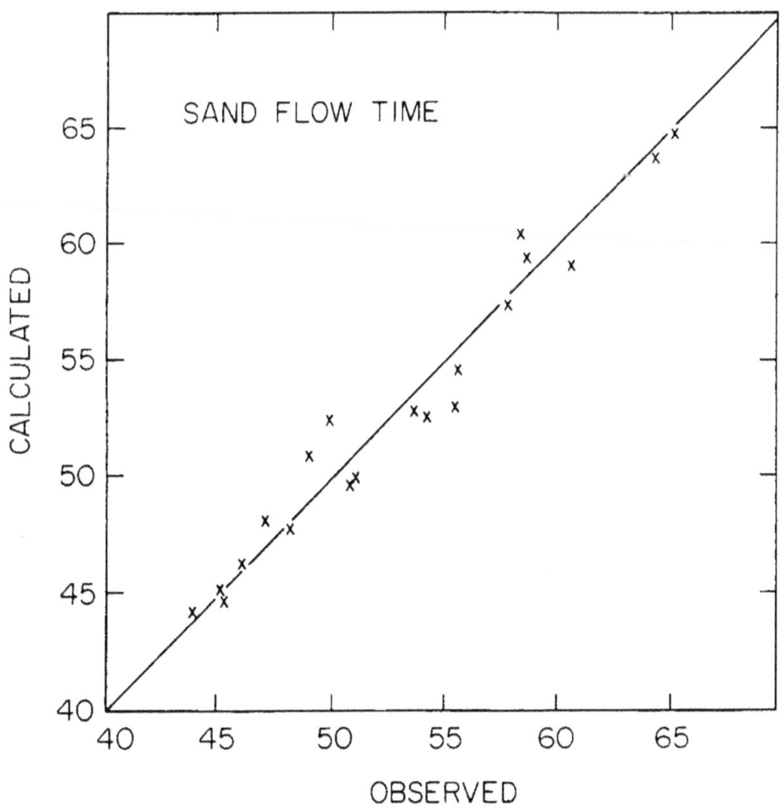


FIGURE 14. Plot of calculated value of flow time vs. observed.

## V. DISCUSSION AND CONCLUSION

This is the first attempt ever to quantitatively relate the shape terms derived from R-θ method to bulk properties. The equations presented are empirical, and the promising results indicate the potential of morphological analysis. However, further studies aimed at the theoretical analysis of each property and the establishment of theoretical relationships are highly desirable.

In closing, some remarks are worth emphasizing:

1. A broad size distribution will further complicate the study of shape since how the size distribution will affect the bulk properties quantitatively is not well understood. This may clearly be seen from the comparison of the results of plastic powders and that of sands.
2. The surface roughness of the particle also may affect the bulk properties. This is well demonstrated in the friction angle of agricultural grains.
3. Shape distribution of the particles may as well affect the bulk properties.
4. It should also be pointed out that although there are four or more morphic terms used in the correlation equations presented here, the number of morphic terms used may well be reduced. In most cases, two or three morphic terms are enough.[5]

## LIST OF SYMBOLS

| | |
|---|---|
| A,B,C,M: | Correlation constants |
| $A_0$: | Mean radius; defined as $A_0 = \frac{1}{2\pi}\int_0^{2\pi} R(\theta)\, d\theta$ |
| F.T.: | Flow time of sand |
| H.R.: | Hausner ratio; (tap density/apparent density) |
| IRR: | Irregularity, defined as $\left(\frac{R_0}{A_0}\right)^2$ |
| $L_0$: | Defined as $\frac{A_0}{R_0}$ |
| P: | Processing parameter of PVC powders |
| PER: | Normalized perimeter; $PER = \frac{\text{perimeter of particle}}{2\pi R_0}$ |
| $R_0$: | Equivalent radius: is the radius of the circle which has the same area as the projection area of the particle. Also, the mean sieve opening in the cases of PVC and PE powders. |
| RDC: | Radance; defined as $RDC = \Sigma L_2(n)$ |
| SKE: | Skewness; defined as $SKE = \lvert\Sigma\,\Sigma\, L_3(m,n)\rvert$ |
| V.R.: | void ratio; $V.R. = \frac{\text{particle density} - \text{bulk density}}{\text{bulk density}}$ |
| $k,k_1,k_2,k_3$: | Correlation constants |
| x,y,z,u,v: | Correlation constants |
| $\phi$ | Angle of internal friction |
| $\sigma$: | Standard deviation of the size distribution |

## REFERENCES

1. **Reisner, W. and Rothe, M. V. E.,** Bins and Bunkers for Handling Bulk Materials, Trans. Tech. Publishing, Rockport, Mass., 1971; Molerus, O., *Powder Technol.,* 12, 25, 1975.
2. **ASTM,** *Book of ASTM Standards,* Part 9, American Society for Testing Materials, Philadelphia, 1978, p.43 and 196.
3. **Beddow, J. K. and Kostelnik, M.,** Tap density — a new test for the characterization of metal powders, in *Advances in Powder Metallurgy,* Vol. 4, Hausner, H. H., Ed., Plenum Press, New York, 1971, 28.
4. **ASTM,** *Book of ASTM Standards,* Part 9, American Society for Testing and Materials, Philadelphia, 1978, 45.
5. **Chang, C. R.,** Particle Morphology and Its Effect on Bulk Property, Ph.D. dissertation, University of Iowa, Iowa City, 1982.

Chapter 18

# THE APPLICATION OF MORPHOLOGICAL ANALYSIS IN KINETIC MODELING

**Louise Hua, John Keith Beddow, and Arthur F. Vetter**

## TABLE OF CONTENTS

## I. INTRODUCTION

The morphological changes of a crystal during kinetic processes are strongly dependent upon process variables[1] such as temperature, supersaturation, agitation, type of solvent, impurities, surface active agents, and so on. The size and shape of a crystal is controlled primarily by the rates of growth of its individual faces. A correct kinetic model should enable us to estimate the effect of process variables[2] on the facial growth rates and thus to control the size and shape of crystals.

The study report has had the following objectives:

1. To develop a model which more closely follows the growth of real crystals
2. To achieve the possibility of controlling crystal growth processes as a result of increased understanding gained from the kinetic model
3. To eventually develop the capability for producing crystal growths of a predetermined morphology

## II. MODELING

The use of morphological analysis in kinetic modeling has been studied by using $(R,\theta)$ space, in which the radial growth of a crystal (R) is assumed to be isotropic along $\theta$-direction.[3] A more realistic kinetic model of crystal growth, with the direction of growth normal to the individual crystal faces, is now studied. In this model, facial growth rates may vary from face to face while interfacial angles stay constant, so that the successive displacements of a face during growth are parallel to each other. The fast-growing faces may disappear and this has been known as "overlapping" as shown in Figures 2A, 3A (center), 3A (right), 4A (center), 4A (right), 5A (center), 5A (right), 8A, and 9A. On the other hand, crystals may maintain their basic geometric pattern as they grow, and this has been called "invariant" as shown in Figures 3A (left), 4A (left), 5A (left), and 6A. A criterion for eliminating crystal faces is discussed later.

A computer program is developed using numerical method to simulate the growth of a crystal and to estimate the size term ($R_0$) and other morphic terms[4] ($L_0$, variance, skewness) as functions of time. Changes with time in $R_0$ and other morphic features are shown in a series of graphs. These data are not interpreted herein.

## III. RESULTS

The growths of an equilateral hexagon and octagon are considered in this study.

### 1. Growth of an equilateral hexagon: (five cases are considered)

The six faces are labeled from 1 to 6 as shown in Figure 1. The simplest case of growing a hexagon will be the case that all the six faces grow with the same rate. In this case, the shape is retained always as an equilateral hexagon. However, the situation becomes more complicated when the facial growth rates are different from one another. Some of the simpler cases are studied, in which only two different facial growth rates are considered, with the ratio 1:5.

*Case 1* — with one of the six faces (i.e., face 2 in Figure 2A) growing faster than the others. The time dependence of the shape, size, and morphic terms refer to Figures 2A to 2E.

*Case 2* — with two of the six faces growing faster than the rest. Three subcases are involved in this case, with the two fast-growing faces ortho (i.e., faces 2 and 3 in Figure 3A (left)), meta (i.e., faces 2 and 4 in Figure 3A (center)), or para (i.e., faces 2 and 5 in

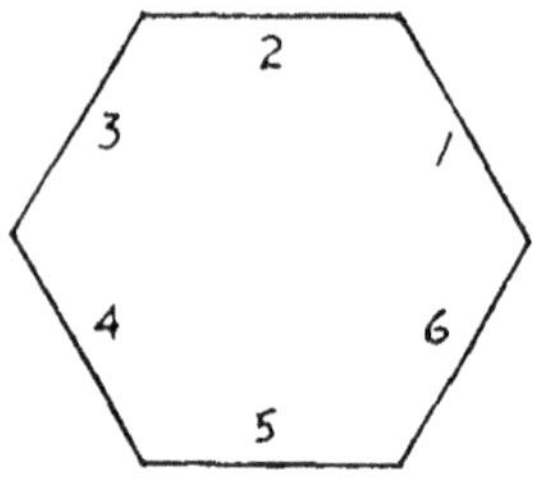


FIGURE 1. Labels of hexagon faces.

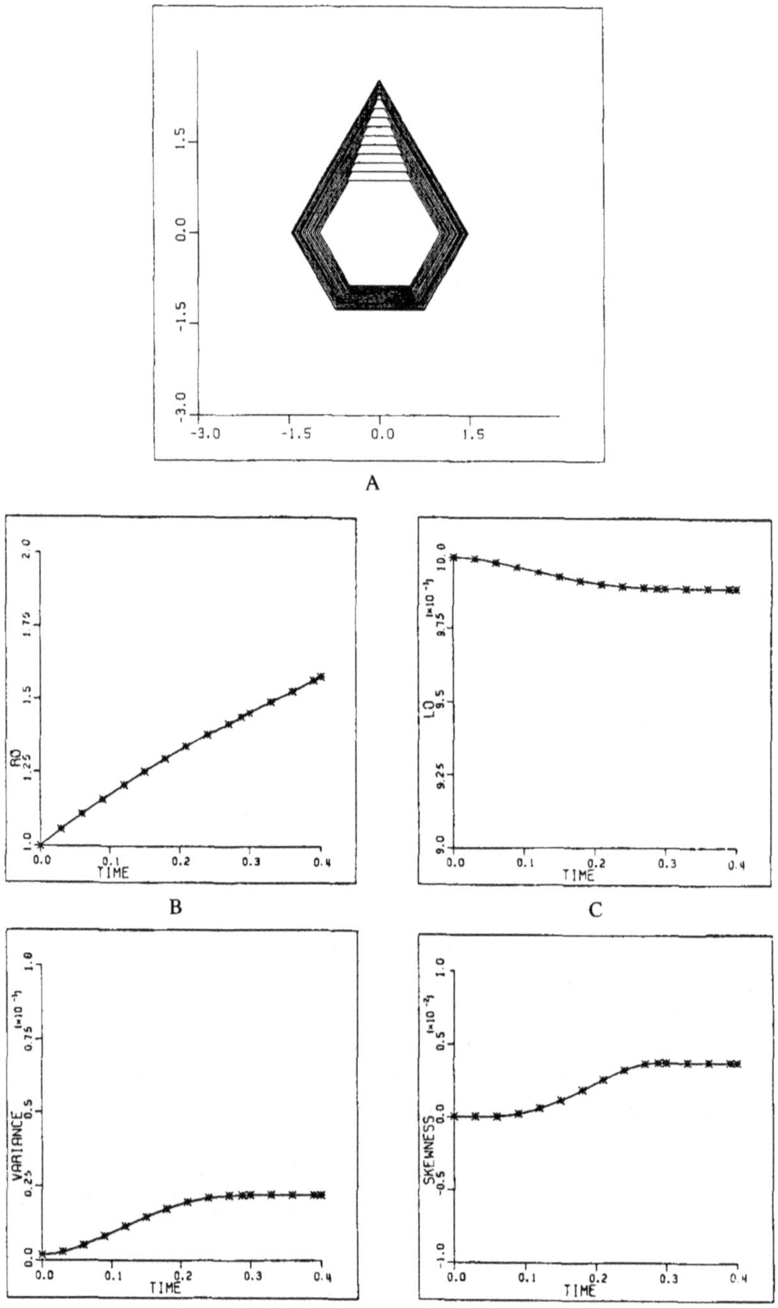


FIGURE 2. Shape, size, and morphic terms vs. time. Fast growing face: 2.

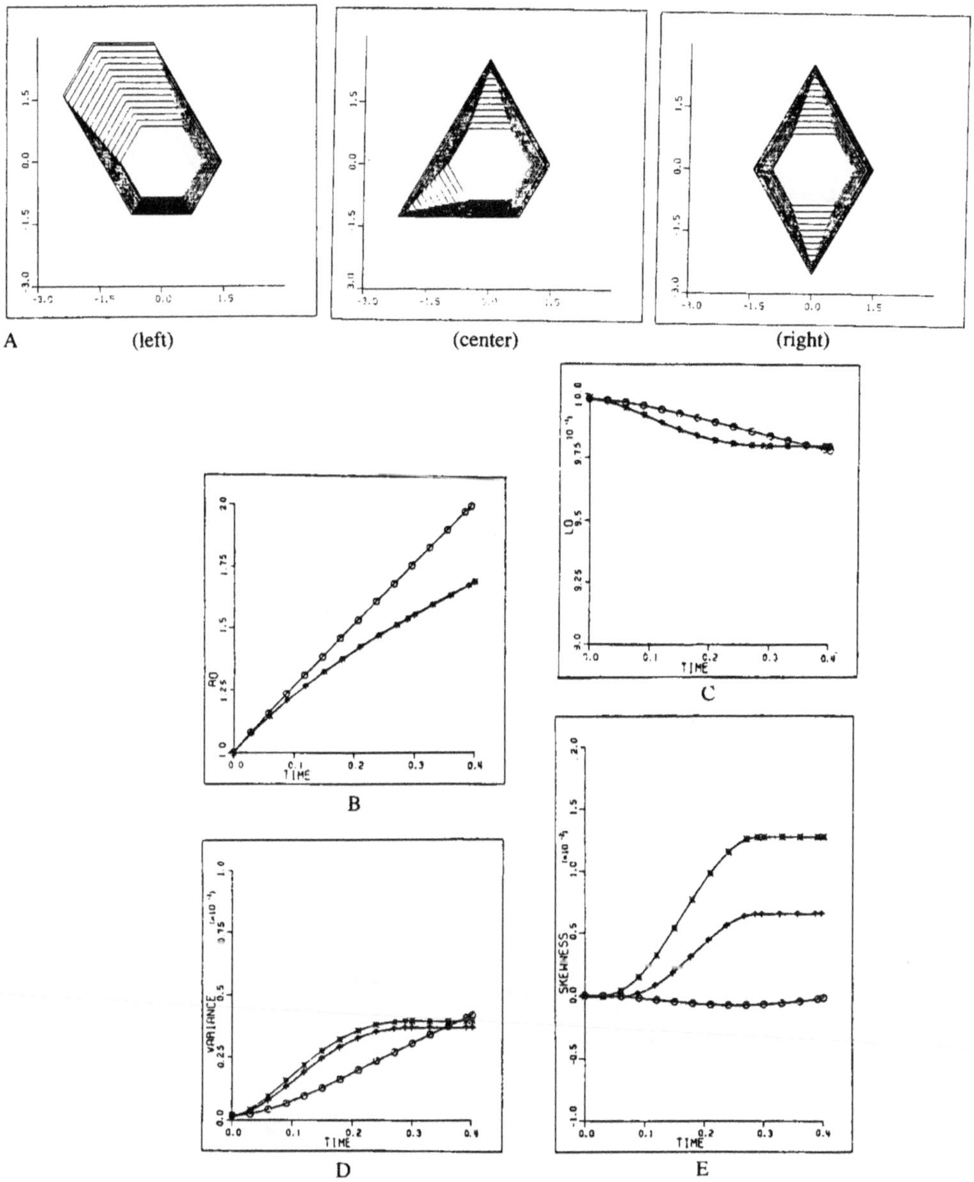


FIGURE 3. Shape, size, and morphic terms vs. time. Fast growing faces: O-2 and 3, ↑-2 and 4, and *-2 and 5.

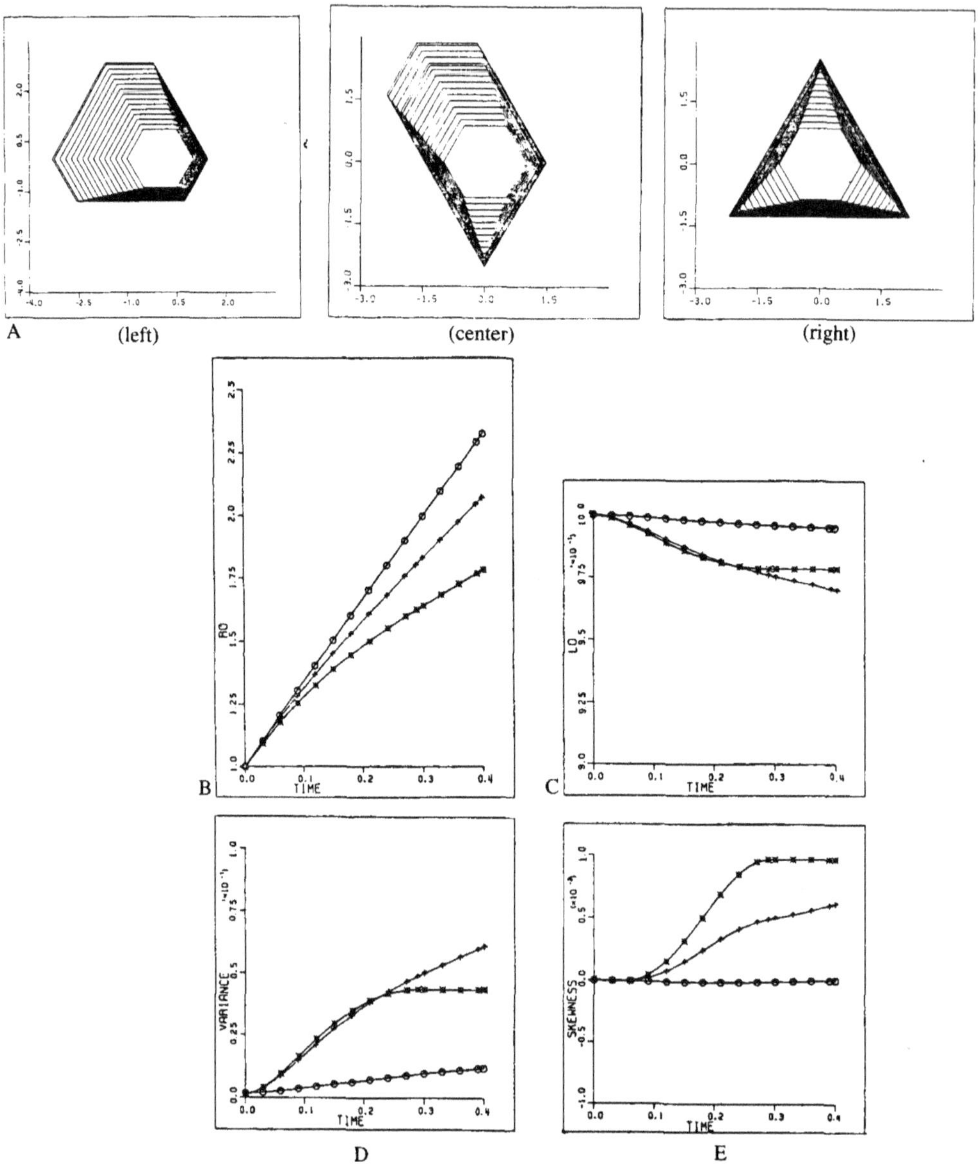


FIGURE 4. Shape, size, and morphic terms vs. time. Fast growing faces: O-2,3, and 4; ↑ -2, 3, and 5; and *-2, 4, and 6.

Figure 3A (right)) to each other. The size and morphic terms as a function of time are shown in Figures 3B to 3E.

*Case 3* — with three of the six faces growing faster than the others. Again, this case has three subcases faces 2, 3, and 4 (Figure 4A (left)), faces 2, 3, and 5 (Figure 4A (center)), or faces 2, 4, and 6 (Figure 4A (right)) to be fast-growing faces. Figures 4B to 4E show how the size and morphic terms change with time.

*Case 4* — with four fast growing faces. Similarly, there are three subcases involved. With the two slow-growing faces ortho (i.e., faces 1 and 6 in Figure 5A (left)), meta (i.e., faces 1 and 5 in Figure 5A (center)), or para (i.e., faces 1 and 4 in Figure 5A (right)) to each other. The size and morphic terms, as functions of time, are shown in Figures 5B to 5E.

*Case 5* — with only one slow-growing face (i.e., face 1 in Figure 6A). The time dependence of the shape and size and morphic terms are shown in Figures 6A to 6E.

**2. Growth of an equilateral octagon: (two cases are considered)**

The eight faces are indexed from 1 to 8 as shown in Figure 7. Growths of an octagon into a square and rectangle are studied. In the former case, the octagon grows with two different rates in alternate faces, i.e., with faces 2, 4, 6, and 8 growing three times faster than the rest. In the latter case, three different rates are considered. Faces 2, 4, 6, and 8 are the fastest growing faces with rates six times faster than the slowest growing faces — faces 1 and 5 — while faces 3 and 7 grow three times faster than faces 1 and 5. The time dependence of the shape and size and morphic terms for both cases are shown in Figures 8A to E and 9A to E, respectively.

## IV. DISCUSSION

Advantages of this model include:

1. Anisotropy is incorporated in a kinetic model for crystal growths for the first time.
2. For an asymmetrical growth, the CG (center of gravity) changes with time. The movement of CG is calculated from the new shape, and all morphic terms are relative to the updated CG. This is in contrast to the (R,θ) method in which the CG cannot be updated so easily.

The position of the fast-growing faces dramatically affect the final shape of a crystal, as well as the size, even if the number of fast-growing faces are the same. The effects of this growth rates selectively on the size and morphic terms are shown in the hexagon case 2, case 3, and case 4. Case 2 is more interesting than case 3 or case 4. As may be seen from Figures 3C to E, each morphic term does not vary much in all the three subcases except for skewness. This indicates that skewness is an important morphic feature.

## V. CRITERION FOR FACE ELIMINATION

In general, whether a face of a crystal is to disappear or not during crystal growth depends upon the relative growth rates of that face and its two neighboring faces, and also the angles between that face and its neighbors. If the environmental conditions (i.e., concentration gradient, hydrodynamics, temperature, etc.) are the same for all faces, and the individual growth rate of each face is constant with respect to time, then a critical condition for a face to disappear may be derived. Establishing this critical condition is a necessary step in achieving the possibility of controlling crystal growth processes.

Consider Figure 10. Suppose the facial growth rates of faces 1, 2, and 3 are $k_1$, $k_2$, and $k_3$, respectively. Let $\alpha$ be the angle made by faces 1, 2, and $\beta$, the angle made by faces 2, 3. AA'c and BFB' are two faces perpendicular to face 2. After time interval $\Delta t$, faces 1, 2, and 3 move distances of $k_1\Delta t$, $k_2\Delta t$, and $k_3\Delta t$, respectively. The new face 2 is denoted by DG. The new face 1 intercepts new face 2 and D at face AA'C at A' with angle $\theta_1$ ($0 < \theta_1 < 90°$). The new face 3 intercepts new face 2 at G and face BFB' at B' with angle $\theta_2$ ($0 < \theta_2 < 90°$). Then

$$\theta_1 = \alpha - 90° \quad (1)$$

$$\theta_2 = \beta - 90° \quad (2)$$

and CD, FG will be the increments of face 2 at the end near face 1, face 3, respectively.

Define $k_{c21}$ to be the outward displacement of face 2 per unit time if there is no increment of face 2 near face 1, i.e.,

$$k_{c21} = \frac{AA'}{\Delta t} \quad (3)$$

If the displacement is inward then it is set to be negative. According to this definition, we make the following conclusion:

$$k_{c21} < 0 \text{ if } \alpha < 90°$$
$$k_{c21} > 0 \text{ if } \alpha > 90°$$
$$k_{c21} = \infty \text{ if } \alpha = 90°$$

$$\text{Since } \sin\theta_1 = \frac{k_1\Delta t}{|AA'|} = \frac{k_1\Delta t}{AA'} \quad (4)$$

$$\text{and } \tan\theta_1 = \frac{|CD|}{|A'C|}$$

$$= \frac{-CD}{AC - AA'} \text{ (negative increment)}$$

$$= \frac{CD}{AA' - k_2\Delta t} \quad (5)$$

combine Equations 1, 3, and 4, we have

$$k_{c21} = \frac{k_1}{\sin(\alpha - 90°)} \quad (6)$$

substitution of Equations 1 and 3 into 5 and rearranging to yield

$$CD = \Delta t\,(k_{c21} - k_2)\tan(\alpha - 90°) \quad (7)$$

For $\alpha < 90°$, the increment is always positive. For $\alpha > 90°$, if $k_{c21} > k_2$ then the increment is positive, otherwise it is negative.

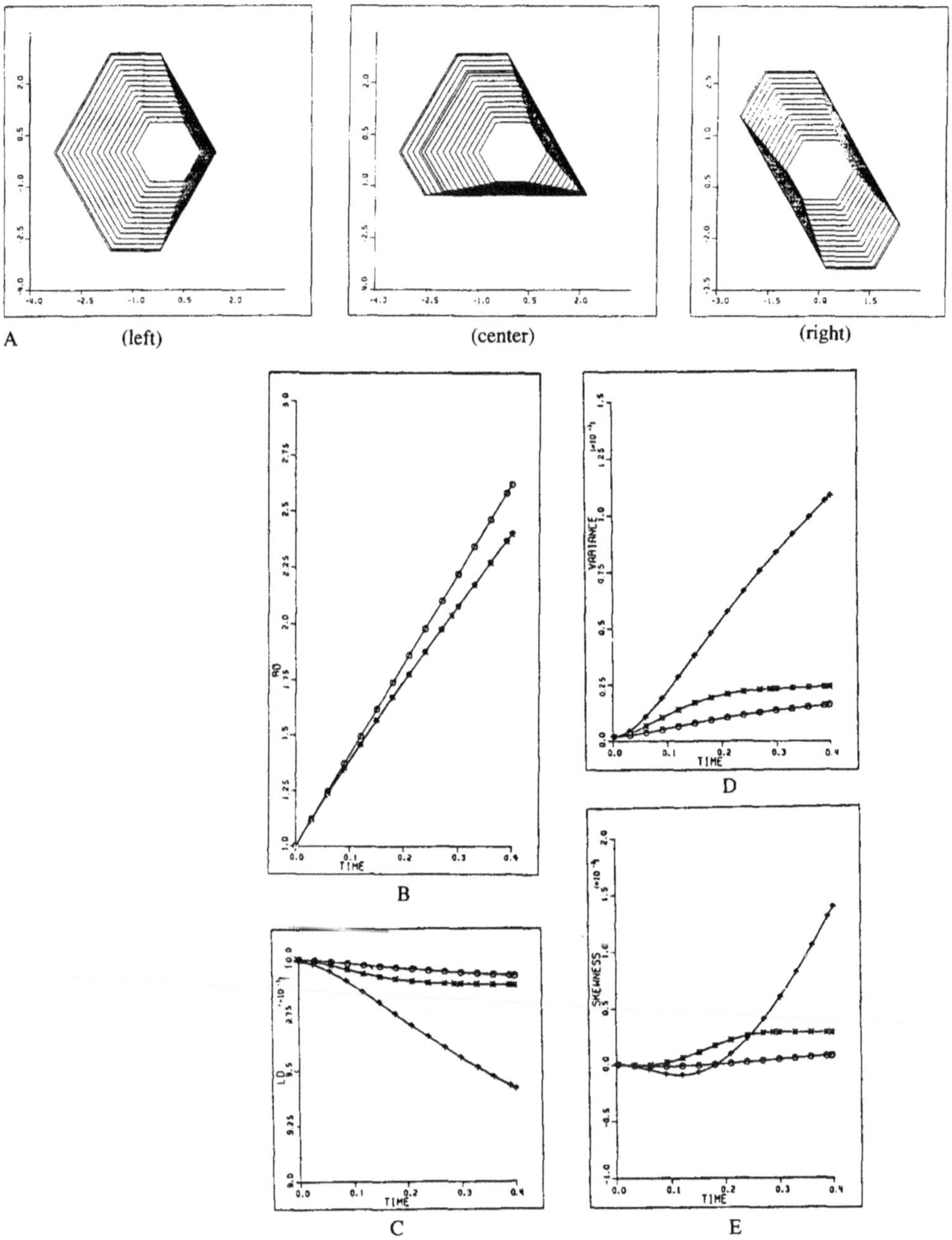


FIGURE 5. Shape, size, and morphic terms vs. time. Slow growing faces: 0-1 and 6, ↑-1 and 4, and *-1 and 5.

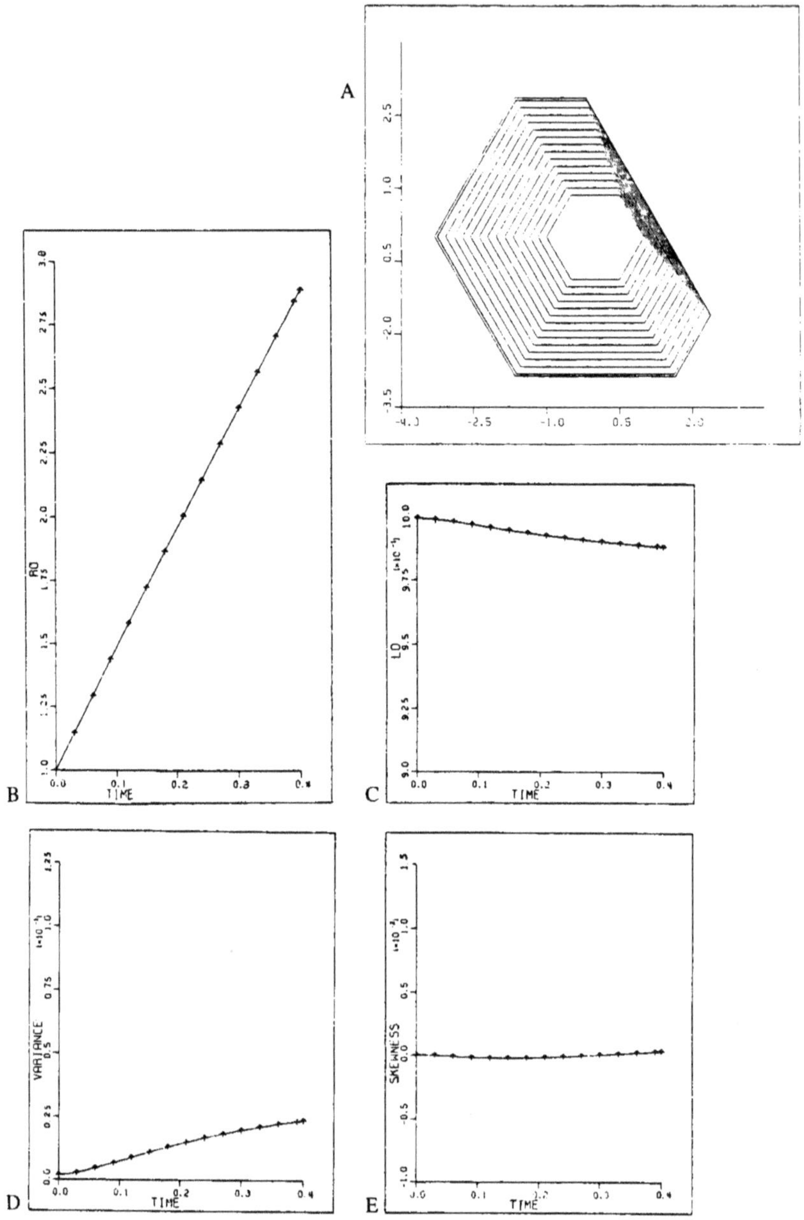


FIGURE 6. Shape, size, and morphic terms vs. time. Slow growing face: 1.

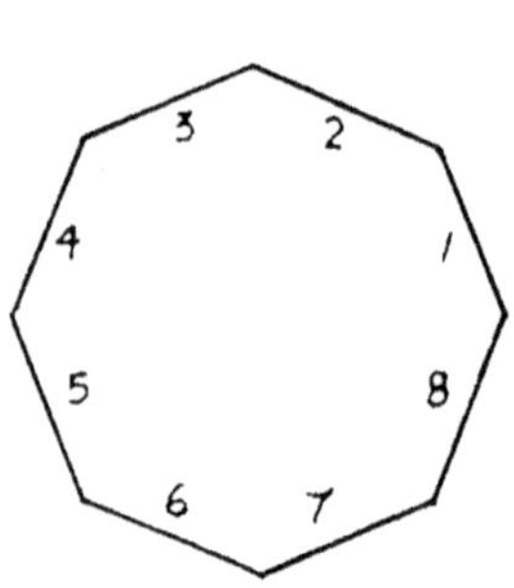


FIGURE 7. Labels of octagon faces.

Substitute Equation 6 into Equation 7, we have

$$CD = \Delta t \left(\frac{k_1}{\sin(\alpha - 90°)} - k_2\right) \tan(\alpha - 90°) \tag{8}$$

Similarly, we can get

$$FG = \Delta t \left(\frac{k_3}{\sin(\beta - 90°)} - k_2\right) \tan(\beta - 90°) \tag{9}$$

Now, for eliminating a face the net increment must be negative, i.e., $CD + FG < 0$. Therefore, we have

$$\Delta t \left(\frac{k_1}{\sin(\alpha - 90°)} - k_2\right) \tan(\alpha - 90°) + \Delta t \left(\frac{k_3}{\sin(\beta - 90°)} - k_2\right) \tan(\beta - 90°) < 0 \tag{10}$$

Equation 10 can be simplified to

$$-k_2[\tan(\alpha - 90°) + \tan(\beta - 90°)] + \left[\frac{k_1}{\cos(\alpha - 90°)} + \frac{k_3}{\cos(\beta - 90°)}\right] < 0 \tag{11}$$

therefore

$$k_2 > \frac{\dfrac{k_1}{\cos(\alpha - 90°)} + \dfrac{k_3}{\cos(\beta - 90°)}}{\tan(\alpha - 90°) + \tan(\beta - 90°)} \tag{12}$$

and Equation 12 is a necessary and sufficient condition for face 2 to disappear.
If we define

$$k_{2c} = \frac{\dfrac{k_1}{\cos(\alpha - 90°)} + \dfrac{k_2}{\cos(\beta - 90°)}}{\tan(\alpha - 90°) + \tan(\beta - 90°)}$$

then we can conclude the criterion:

1. If $k_2 > k_{2c}$ then face 2 will eventually disappear
2. If $k_2 < k_{2c}$ then face 2 will not disappear

## VI. EXPERIMENTAL CONFIRMATION OF CRITERION

The study of a single crystal growth of potash alum by Mullin and Garside[5] confirms this criterion. In that report, the effects of supersaturation ($\Delta C$) and solution velocity (U) on the (111) facial growth rate was characterized as follows:

$$k_{(111)} = 6.24 \times 10^{-4} U^{0.65} \Delta C^n$$

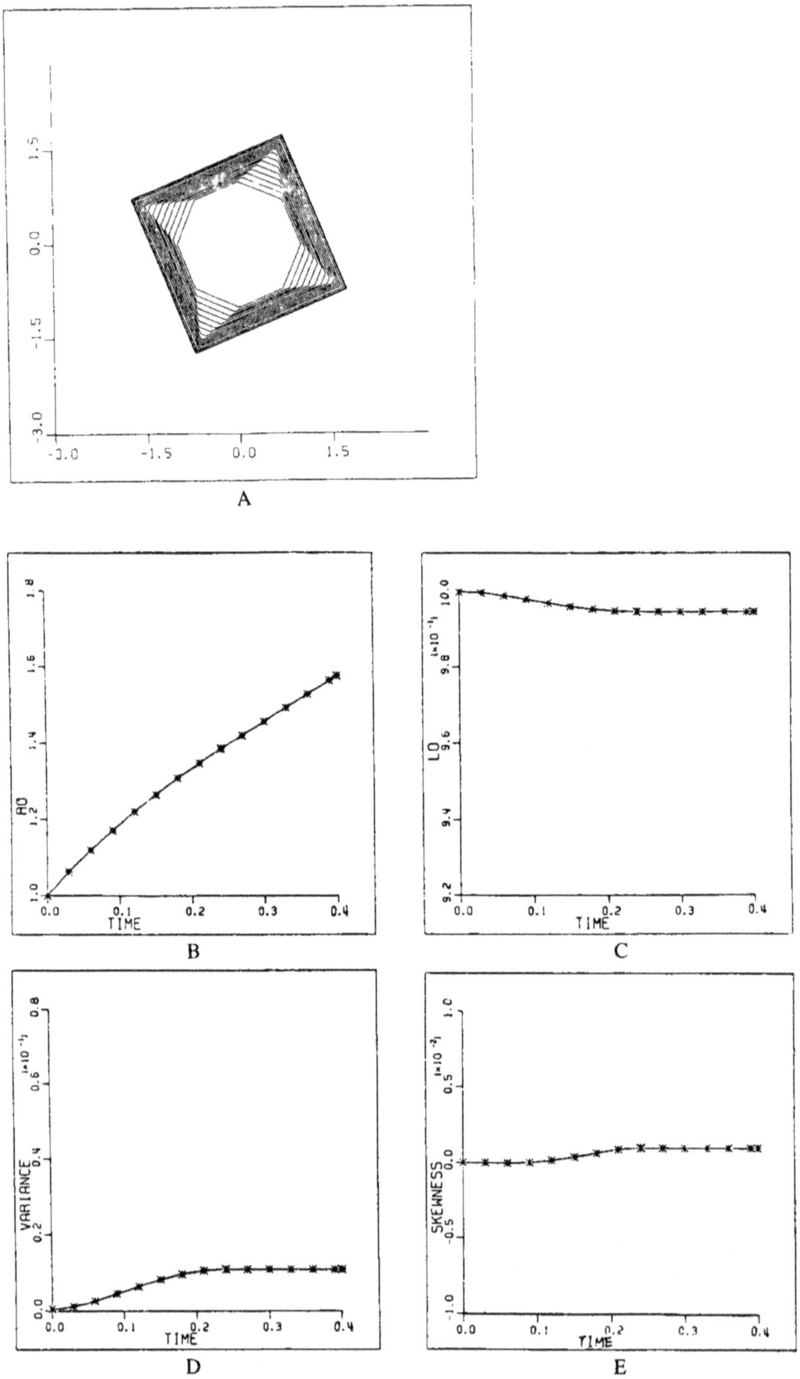


FIGURE 8. Shape, size, and morphic terms vs. time. Fast growing faces: 2, 4, 6, and 8.

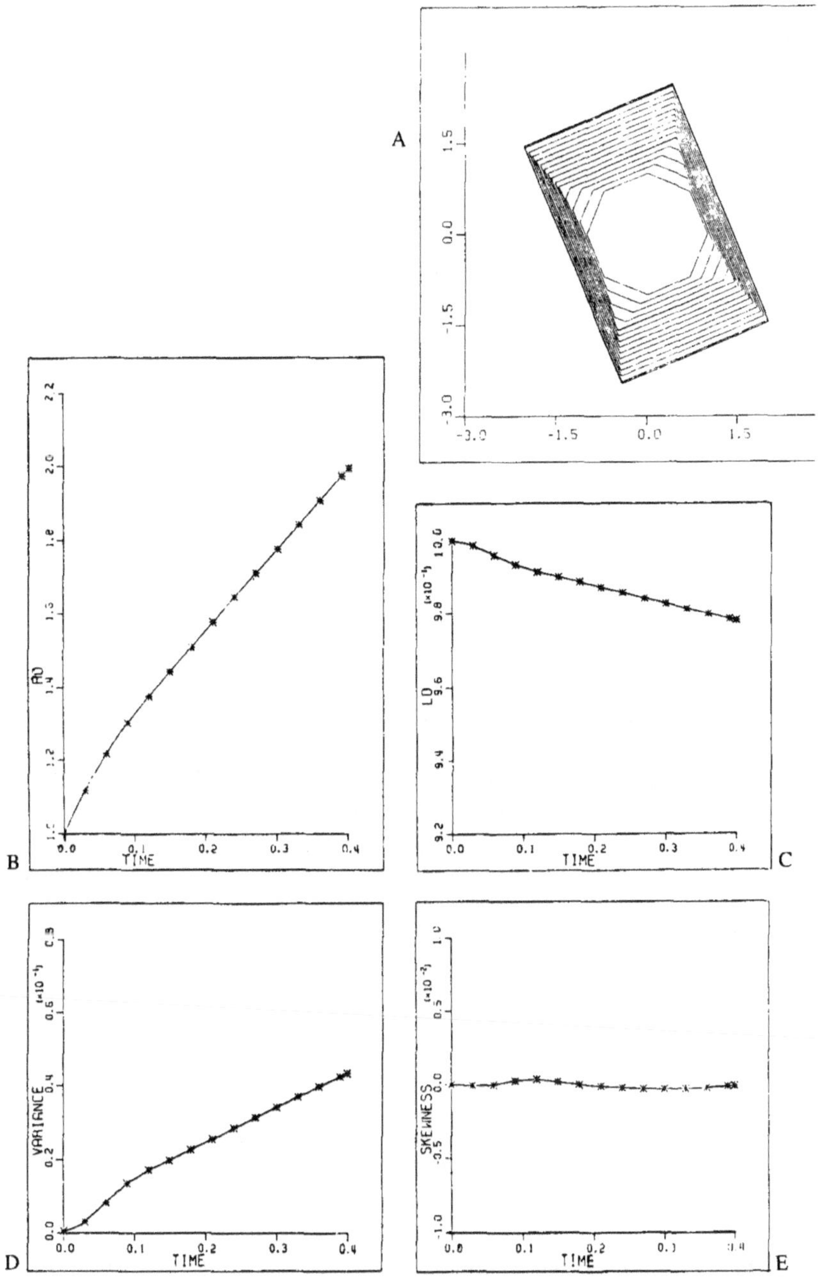


FIGURE 9. Shape, size, and morphic terms vs. time. Fast growing faces: 2, 4, 6, and 8. Slow growing faces: 1 and 5.

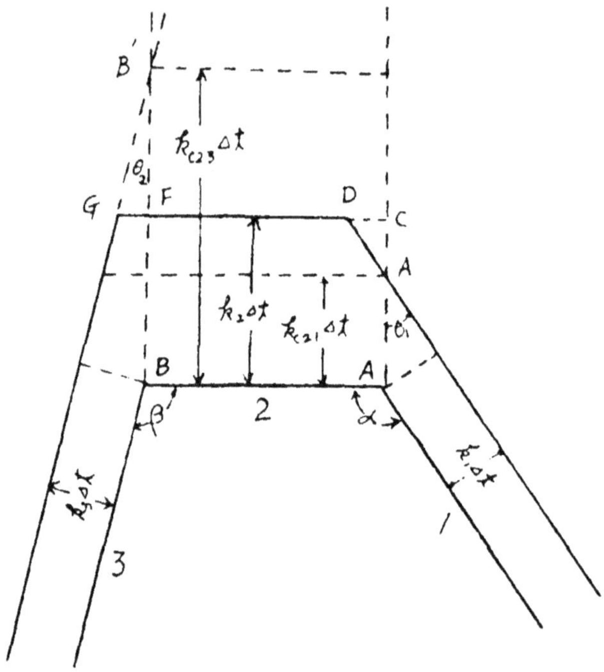


FIGURE 10. Geometrical construction in criterion for face elimination.

where $1.25 < n < 1.62$ over the velocity range $0.003 < U < 0.15$ m/sec. The (100) facial growth rate was also measured under various supersaturations and solution velocities. It was found that the (100) face eventually disappeared if $k_{(100)}/k_{(111)} > 1.732$, leaving only (111) faces, or the (100) face began increasing in size if $k_{(100)}/k_{(111)} < 1.732$. The crystal face has a truncated octahedron shape if a (100) face was present, and $k_1 = k_3 = k_{(111)}$, $k_2 = k_{(100)}$, $\alpha = \beta = 125°16'$,

$$k_{2c} = \frac{\dfrac{k_{(111)}}{\cos(125°16' - 90°)}}{\tan(125°16' - 90°)}$$

$$= \frac{k_{(111)}}{\sin(35°16')}$$

$$= 1.732\ k_{(111)}$$

Therefore, the experimental result is in excellent agreement with the criterion.

## VII. APPLICATION

The morphological changes of a crystal as it grows can be determined experimentally. By analyzing the time-sequence morphologies under different environmental conditions, it is possible to characterize the influence of various process variables on the facial growth rates. Once the effects of process variables are characterized, and by applying the critical condition for a face to disappear, one should be able to control the size and shape of a crystal by controlling the process variables in the manner indicated by model analysis.

* All symbols refer to Figure 10.

## LIST OF SYMBOLS*

$k_1$ = Facial growth rate of face 1
$k_2$ = Facial growth rate of face 2
$k_3$ = Facial growth rate of face 3
$\alpha$ = Angle made by faces 1 and 2
$\beta$ = Angle made by faces 2 and 3
$\theta_1$ = Angle made by faces A′D and A′A
$\theta_2$ = Angle made by faces B′G and B′F
$\Delta t$ = Time interval
CD = Increment of face 2 at the end near face 1
FG = Increment of face 2 at the end near face 3
AA′ = Outward displacement of face 2 with no increment at the end near face 1
$k_{2c1}$ = Outward displacement of face 2 per unit time with no increment at the end near face 1

## REFERENCES

1. **Nyvlt, J.,** *Industrial Crystallization from Solutions,* Butterworths, London, 1971.
2. **Breckner, E., Day, J., and Carmichael, G. R.,** The effect of particulate shape, solvent, concentration, and agitation on dissolution, submitted.
3. **Luerkens, D. W.,** Morphological Analysis of Fine Particles Using the Fourier Method, Ph.D. thesis, University of Iowa, Iowa City, 1980.
4. **Luerkens, D. W., Beddow, J. K., and Vetter, A. F.,** Morphological Fourier descriptors, *Powder Technol.*, 31(2), 209, 1982.
5. **Mullin, J. W. and Garside, J.,** The crystallization of aluminum potassium sulphate: a study in the assessment of crystallizer design data. I. Single crystal growth rates, *Trans. Instr. Chem. Eng.*, 45, T285, 1967.

# *Index*

# INDEX

## N



## O



## P

## Q



## R



## S

T



U



V



W

For Product Safety Concerns and Information please contact our EU representative GPSR@taylorandfrancis.com Taylor & Francis Verlag GmbH, Kaufingerstraße 24, 80331 München, Germany

**Batch number: 10731074**

Printed by Printforce, the Netherlands